Nanotechnology and Plant Disease Management

Nanotechnology and Plant Disease Management explores the intersection of nanotechnology and agriculture. This book serves as a comprehensive exploration of the current state and future potential of nanoparticles in revolutionizing plant disease management within the realm of agriculture. This book elucidates the synthesis, characterization and judicious application of nanoparticles, providing a clear and accessible explanation of what nanomaterials are, how they are characterized, and their pivotal role in reshaping the plant disease management systems. It scrutinizes innovative strategies that influence the unique properties of nanoparticles to identify and monitor the presence of pathogens at early stages. The book also examines the limitations inherent in the use of nanomaterials for disease management in plants by critically evaluating both sides of the spectrum. This aims to provide a candid overview of the hurdles that must be overcome to unlock the full benefits of nanotechnology in this field. By pinpointing and understanding these obstacles, the current work aims to pave the way for effective strategies and solutions, ensuring the responsible and optimized application of nanomaterials for enhanced plant disease management. This in-depth examination serves as a cornerstone, providing readers with a profound understanding of the intricate processes involved in synthesis, characterization and utilizing nanomaterials for disease control. *Nanotechnology and Plant Disease Management* is a testament to the transformative potential of nanotechnology in agriculture. The authors invite readers to embark on this enlightening journey, exploring the intricate world of nanomaterials and their application in safeguarding the health and vitality of plants.

Nanotechnology and Plant Disease Management

Edited by

Irshad Mahmood, Rizwan Ali Ansari and Rose Rizvi

CRC Press
Taylor & Francis Group
Boca Raton London New York

CRC Press is an imprint of the
Taylor & Francis Group, an **informa** business

Designed cover image: ShutterStock Photo ID: 2203319013

First edition published 2025
by CRC Press
2385 NW Executive Center Drive, Suite 320, Boca Raton FL 33431

and by CRC Press
4 Park Square, Milton Park, Abingdon, Oxon, OX14 4RN

CRC Press is an imprint of Taylor & Francis Group, LLC

Library of Congress Cataloging-in-Publication Data
Names: Mahmood, Irshad, editor. | Ansari, Rizwan Ali, editor. | Rizvi, Rose, editor.
Title: Nanotechnology and plant disease management / Irshad Mahmood, Rizwan Ali Ansari, Rose Rizvi.
Description: First edition. | Boca Raton, FL : CRC Press, 2025 |
Includes bibliographical references and index. |
Identifiers: LCCN 2024024986 (print) | LCCN 2024024987 (ebook) |
ISBN 9781032188829 (hardback) | ISBN 9781032188843 (paperback) |
ISBN 9781003256762 (ebook)
Subjects: LCSH: Agricultural innovations. | Nanobiotechnology. |
Plant diseases. | Plants, Protection of.
Classification: LCC S494.5.I5 N3444 2025 (print) | LCC S494.5.I5 (ebook) |
DDC 338.1/6—dc23/eng/20241010
LC record available at https://lccn.loc.gov/2024024986
LC ebook record available at https://lccn.loc.gov/2024024987

ISBN: 9781032188829 (hbk)
ISBN: 9781032188843 (pbk)
ISBN: 9781003256762 (ebk)

DOI: 10.1201/9781003256762

Typeset in Times
by Newgen Publishing UK

Contents

Preface

It is with great pleasure and enthusiasm that we present *Nanotechnology and Plant Disease Management*. This book serves as a comprehensive exploration of the current state and future potential of nanoparticles in revolutionizing plant disease management within the agriculture.

In the pursuit of clarity, each chapter meticulously addresses essential aspects of nanotechnology in plant disease management. To facilitate an accessible journey through this dynamic field, we have structured the book to provide a well-balanced blend of theory, research findings, and practical examples. The chapters are crafted with a commitment to clarity, ensuring that even readers less acquainted with the subject matter can glean valuable insights. A particular emphasis is placed on elucidating the synthesis, characterization and application of nanoparticles. This foundational chapter sets the stage for a profound exploration into the advantages, disadvantages, and limitations inherent in the use of nanomaterials. The initial chapters lay a solid foundation by explaining the nature of nanomaterials. We provide a clear and accessible explanation of what nanomaterials are, how they are characterized, and their pivotal role in reshaping the plant disease management systems. This proactive approach is central to minimizing the impact of diseases on crop yields and overall agricultural productivity. The book also examines the limitations inherent in the use of nanomaterials for disease management in plants by critically evaluating both sides of the spectrum. We have tried to gather the information on various challenges and limitations that might hinder the complete realization of nanomaterials' potential in plant disease management. This scrutiny aims to provide a candid overview of the hurdles that must be overcome to unlock the full benefits of nanotechnology in this field. By pinpointing and understanding these obstacles, we aim to pave the way for effective strategies and solutions, ensuring the responsible and optimized application of nanomaterials for enhanced plant disease management. This in-depth examination serves as a cornerstone, providing readers with a profound understanding of the intricate processes involved in synthesis, characterization and utilizing nanomaterials for disease control. To bridge the theoretical concepts with practical applications, the book incorporates well-researched examples and case studies. These real-world scenarios showcase the tangible impact of nanotechnology on mitigating the effects of plant diseases, offering readers a glimpse into the transformative potential of this burgeoning field. "Nanotechnology and Plant Disease Management" is a testament to the transformative potential of nanotechnology in agriculture.

We express our profound appreciation to Prof. Naima Khatoon, the esteemed Vice Chancellor of Aligarh Muslim University, for her exceptional support and encouragement in facilitating our research and development endeavours. Under her visionary leadership, the university has made significant advancements in plant sciences. Her steadfast commitment to fostering academic excellence and a robust research culture serves as a perennial source of inspiration.

Our sincere gratitude extends to Prof. Akram Ahmad Khan, Dean, Faculty of Agricultural Sciences whose unwavering support has been a constant throughout this project. We also acknowledge with indebtedness to Prof. M. Badruzzaman Siddiqui, Chairperson of the Department of Botany, and Prof. Mujeebur Rahman Khan, Chairperson of the Department of Plant Protection whose insightful suggestions have played a pivotal role in shaping the content of this book. Prof. Saghir Ahmad Ansari, with his extensive knowledge and research acumen in agricultural sciences, merits our deepest appreciation for providing critical insights to enhance this work.

Special acknowledgment is due to esteemed professors, including Prof. Zaki Anwer Siddiqui, Prof. Iqbal Ahmad, Prof. Malik Ahmad, Prof. Saghir Ahmad Khan, Prof. Anwar Shahzad, Prof. Abrar Ahmad Khan, and Prof. Masroor Akhtar Khan. Their valuable feedback, suggestions, and unwavering support have been instrumental in refining the content of this book, leveraging their expertise and knowledge in the field.

We extend our heartfelt gratitude to Dr. Aisha Sumbul, a distinguished scholar whose invaluable contributions considerably enriched the writing process of this book. Dr. Sumbul's profound insights and prolific suggestions have been pivotal in shaping the content. Her unwavering support and commitment to excellence have been instrumental in translating our vision into reality.

Additionally, our gratitude extends to dedicated researchers, Dr. Sartaj Ali Tiyagi, Dr. Safiuddin, Dr. Raees Ullah Khan, Dr. Athar Ali Khan, Dr. Sana Chaudhary, Dr. Faheem Ahmad and Dr. Tahir Mohammad Chauhan whose considerable motivations and suggestions have enriched the content of this book.

Finally, we recognize the untiring support and inspiration from our families without whom the realization of this book would not have been possible. Their understanding and encouragement have been a constant driving force. We extend our heartfelt thanks with sincere apologies for any inadvertent disruptions during the writing process.

With warm regards

Prof. Irshad Mahmood, Aligarh, India
Dr. Rizwan Ali Ansari, Aligarh, India
Dr. Rose Rizvi, Aligarh, India

Editors' Bios

Irshad Mahmood, an erudite Professor of Plant Pathology and Nematology in the Department of Botany, Aligarh Muslim University, Aligarh. He obtained his Ph.D. degree from Aligarh Muslim University in the field of Plant Pathology and Nematology. He has been engaged with the teaching programme of undergraduate and post-graduate level students for the last 30 years and has made many overseas visits, including to the United States, France and the UK. He has attended a significant number of national and international conferences on a wide area of agricultural sciences and has published around 150 original research papers, review articles and book chapters in various refereed national and international publications. He has successfully completed many training courses in various ICAR sponsored research institutes in India and also at North Carolina State University, Raleigh, USA. He is also an active member of national and international scientific organizations, served as an expert for selection committees, reviewer of journals, doctoral theses and funding agencies, and is a recipient of a scientist of the year award in the field of Plant Pathology and Nematology. He has guided ten Ph.D., several M. Phil. students and a large number of M.Sc. dissertations.

Rizwan Ali Ansari, Assistant Professor in the Department of Plant Protection, Faculty of Agricultural Sciences at Aligarh Muslim University in India, is a highly accomplished scholar in the field of Plant Pathology/Nematology. His academic credentials include a Ph.D. from the same institution, where he has actively engaged in cutting-edge research and development strategies that have revolutionized the field. Dr. Ansari has formulated innovative management modules that leverage a wide array of microorganisms, antagonists and organic additives to combat a diverse range of soil-borne plant pathogens that infest numerous agricultural crops. He has also attended several prestigious national and international conferences, where his contributions have earned him coveted accolades from renowned scientific societies such as the Society of Plant Protection Sciences (SPSS), Indian Phytopathological Society (IPS) and the Nematological Society of India (NSI). As a prolific scholar, he has authored several highly acclaimed book chapters, research papers and review articles that delve into the utility of organic additives, mycorrhizal fungi and plant growth-promoting bacteria in the sustainable management of plant pathogens. His research has been published in several reputable journals and books. Dr. Ansari has edited many books which have garnered global appreciation. His current research domains include the judicious application of organic additives, synthesis of nanoparticles and their application and biological agents in the management of plant diseases, improving soil and crop health, and conducting surveys on disease prevalence caused by soil-borne pathogens on various economically significant crops. His pioneering research has the potential to significantly impact the sustainable management of plant pathogens and improve soil and crop health, thereby advancing agricultural productivity and food security.

Rose Rizvi is currently working as Assistant Professor in the Department of Botany, Aligarh Muslim University, Aligarh, India. She earned her Ph.D. degree from the same department in Plant Pathology/Nematology. She has thus far published a good number of research and review articles in reputable journals. Dr. Rizvi has also participated in various national and international symposia/conferences/workshops and successfully responded to queries from the delegates pertaining to the area of her research. She has been the recipient of various awards such as UGC-BSR Research Fellow from the University Grants Commission (UGC); Norman Ernest Borlaug Research Award from Plant Pathology (Photon Foundation); Junior Scientist Award from the National Environmental Science Academy (NESA), New Delhi, India; CSIR-Research Associateship from MHRD, New Delhi; Scientist of the Year Award 2018 from the National Environmental Science Academy (NESA), New Delhi, India; Young Scientist of the Year Award 2018 from the International Foundation for Environment and Ecology (IFEE), Kolkata, India. Her area of research is the integrated disease management of phytoparasitic nematodes by the application of various means like biological control fungi, rhizobacteria, organic additives and nanoparticles.

Contributors

Mahfouz M. M. Abd-Elgawad
Pests & Plant Protection Department,
Agriculture and Biological Division-National Research Centre,
33rd ElBehouse St. (El-Tahrir Street) Dokki,
Giza, Egypt. (P.O. Box 12622), Fax: 002/ 33370931

Taofeek T. Adegboyega
Food Security and Safety Niche Area,
Faculty of Natural and Agricultural Sciences,
North-West University, Private Bag X2046,
Mmabatho 2735,
South Africa & Department of Microbiology & Parasitology,
School of Medicine & Pharmacy,
University of Rwanda

Rizwan Ali Ansari
Department of Plant Protection,
Faculty of Agricultural Sciences,
Aligarh Muslim University,
Aligarh 202002

Archana A.M.
University of Horticultural Sciences,
Bagalkot -587104, India

Archana T. S.
Lovely Professional University,
Phagwara, Punjab -144411,
India

Tesleem Taye Bello
Department of Agricultural Science Education,
Federal College of Education,
PMB 2096 Abeokuta,
Ogun State, Nigeria

Jamila Boudaden
Fraunhofer EMFT,
Research Institution for Microsystems and Solid State Technologies EMFT,
HansastraBe 27d, D-80686 Munich,
Germany

Gökhan Boyno
Department of Plant Protection,
Faculty of Agriculture,
Van Yuzuncu Yil University,
Van, Turkiye

Younes Rezaee Danesh
Department of Plant Protection,
Faculty of Agriculture,
Van Yuzuncu Yil University,
Van, Turkiye

Semra Demir
Department of Plant Protection,
Faculty of Agriculture,
Van Yuzuncu Yil University,
Van, Turkiye

Emre Demirer Durak
Department of Plant Protection,
Faculty of Agriculture,
Van Yuzuncu Yil University,
Van, Turkiye

Sozan E. El-Abeid
Mycology and survey research department and Nanotechnology and advanced Nano-materials Lab (NANML),
Plant Pathology Research Institute,
ARC, Giza, Egypt, 12619

Oluwatoyin Adenike Fabiyi
Department of Crop Protection,
Faculty of Agriculture,
University of Ilorin,
PMB 1515 Ilorin,
Nigeria.

Noor Fatima
Section of Plant Pathology and Nematology,
Department of Botany,
Aligarh Muslim University,
Aligarh 202002

Ahmed Ghannouchi
Centre International de Hautes Etudes Agronomiques Méditerranéennes,
Mediterranean Agronomic Institute of Bari,
Via Ceglie 9,
70010 Valenzano, Italy

Manjunath Hubballi
University of Horticultural Sciences,
Bagalkot -587104,
India

Dina S. S. Ibrahim
Nematology and Biotechnology Department of Nematodes diseases and central Lab of biotechnology;
Plant Pathology Research Institute,
ARC, 9 Cairo University St.,
Giza 12619, Egypt

Chara Preet Kaur
University of Delaware,
Newark -19716, USA

Irshad Mahmood
Section of Plant Pathology and Nematology,
Department of Botany,
Aligarh Muslim University,
Aligarh 202002

Mohamed A. Mosa
Nanotechnology & Advanced Nano-Materials Laboratory (NANML);
Plant Pathology Research Institute; Agricultural Research Center (ARC); Giza 12619; Egypt. &
Mycology and Disease Survey Research Department, Plant Pathology Research Institute, ARC. 12619, Egypt

Manal Mostafa
Nematodes diseases department and Central Lab of Biotechnology,
Plant Pathology Research Institute,
ARC, Giza,
Egypt, 12619

Uma Shankar Rajput
Department of Plant Protection,
Faculty of Agricultural Sciences,
Aligarh Muslim University,
Aligarh 202002

Geeta Rautela
Section of Plant Pathology,
Department of Botany,
Aligarh Muslim University,
Aligarh, India

Mehmet Ramazan Rişvanlı
Department of Plant Protection,
Faculty of Agriculture,
Van Yuzuncu Yil University,
Van, Türkiye

Rose Rizvi
Section of Plant Pathology and Nematology,
Department of Botany,
Aligarh Muslim University,
Aligarh 202002

Magda Mahmoud Amin Sabbour
Pests & Plant Protection Department,
Agriculture and Biological Division-
National Research Centre,
33rd ElBehouse St. (El-Tahrir Street) Dokki,
Giza, Egypt. (P.O. Box 12622), Fax: 002/33370931

Fayrouz M. Saeed
Euro-Mediterranean master's program in Neuroscience and Biotechnology,
Faculty of Science,
Alexandria University,
Alexandria,
Egypt, 21568

Safiuddin Safiuddin
Section of Plant Pathology and Nematology,
Department of Botany,
Aligarh Muslim University,
Aligarh 202002

Ahmed M. Saleh
Department of Pharmaceutical Chemistry;
Faculty of pharmacy;
Horus University;34518,
Egypt

Arsh Rehan Siddiqui
Gems Founder School,
Dubai,
United Arab Emirates

Ahmed G. Soliman
Biotechnology program,
Faculty of Agriculture,
Ain Shams- university,
Cairo, Egypt

Aisha Sumbul
Section of Plant Pathology and Nematology,
Department of Botany,
Aligarh Muslim University,
Aligarh 202002

Necmettin Teniz
Department of Agricultural Biotechnology,
Faculty of Agriculture,
Van Yuzuncu Yil University,
Van, Turkiye

Noah Totsline
University of Delaware,
Newark, DE 19716,
USA

Atul Upadhyay
Department of Plant Protection,
Faculty of Agricultural Sciences,
Aligarh Muslim University,
Aligarh 202002

Abdulmujib Gboyega Yusuf
Plant Protection Department,
College of Food and Agricultural Sciences,
King Saud University,
P.O. Box 2460, Riyadh 11451,
Saudi Arabia

Aiman Zafar
Section of Plant Pathology,
Department of Botany,
Aligarh Muslim University,
Aligarh, India

1 Nanotechnology

A New Frontier in Sustainable Agricultural Pest and Disease Management

Magda Mahmoud Amin Sabbour, Mehmet Ramazan Rişvanlı, and Gökhan Boyno

1.1 INTRODUCTION

The reduction of crop productivity is caused by insects (14%), pathogens (13%), and weeds (13%) (Oerke, 2006). According to the United Nations Food and Agriculture Organization, annual pest losses to crop production range from 20% to 40% (FAO, 2019). Invasive insect pest species cost the global economy about $70 billion annually, while plant diseases cost about $220 billion (FAO, 2020). Pesticides are used in agriculture to control weeds, insect infestation, and diseases. Properly applying pesticides can result in significant socioeconomic and environmental benefits, including producing safe, healthy, and affordable food, stable farm incomes, and managing natural resources more sustainably by using soil, water, and general land use more effectively (Aktar, et al., 2009). However, achieving responsible pesticide application practices can be very challenging. This situation has led us to search for alternative ways to reduce the using chemicals. Nanotechnology is a field in rapid development based on producing many nanomaterials (NMs), manipulating their physical and physicochemical features such as size, shape, crystalline nature, and interaction with biological systems (Pradeep, 2007).

Professor Kerie E. Drexler, a lecturer and researcher in nanotechnology, is known for coining the word nanotechnology and the word nano, which is derived from the Greek word meaning dwarf. Richard Feynman first presented nanotechnology as a concept during his lecture There is Plenty of Room at the Bottom at the American Institute of Technology in 1959 during the annual meeting of the American Physical Society (Feynman, 1959). While, in 1974, the word of Nanotechnology was introduced by Norio Taniguchi at Tokyo Science University (Hulkoti & Taranath, 2014).

The Nanotechnology has become one of the most crucial technology among various technologies in the recent years (Goswami et al., 2010). Numerous efforts have been made to advance agriculture using nanotechnology, called the next industrial revolution. The next revolution in agriculture and agrotechnology is now possible thanks to this intriguing and promising technology (Pan et al., 2023). The nanotechnology used in agriculture, like the synthesis of nano-fertilizers produces nano-pesticides, nano-chemical insecticides, and nano-oils. It reduces the amount of substances, decreases the economic cost and toxicity in soil and groundwater, and increases the quality of treatments, enhancing properties such as efficacy, quality, and increased yield (Kumar et al., 2019). Nanosensors and innovative nano-based delivery systems help use natural resources, chemicals and nutrients during cultivation (Johnson et al., 2021; Zhu et al., 2022). Nanotechnology has shown promise in making agricultural systems more sustainable and effective to help feed the world (Figure 1.1).

DOI: 10.1201/9781003256762-1

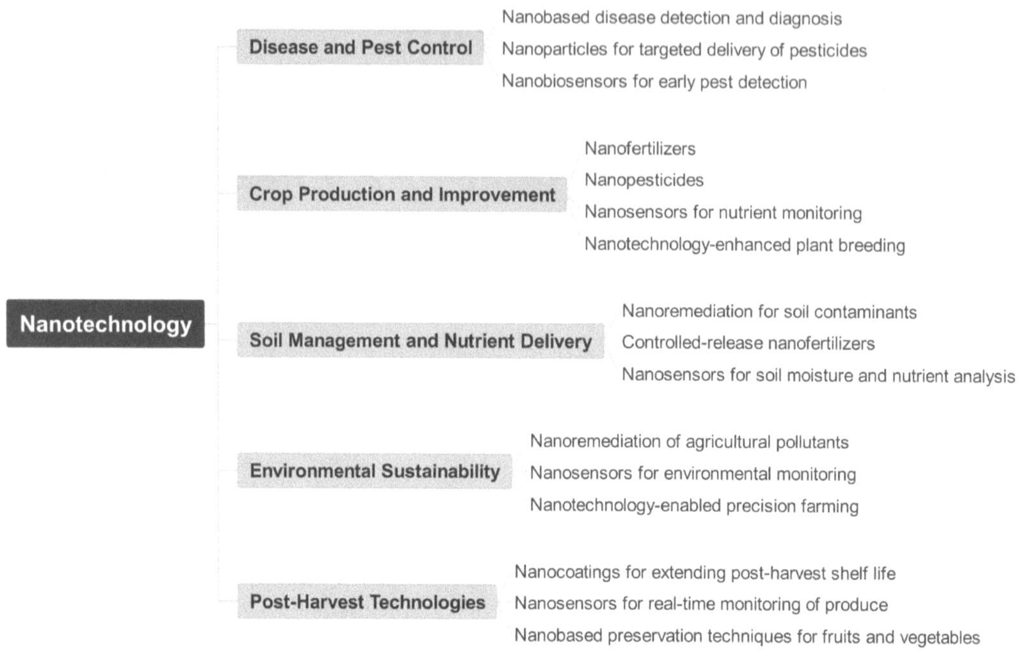

FIGURE 1.1 Applications of nanotechnology in various agricultural fields: (1) Disease and pest control: Nanotechnology helps in targeted and efficient control of agricultural diseases and pests. (2) Crop production and improvement: Nanoparticles enhance crop growth and increase yield through improved nutrient uptake. (3) Soil management and nutrient delivery: Nanoscale materials assist in precise nutrient delivery and soil health monitoring. (4) Environmental sustainability: Nanotechnology promotes eco-friendly practices in agriculture, reducing environmental impact. (5) Post-harvest technologies: Nanomaterials can help preserve the quality and extend the shelf-life of harvested produce by using innovative post-harvest technologies (Modified from: Prasad et al., 2014; Das et al., 2020; Ahmed et al., 2021; Khan et al., 2022; Pan et al., 2023).

Nanotechnology helps producing many nanostructures due to radical changes in the material size (Gao et al., 2012). These structures have gained importance in advanced technologies due to changes in their physical and chemical characteristics as size, shape, melting point, solubility, thermal conductivity, hardness, light absorption, scattering, and permeability leading to enhancement in the performance over their bulk because of the high surface area of the nanoparticles, so, this technology helps in development of almost all applied sciences (Jeevanandam et al., 2018; Delshadi et al., 2020).

An international system of units called nanometers (nm) or metric units of length; that is equivalent to a billionth of a meter (10^9 meters in length) equal to 0.001 micrometers. The DNA helix is about 2.5 nm in diameter, while human hair is about 80000 to100000 nm wide (Nithya, 2011; Hulkoti & Taranath, 2014). Nanotechnology is an applied science that explains and studies the behaviour or features of materials and structures with dimensions ranging from 1to100 nm, where these materials are known as NMs (Saleem & Ziadi, 2020). The Food and Drug Administration (FDA) also refers to NMs as a material with at least one dimension in the range of approximately 1 to 100 nm (Esquivel et al., 2007; FDA, 2011). Despite the apparent benefits of NMs, there is concern about their safety margin for the environment and humans; hence new safety rules are needed for this technology (Nikmaram et al., 2017). This chapter aims to focus on the opportunities and the future research perspectives for the use of NMs in pest and disease management.

1.2 NANOPESTICIDES

Farmers and growers are concerned about diseases, pests, and weed species because they can negatively affect plants throughout production and storage. Pesticides comprise any substance or combination of substances used to manage these pests and diseases (Zacharia, 2011). Studies have shown that only 0.1% of pesticide exposure reaches its target, while 99.9% remains in the environment as residual pesticides (Horrigan et al., 2002; Carriger et al., 2006). Insecticides, herbicides, fungicides, and rodenticides etc. are all referred to as pesticides and can be used to kill various organisms. However, because they are highly toxic and may not be biodegradable, chemical pesticides contribute to various environmental pollution (Abubakar et al., 2020).

Utilizing nanopesticides to control diseases and pests is a potential application of nanotechnology in agriculture. To get beyond the limits of traditional pesticides, researchers have created nano-pesticides. By employing NMs to encase the chemical components, these nano-pesticides were created. They have a number of advantages over conventional pesticides, including improved stability, controlled release of the active ingredients, superior efficacy, a need for lower doses, good dispersion, and reduced residue (Li et al., 2019). Nanotechnological methods can further enhance the controlled release of the active components of bio-inoculants by using nanoparticles as carriers for these chemicals. Such formulations can significantly reduce pesticide use and associated environmental risks. Because there is at least a 10 to 15 fold reduction in the amount of chemicals needed to achieve an impact compared to traditional formulations, nanopesticides will require fewer chemicals to achieve the same effect (Sharma et al., 2018). However, nano pesticides may also have risks, such as toxicity to organisms other than their intended targets, transportation, bioaccumulation, and interactions with other environmental pollutants (Xu et al., 2022).

Developing new nanopesticide formulations using nanotechnology would reduce the risks caused by chemical pesticides and increase the effectiveness of biopesticides. There is a lot of interest in developing safe and efficient insecticides thanks to environmentally friendly NMs with biodegradable, responsive, and biocompatible qualities. Additionally, it may help in the development of less toxic biopesticides with improved biological safety, increased active ingredient stability, and improved activity against target pests (Pan et al., 2023).

1.2.1 Potential Environmental Concerns of Nanopesticides

Nanopesticides are still in their infancy and require a great deal of study to determine their safety and efficacy. The advantages and risks of using nanopesticides cannot yet be reliably assessed given the existing state of knowledge. There is an urgent need for a more reliable methodology for risk assessments of nanopesticides (Anonymous, 2010; Kah et al., 2021). According to researchers, using nanotechnology with pesticides may improve active ingredient delivery and control, reduce ecological drift, create formulations that will reach the target most effectively organism, and possibly improve agricultural workers protection (Li et al., 2019). However, there are still concerns about the safety of nanopesticides. The Office of Pesticide Programs (OPP) is concerned about the potential human health effects of nanopesticides. It is necessary to take into account the potential for particular dangers posed by the nanoscale in an agricultural context, even if crops and fruit trees are currently treated with bulk forms of metal-based pesticides. For instance, the potential for NMs persistence and bioaccumulation in the environment and in plants and the long-term effects of NMs on the soil microbiota have not been determined (Su et al., 2019). Furthermore, it is unclear whether NMs may move between different plants and between edible and inedible parts of those plants. If NMs are discovered to translocate to and persist in the edible part of a crop, ingestion is a possible risk (Gilbertson et al., 2020). One of the concerns is that nanoparticles can be absorbed through the skin and may pass through cell membranes. Another concern is that nanoparticles can go to the deep lung and may translocate to the brain, which means they could cross the blood-brain barrier (Graily-Moradi & Lajayer, 2021; Shekhar et al., 2021; Pan et al., 2023). There is an urgent need for a more reliable methodology for risk assessments of nano-pesticides. Research priorities in this area include identifying the presumptions currently used that do not hold true for nanopesticides, assessing the areas or circumstances where differences may significantly impact exposure assessment results, and improving or adapting current assessment protocols as needed.

1.3 TYPES OF ENGINEERED NANOPARTICLES

By creating nanoscaled agrochemicals, engineered nanomaterials have demonstrated promising sustainable applications in agricultural practices (Gilbertson et al., 2020). The morphology, size, and chemical characteristics of NMs are used to categorize them broadly into several groups. Metal nanoparticles (MNPs), carbon-based nanoparticles (CNMs), polymeric nanoparticles (PNMs), and hybrid nanoparticles are the four primary categories of NMs (Jeevanandam et al., 2018).

1.3.1 Metal-Based Nanoparticles

MNPs are produced by high-pressure homogenization in various sizes (10 to100 nm), shapes spherical or polyhedral. MNPs are one of the types of engineered nanoparticles that have shown promise in preventing plant diseases (Servin et al., 2015; Dutta et al., 2023). They are utilized in the production of insecticides, herbicides, and insect repellents because they have been shown to be highly efficient against plant diseases (Barik et al., 2008; Yadav et al., 2023). It has been discussed how plant disease diagnostic use of metal and metal oxide nanoparticles to control crop diseases. These NPs are novel tools in the control plant pathogens, making it easier to diagnose and manage plant diseases early on, especially when resources are scarce (Elmer et al., 2018). Most metals used in synthesizing metal-based insecticides include silver, zinc, titanium, etc. (Graily-Moradi & Lajayer, 2021). Also, an enormous range of natural and synthetic, organic and inorganic nanomaterials can be used for nano-insecticide formulations, for example, silver nanoparticles (AgNPs) (Rouhani et al., 2013), nano-chitosan is not a metal, however, it is frequently used to incorporate metal ions into metal-based nanoparticles (Kumar et al., 2000; Sabbour, 2016; Hernandez-Hernandez et al., 2018), silicon dioxide nanoparticles (El-Naggar et al., 2020), zinc oxide nanoparticles (Khooshe-Bast et al., 2016), and more.

1.3.2 Carbon-Based Nanomaterials

There is great interest in agriculture in a class of nanoparticles called carbon-based nanomaterials. because of their distinctive physicochemical characteristics and possible uses. Carbon nanotubes (CNTs), graphene, carbon dots, fullerenes, and nanodiamonds are only a few examples of CNMs. These nanoparticles can increase crop growth, protection, and production by acting as light converters, nanosensors, and delivery tools (Mukherjee et al., 2016; Zhu et al., 2022).

Detecting and measuring physical, chemical, or biological signals at the nanoscale is the function of nanosensors. The soil moisture, nutritional status, pH, temperature, salinity, pathogens, and pollutants are only a few of the characteristics that CNMs can monitor as nanosensors for plants. For instance, graphene can be utilized as a nanosensor to track changes in electrical conductivity to find nitrate levels in soil (Johnson et al., 2021; Zhu et al., 2022). Fertilizers, insecticides, hormones, DNA, and other compounds can be delivered to plants or plant cells using CNMs as delivery ways. For instance, by penetrating the cell wall and membrane and releasing nitrogen inside the cells, carbon dots can be employed as methods of delivery to provide nitrogen fertilizer to rice plants (Li et al., 2020).

In numerous studies, CNMs have demonstrated promising gains in plant protection, productivity, and growth. However, using them in agriculture also comes with some challenges and risks. The lack of standardized characterization techniques, the possible toxicity and environmental impact of CNMs, the ethical and social ramifications of nanotechnology, and the regulatory and legal concerns with CNM-based goods are a few of these (Li et al., 2020; Zhu et al., 2022) More study is therefore required to solve these issues and guarantee the sustainable and safe use of CNMs in agriculture.

1.3.3 Polymeric Nanoparticles

Agricultural chemicals such as herbicides, fertilizers, insecticides, and plant growth regulators can be delivered via PNMs, which are nanoscale particles made of organic or synthetic polymers. In comparison to traditional pesticide formulations, PNMs have improved stability, solubility, homogeneous distribution, controlled release, and less environmental impact. Additionally, PNMs can enhance the effectiveness and uptake of the active ingredients by plants, resulting in increased agricultural yields and reduced losses (Vemula et al., 2023).

Several PNMs, including chitosan, polyvinyl alcohol (PVA), lipids, and poly (lactic-co-glycolic acid) (PLGA), have been developed and utilized in agriculture. Depending on their composition, structure, size, and surface modification, these NMs have a variety of diverse characteristics and uses. For instance, the natural polysaccharide known as chitosan, which is generated from crustacean shells, possesses antibacterial, antifungal, and biodegradable capabilities. Using chitosan nanoparticles, fungicides, insecticides, and plant growth regulators can be delivered to the plants. The synthetic polymer PVA is highly soluble in water, biocompatible, and capable of film-forming (Kose et al., 2023). Herbicides can be encapsulated and released in a regulated way using PVA nanoparticles (Pascoli et al., 2018). Lipids are natural or synthetic fatty acids that can form vesicles or micelles to carry hydrophobic or hydrophilic compounds. Lipid nanoparticles can be utilized to deliver to plants plant hormones, nutrients, and insecticides (Mittal et al., 2020; An et al., 2022). PLGA is a synthetic copolymer that has biodegradable and biocompatible properties. PLGA nanoparticles can be used to protect and deliver insecticides, fungicides and herbicides to the plants (Hernandez-Giottonini et al., 2020). The polymeric nanocapsule is mainly prepared for the active compounds-controlled release due to polymeric support systems for the steric (strong) stabilization. Many natural and synthetic polymers, including protein (albumin, legumin, lecithin, and gelatin), polysaccharides (alginates, chitosan, dextrin, gums, and starch), and synthetic polymers (as polyethylene glycol and poly-E -caprolactone) have been developed as the main components for the nanoencapsulation process (Jeevanandam et al. 2018; Ngwuluka et al., 2021). It has been found that different polymeric nanoformulations give highly reductions in active ingredients consumption with maintaining nontoxic behavior, low cost, and biodegradable properties (Jeevanandam et al. 2018; Vemula & Reddy, 2023). Meyer et al. (2015) mentioned that the role of nano-polymer in minimize the entrapped insecticide concentrations where the encapsulation of imidacloprid (systemic insecticide) was about 200 times lower than commercial formulations in term of the used concentration. Thus, polymeric nanocapsules can help pesticides to be used more efficiently while also decreasing their adverse outcomes for the human health and environment (Kumar et al., 2000; Hernandez-Hernandez et al., 2018; Vemula & Reddy, 2023).

The sustainability and productivity of agriculture can be improved by using PNMs as agrochemical carrier systems. Nevertheless, this area still has several difficulties and restrictions, such as the large-scale production, cost-effectiveness, standardization, regulation, and safety assessment of these NMs (Vemula & Reddy, 2023). More research is required to fully comprehend how PNMs operate in the environment and in plants and how they are transported, transformed, and toxic.

1.3.4 Hybrid Nanoparticles

NMs called hybrid nanoparticles consist of two or more different components, such as metals, oxides, polymers, etc. Their unique and adaptable qualities, which can be used for a wide range of applications, including agriculture, have become increasingly popular in recent years. Due to their distinctive and adaptable physicochemical features caused by their small sizes, NMs have demonstrated enormous potential for their numerous applications in a range of industries, including medicine, healthcare, the food industry, the environment, agriculture, electronics, and pharmaceutics (Xin et al., 2021; Singh et al., 2022).

Through the combination of several functionalities and synergistic effects, hybrid nanoparticles have been employed to improve the efficacy of nanopesticides, nanofertilizers, and nanosensors. By encapsulating urea and releasing it slowly in response to soil pH, for instance, hybrid nanoparticles made of metal oxides and polymers have improved nitrogen availability and decreased leaching losses (Xue & Luo, 2021). Inhibiting fungal growth and improving plant tolerance to Fusarium wilt have both been accomplished using hybrid nanoparticles made of silver and chitosan. With the use of surface-enhanced fluorescence, nitrate levels in soil have been detected using hybrid nanoparticles made of gold and graphene oxide (Chen et al., 2016; Kumar et al., 2019; Akpınar et al., 2021).

The use of hybrid nanoparticles in agriculture is also limited due to a number of other factors, including the lack of standardized characterization methods, potential toxicity and eco-toxicity effects, regulatory and ethical issues, etc. To better understand how hybrid nanoparticles interact with plants, soil, and other biota and to improve their design, synthesis, use, and disposal, more research is required (Mittal et al., 2020; Xin et al., 2021; Singh et al., 2022).

1.4 NANOPARTICLES FOR PLANT DISEASE MANAGEMENT

Nanotechnology offers a promising new technique for managing plant diseases compared to traditional treatments and methods. Nanotechnology will make it possible to develop numerous new techniques for controlling diseases in greenhouses and arable land, improve disease diagnostics, and make a variety of new tools for molecularly manipulating pathogens and plants (Elmer et al., 2018). These advantages include improved efficacy, fewer input requirements, and lower eco-toxicity. Due to their incredibly tiny, even tinier than a virus particle and great reactivity, nanoparticles may have an impact on the pathogenicity of microbes. The findings of studies on crop plants using different NMs as protective agents are encouraging. At the right dosage, NMs showed the ability to reduce pathogen infection and boost plant development (Sharma et al., 2018; Fu et al., 2020; Younas et al., 2020). Additionally, efforts are required to commercialize applications based on NMs. Screening and crop-specific nanomaterial optimization can help with this. Modifying the characteristics and consistency of such developed NMs can increase their effectiveness. Therefore, it is necessary to create some unique nano-approaches that are very effective in absorbing agriculturally active substances, have better stability, and transport pesticides, fungicides, and fertilizers in a targeted manner. (Yadav et al., 2023).

Numerous researches have investigated the potential application of NMs as control agents for crop plants. By altering the physiological and biochemical functions of plants, NMs can improve plant development, yield, and stress resistance. For instance, *Bipolaris sorokiniana* infection in wheat plants was effectively controlled by the AgNPs potent antifungal activity. It was discovered under both in vitro and in vivo conditions that bsAgNPs had a promising capacity for biocontrol. (Mishra et al., 2014).

Another potential use of NMs is to increase agricultural yields by genetically modifying plants and delivering pesticides and genes to precise cell-level locations. The plant defense system can be strengthened using nanoarray and genomic technologies to defend against disease or manage stress (Younas et al., 2020). Forecasting and diagnosing pathogen epidemics, which are essential for controlling plant diseases, can be made more accessible by nanotechnology (Kashyap et al.,

2016). Researchers have made significant progress in this direction by developing nano-robots for use in a range of agricultural practices. These nano-robots have proven to be beneficial in diagnostic applications within crop fields (Pandey, 2018). It is necessary to investigate both the impact of nanosized particles on microbes and their use in the synthesis of pesticides to fully leverage the benefit of nanotechnology to protect and manage plant diseases. Due to their small size and strong reactivity, nanoparticles may also prove to be very useful diagnostic tools for plant diseases and infections as well as for analyzing pesticide residues (Sharma et al., 2018).

The data mentioned above unmistakably demonstrates the broad application of NPs in agriculture, particularly for controlling plant diseases.

1.4.1 Mechanisms of Action of Nanoparticles in Plant Disease Management

1.4.1.1 Antifungal Activity

Nanoparticles have many uses in agriculture, including their antifungal action against phytopathogenic fungi that significantly reduce crop output and quality (Akpinar et al 2021). The size, shape, concentration, chemical composition, and solubility of nanoparticles are only a few of the variables that can affect their antifungal effectiveness (Singh et al., 2015). The interaction of nanoparticles with the fungal cell wall and cell membrane, which are their main action targets, is influenced by all of these factors.

The fungal cell membrane and cell wall can be damaged by nanoparticles through a number of methods, including:

> Nanoparticles can penetrate into the fungal cell wall and cell membrane due to their high surface energy and small size, causing mechanical damage and leakage of cellular contents (Gajbhiye et al., 2009; Priyadarshini et al., 2021). Nanoparticles can generate reactive oxygen species (ROS) that can oxidize and degrade the fungal cell wall and cell membrane components, leading to loss of function and viability (Kim et al., 2012; Priyadarshini et al., 2021). Nanoparticles can release metal ions that can interact with the fungal cell wall and cell membrane components, altering their structure and function (Nisar et al., 2019; Priyadarshini et al., 2021). Nanoparticles can bind to the fungal cell wall and cell membrane components, such as proteins and enzymes, inhibiting their activity and disrupting their metabolic pathways (Jo et al., 2009).

The antifungal activity of nanoparticles can change depending on the type of nanoparticle and the type of fungus. Fungal hyphae and conidial germination have been found to be adversely affected by the use of AgNPs as antimicrobial agents (Kim et al., 2009). In comparison to synthetic fungicides, silver may be used to treat various plant infections with the greatest degree of safety due to its multiple mechanisms of blocking action against microbes (Min et al., 2009; Sharma et al., 2018). For instance, AgNPs have been found to have strong antifungal activity against the soil-borne fungus *Fusarium oxysporum*, which causes tomato wilt (Kim et al., 2012; Akpinar et al., 2021). By impairing the cell wall and membrane integrity of *F. oxysporum*, AgNPs are able to inhibit mycelial development, spore germination, biomass production, and conidial morphology.

Additionally, AgNPs can cause oxidative stress and produce silver ions that can disrupt the metabolism of fungi (Akpinar et al 2021). Also, the fungicidal activity of the AgNPs was evaluated against a variety of yeasts and molds, including *Candida krusei*, *C. albicans*, *C. glabrata*, *C. tropicalis,* and *Aspergillus brasiliensis*. The hybrid materials confirmed the powerful antifungal activities against these microorganisms. It has been demonstrated that *Botrytis cinerea* and *Penicillium expansum* are two fungal infections that zinc oxide nanoparticles (ZnO) have an antifungal effect. By suppressing the growth of conidiophores and conidia in *P. expansum*, the treatment with nanoparticles also caused deformation in the hyphae of *B. cinerea*, which further contributed to the destruction of fungal hyphae (Sharma et al., 2018).

1.4.1.2 Antibacterial Activity

The size, shape, surface charge, composition, and coating of nanoparticles are just a few of the variables that affect their antibacterial effectiveness. These factors influence how nanoparticles interact with bacterial cells and their biological components.

Nanoparticles have the remarkable ability to penetrate the membranes and cell walls of pathogens, causing significant damage to their internal structures and leading to the release of cytoplasmic contents. As a consequence of this disruption, the affected cells may experience either death or a loss of viability. For instance, *Escherichia coli* and *Pseudomonas aeruginosa*, have been found to have their cell walls and membranes damaged by AgNPs (Ramalingam et al., 2016). Ocsoy et al. (2013) conducted a study on Bacterial Spot, a severe disease affecting tomatoes caused by *Xanthomonas perforans*. They developed DNA-directed AgNPs grown on graphene oxide (GO) to target *X. perforans* cells. These composites displayed excellent antibacterial properties at very low concentrations (16 ppm) in culture, showing enhanced stability, stronger antibacterial activity, and improved adsorption properties. Min et al. (2009) also examined how AgNPs affected phytopathogenic fungus development. Tests on sclerotial germination growth demonstrated that the nanoparticles had a sizable amount of potential for suppression. In example, modest quantities of AgNPs most effectively reduced the sclerotial germination development of *S. sclerotiorum*. A microscopic examination demonstrates that hyphae exposed to AgNPs were severely compromised, causing the separation of hyphal wall layers and hyphal collapse. This experimental investigation suggested using AgNPs instead of insecticides to suppress scleotium-forming phytopathogenic fungi. Gupta & Chauhan (2016) conducted a study focusing on the antifungal properties of AgNPs against *Alternaria brassicicola*, the causative agent of Black Spot disease in cauliflower, radish, cabbage, and kale, leading to significant agricultural losses. They experimented with *A. brassicicola* grown on PDA in petri dishes using different quantities of produced AgNPs. AgNPs have the potential to be used as a fungicide to control plant diseases, as the study showed that they effectively slowed the growth of *A. brassicicola.*

Nanoparticles can produce reactive oxygen species (ROS), including hydroxyl radicals, hydrogen peroxide and superoxide anions, which can harm bacterial DNA, proteins, lipids, and enzymes (Holt & Bard, 2005; Radzig et al 2013). Nanoparticles can release metal ions such as zinc, copper, or silver ions. These ions can interact with bacterial proteins and impair their functionality. For instance, silver ions can bind to bacterial DNA and prevent it from being replicated and being translated (Rai et al., 2009; Hamad et al. 2020). Excellent antibacterial efficacy of AgNPs against methicillin-resistant gram + bacteria such *E. coli*, *P. aureginosa*, and *S. aureus* was described by Guzman et al., (2009). More research and development are required to successfully incorporate this technology into strategies for managing plant diseases.

1.4.1.3 Antiviral Activity

Controlling plant viral infections, which result in considerable losses in crop quality and yield, is one of the most significant difficulties in agriculture (Kang et al., 2005). By preventing the infection cycle, development, or transmission of plant viruses, NPs can act as antiviral agents. Pesticides or RNA-interference compounds that can target particular viral genes or proteins can also be delivered using NPs as nanocarriers. Additionally, NPs can operate as biostimulants or nanofertilizers to improve plant immunity and growth (Galdiero et al., 2011; Mishra et al., 2022; Rajani et al., 2022). Additionally, NPs can function as nanofertilizers or biostimulants to improve plant immunity and growth (Dutta et al., 2023).

Depending on their nature, size, shape, surface charge, and concentration, NPs can defend plants from viral infections through various ways. Several of the potential mechanisms include:

NPs have the ability to bind to viral particles and prevent them from entering or attaching to host cells. AgNPs antiviral efficacy against an RNA-based plant virus shown strong impacts.

For instance, tobacco mosaic virus (TMV) infection can be prevented by AgNPs (Ahsan, 2020; Al-Askar et al., 2023). Also, the antiviral activity of SiO2NPs and ZnONPs against the TMV was reported by Vargas-Hernandez et al., (2020).

NPs can impact viral nucleic acids or enzymes, which can prevent the viral genome from being replicated or translated (Cai et al., 2019; Mishra et al., 2022; Rajani et al., 2022) The antiviral activity of NPs and their capacity to activate plant defense mechanisms may be directly related to the efficacy of NPs. (Romero et al., 2011).

Plant viral infections like TMV, Potato virus Y (PVY), Sunhemp rosette virus (SHRV), Cauliflower mosaic virus (CaMV), Tomato yellow leaf curl virus (TYLCV), cucumber mosaic virus (CMV), and banana bunchy top virus (BBTV) have all been successfully controlled by NPs (Galdiero et al., 2011; Elbeshehy et al., 2015; El-Sawy et al., 2017; Vargas-Hernandez et al., 2020; Abdelkhalek et al., 2022; Dutta et al., 2023) The possible toxicity, environmental impact, stability, and control of NPs as a tool for managing plant diseases remain some of the challenges and restrictions. In order to optimize the synthesis, characterization, application, and evaluation of NPs for plant disease management, more study is required.

1.4.1.4 Induction of Plant Defense Responses

Many kinds of nanoparticles have drawn a lot of attention for their potential use in enhancing and protecting plants. The formation of reactive oxygen species, phytoalexins, changes in membrane permeability, stimulation of defense-related genes, and cell lysis are just a few of the plant defense responses that nanoparticles can trigger (Rasheed et al., 2022).

There is a lack of information on the techniques used by NMs to inactivate plant pathogens. Plant hormones such as salicylic acid (SA), jasmonic acid (JA) and ethylene and the expression of disease-related genes are all induced by NPs (Imada et al., 2016). When plants are stressed, their ROS levels rise, which can affect the integrity of pathogens. There is a lack of information regarding the techniques that NMs use to make plant pathogens inactive. Plant hormones like jasmonic acid (JA), salicylic acid (SA), ethylene, reactive oxygen species (ROS), and expression of genes associated to disease are all induced by NPs. When plants are stressed, their ROS levels rise, which can affect the integrity of pathogens (Imada et al., 2016; Cai et al., 2019; Derbalah & Elsharkawy, 2019).

1.4.1.5 Enhancement of Plant Growth and Development

In order to increase plant health and productivity, nanoparticles are potential NMs. Nanoparticles can interact with plants and affect their physiology, growth, and ability to withstand stress. Nanoparticles can increase photosynthesis, nutrient uptake, germination, and biomass accumulation in plants, which can increase plant growth and yield (Ali et al., 2021; Dutta et al., 2022). The results associated with nutritional elements containing NPs (i.e., Fe, Cu, Se, and Co) have demonstrated substantial scientific evidence of their efficacy in improving the plants micronutrients, which has been reflected in improved growth parameters and notable improvements at the physiological level (Ali et al., 2021).

Improvements in metabolite content, higher seed germination, shoot and root length, fruit production, and a significant rise in vegetative biomass in several crops have established the usefulness of nanoparticles in crop plants (Kole et al., 2013) Similar to how they affect plant growth and development, nanoparticles have been shown to increase the efficiency with which nitrogen is used and to increase up photosynthetic rate in a number of significant crops (Prasad et al., 2012)

Although nanoparticles have many positive effects, continuous exposure to them could negatively affect the environment. The effects of NPs on the growth and development of the plant were unclear according to researchers. The size, concentration, physical, and chemical characteristics of NPs and the type of plant all influence how they affect plants (Yadav & Yadav, 2018; Ali et al., 2021; Rajani et al., 2022).

1.4.2 Application Methods of Nanoparticles in Plant Disease Management

1.4.2.1 Foliar Application

The simplest method of protecting plants from pathogen attack is to apply nanoparticles to the soil directly, treat seeds prior to planting or storage, or spray foliar sprays that inhibit infections in a similar way to chemical pesticides (Sharma et al., 2018). The treatment can be coordinated to coincide with the crops particular nutrient requirements to promote growth at various developmental phases or lessen environmental risks (Tripathi et al., 2017; Gilbertson et al., 2020). Foliar-applications of NMs also have a significant potential to improve plant photosynthesis through abiotic stress protection, chlorophyll content enhancement, improved light absorption and photon utilization, scavenging reactive oxygen species, or facilitating electron transfer (Giraldo et al., 2014; Lowry et al., 2019). However, faster photosynthesis might result in less nutrition in the food that is produced (Myers et al., 2014). For the control of plant diseases, foliar application of NPs outperformed other metalloid and MNPs (such as Ag, ZnO, CeO_2, ZnO, MoO_2, Al_2O_3, Fe_2O_3, NiO, SiO_2, and TiO_2 NPs) (Elmer et al., 2016).

As a result, careful consideration of the advantages and disadvantages of foliar application of nanoparticles as well as their compatibility with other plant protection methods is required. More research is required to examine the long-term impacts of nanoparticles on plant health and productivity and improve the formulation and distribution of nanoparticles for foliar application.

1.4.2.2 Seed Treatment

Coating seeds with a variety of chemicals is a traditional strategy for enhancing seed handling and/or germination. Seeds Treatment is one way that nanoparticles are applied in plant disease management. In order to protect seeds from diseases, promote germination, and increase their ability to absorb nutrients, seeds are treated with nanoparticles. The growth of plant pathogens has been shown to be inhibited by nanoparticles with antimicrobial, antifungal, and antibacterial activities, including silver, copper, zinc oxide, titanium dioxide, and carbon nanotubes (Janmohammadi & Sabaghnia, 2015; Tripathi et al., 2017; Mahdy et al., 2020; Mittal et al., 2020). Additionally, nanoparticles can carry insecticides, fungicides, or other bioactive substances to the seeds or seedlings in a controlled manner (Mahdy et al., 2020; Mittal et al., 2020). The usage of chemical pesticides and fertilizers, which are harmful to both human health and the environment, can be decreased by seed treatment with nanoparticles. By modifying the physiological and biochemical processes of the plants, nanoparticles can also improve the seeds vigor, the germination rate, and the plants growth (Djiwanti & Kaushik, 2019; Mittal et al., 2020). Therefore, seed treatment with nanoparticles is a promising nanotechnological approach for sustainable agriculture and plant disease and pest management.

1.4.2.3 Soil Application

One method of using nanoparticles to control plant diseases is by soil application. Nanoparticles applied to the soil can increase nutrient availability, improve soil structure, and protect plants from infections that are carried by the soil (Zhou et al., 2020; Rajput et al., 2021). Conventional agrochemicals can be used more effectively and sustainably by adding nanoparticles to the soil while posing less risk to the environment. AgNPs, copper nanoparticles (CuNPs), zinc oxide nanoparticles (ZnO NPs), titanium dioxide nanoparticles (TiO2 NPs), and iron oxide nanoparticles (Fe3O4 NPs) are some of the kinds of nanoparticles that have been applied to soil to treat plant diseases (Sharma et al., 2018; Das & Pattanayak, 2020; Zhou et al., 2020; Ahmed et al., 2021; Rajput et al., 2021). These nanoparticles have demonstrated antimicrobial activity against a number of soil-borne diseases. However, there are significant potential risks and difficulties associated with applying nanoparticles to soil, including the possibility of toxicity to organisms that are not intended targets, accumulation and transformation in the soil, and unidentified long-term impacts on crop quality and soil health. In order to fully comprehend the fate and behavior of nanoparticles in soil, their interactions with plants and microbes, and their effects on the soil ecosystem and food safety, additional research is required.

1.5 NANOMATERIALS MEDIATED INSECT PEST MANAGEMENT

Agriculture, pharmaceuticals, electronics, and medicine are just a few industries that the development of nanotechnology, a promising area of interdisciplinary research could revolutionize (Hulkoti & Taranath, 2014; Prasad et al., 2014). The uses and potential benefits of formulations are enormous. For example, it is generally known that nanoparticles may effectively manage and control insect pests. They enable the control of insect pests by creating nanomaterial formulations that are based on insecticides and pesticides. Encapsulated nanoparticles are used to expose plants to nutrients continuously, and nanoparticles can edit plants DNA and genes to increase agricultural productivity (Shekhar et al., 2021). These materials can be engineered to have specific characteristics that make them practical for insect pest control. For instance, silica NMs protect the plant from silkworm attack by lowering the lipoprotein found in silkworm larvae. Wu & Liu (2008) have reported the use of Si nanoparticles loaded with validamycin for the controlled protection of host plants against various insecticides. Pesticides have been created using oil and water nanoemulsions, and these pesticides can be utilized to control a variety of insects that reduce crop productivity (Wang et al., 2007; Yadav et al., 2023). According to Patil et al. (2016), latex-fabricated bioactive Au nanoparticles are utilized to control insects by reducing the catalytic potential of trypsin, which acts as an insect protease. In another study, the mosquito species *Culex quinquefasciatus* was used to investigate the neem oil nanoemulsions larvicidal effects. The outcomes confirmed the larvicidal abilities of the nanoemulsion against *Cx. quinquefasciatus.* Additionally, it was shown that the lethal concentration (LC50) of the nanoemulsion against mosquito larvae dropped as the droplet size decreased. (Anjali et al., 2012). The findings of a different study point to the potential use of the green synthesis of AgNPs as the optimum environmentally friendly method for the management of the *Cx. quinquefasciatus* (Mondal et al., 2014). The accompanying figure elucidates and provides a comprehensive exposition of the diverse applications of NMs in pest control (Figure 1.2).

The nanoencapsulation of insecticides enhances their active ingredients stability, solubility, and controlled release. As a result, insecticides can be more effective, their environmental impact can be reduced, and they can be delivered to pests more effectively while minimizing exposure to beneficial microbiota (Chandrashekharaiah et al., 2015; Wani et al., 2019). A nanostructured material can release repellent compounds slowly, protecting against insects for an extended period. These materials can be incorporated into various products such as clothing, bed nets, or surfaces to repel insects and prevent their entry or contact (Abrantes et al., 2021; Pena et al., 2022). Insects use pheromones for communication, including mating and aggregation. Using nanotechnology, pheromone-loaded nanostructures can be developed that release these signals in a controlled manner. As a result, insect mating patterns can be disrupted, pests may become confused, and they may be unable to find suitable hosts or mates (Dantas et al., 2021; Kumar, 2021; Jasrotia et al., 2022). NMs can improve the effectiveness and delivery of biological control agents like nematodes or entomopathogenic fungi. These agents can be shielded from the environment by nanocarriers, which can also increase their stability and increase their ability to reach the intended pests (Kumar et al., 2021; Bihal et al., 2023). Biological control agents or pesticides can be delivered with precision thanks to nanotechnology. The effectiveness of pest management can be increased while limiting off-target impacts by functionalizing nanoparticles to interact precisely with insect pests or their habitats (Duhan et al., 2017; Srujana et al., 2022).

It is possible to create nano-based sensors to track and detect insect infestations. These sensors are capable of picking up damage signals, pest-related biomarkers, or insect pheromones. Farmers and pest management experts can use the collected data to make well-informed decisions about

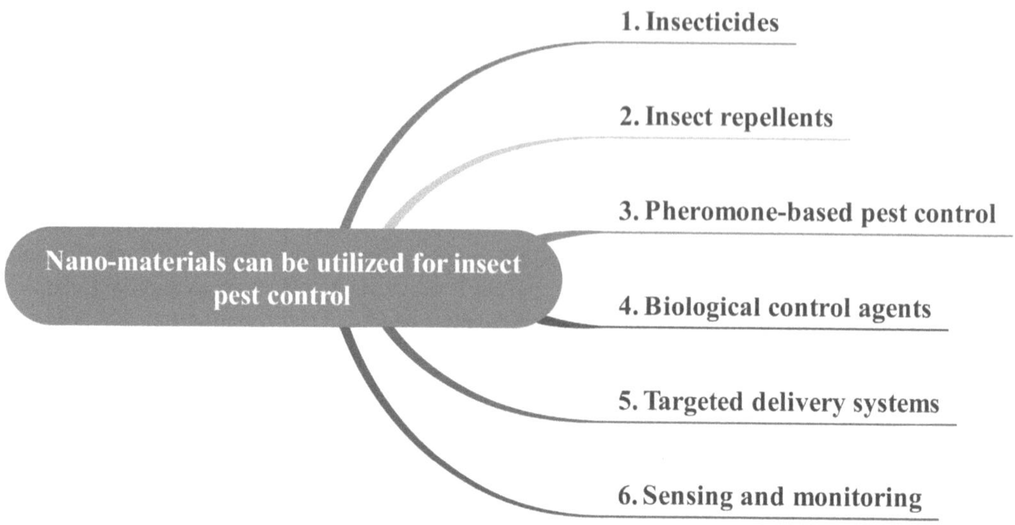

FIGURE 1.2 Nanomaterials can be utilized for insect pest control in several ways: Illustration depicting various applications of nanomaterials in insect pest control. (1) Insecticides: Nanoparticles can act as potent insect-killing agents. (2) Insect repellents: Nanomaterials can repel insects and deter them from the target area. (3) Pheromone-based pest control: Nanotechnology enables the controlled release of insect pheromones to disrupt mating and pest behaviour. (4) Biological control agents: Nanoparticles can enhance the efficacy of biological agents used for pest control. (5) Targeted delivery system: Nanomaterials facilitate precise delivery of pest control substances to the intended locations. (6) Sensing and monitoring: Nanosensors aid in detecting and monitoring pest populations for timely intervention (Modified from: Chandrashekharaiah et al., 2015; Gupta et al., 2021; Jasrotia et al., 2022; Srujana et al., 2022).

pest control measures (Elizabath et al., 2019; Javaid et al., 2023). The use of nanoinsecticides has a number of advantages over traditional insecticides. Higher surface adherence, improved active ingredient release, excellent targeting properties, higher stability and bioavailability, and improved environmental safety are all features of these products. Using nanoinsecticides can reduce the amount of pesticides required to control pests and lower the risk of exposure to people and other non-target species.

The potential of NMs makes precise and effective control of insect pests possible. However, the potential hazards and negative environmental effects that could result from using these materials must be considered. Therefore, using NMs for insect pest management has advantages and disadvantages, which are still being studied and developed to ensure their responsible and safe application (Figure 1.3).

1.5.1 Nano-Formulations Used in Insect Pest Management

Based on their structure, composition, and function, many forms of nano-formulations, such as nanoemulsions, nanoparticles, nano-capsules, nanogels can be used to manage insect pests.

Nanoemulsions: Surfactants stabilize the mixture of water and oil to create nanometer-sized droplets called nanoemulsions. They may improve an active ingredients solubility, stability, and bioavailability, such as insecticides or repellents (Campolo et al., 2020; Lobato et al., 2021). For instance, it has been demonstrated that nanoemulsions-based essential oils are more efficient against mosquitoes than traditional formulations (Lobato et al., 2021).

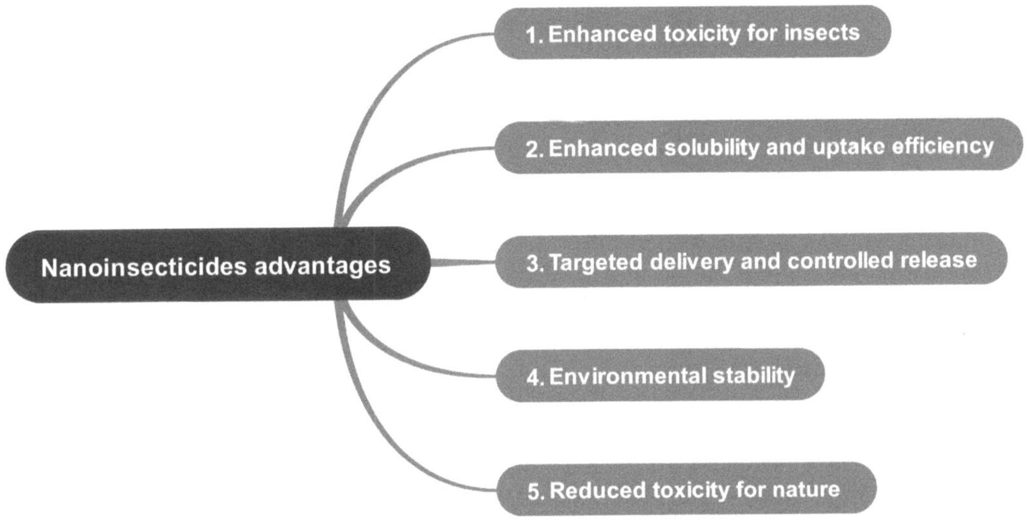

FIGURE 1.3 Advantages of nanoinsecticides: Illustration highlighting the advantages of nanoinsecticides. (1) Enhanced toxicity for insects: Nanomaterials increases the effectiveness of insecticides against pests. (2) Enhanced solubility and uptake efficiency: Nanoinsecticides exhibit improved solubility and are efficiently absorbed by target pests. (3) Targeted and controlled release: The efficacy of pesticides is increased through the exact delivery and controlled release made possible by nanotechnology. (4) Environmental stability: Nanoinsecticides offer increased environmental stability, reducing the risk of contamination. (5) Reduced toxicity for nature: Nanomaterials can be engineered to minimize adverse effects on non-target organisms and the environment (Modified from: Dantas et al., 2021; Gupta et al., 2021; Kumar, 2021; Jasrotia et al., 2022).

Nanoparticles: Solid materials with one or more dimensions in the nanometer range are known as nanoparticles. They may be created from metals, metal oxides, carbon, polymers, or biological components. They can have intrinsic insecticidal qualities or serve as carriers for active substances. For instance, numerous studies have shown that nanoparticles are efficient against a variety of insect pests, including aphids, whiteflies, and beetles (Gupta et al., 2021; Jasrotia et al., 2022).

Nano-capsules: The active substance is contained in the core of a nano-capsules spherical structure, which is shielded from oxidation or release by the shell. The stability, controlled release, and target specificity of active substances can all be improved. Insecticidal activity and persistence against caterpillars have been observed to enhance with the use of nano-capsules of neem oil, for instance (Amin et al., 2019; Dimetry et al., 2019).

Nanogels: Cross-linked polymer networks known as nanogels can expand and contract in response to environmental factors like pH, temperature, and light. They have the ability to encapsulate active substances and release them gradually (Zannat & Afroz, 2021). For instance, Bhagat et al. (2013) investigated the use of nano-gelled pheromone for the management of *Bactrocera dorsalis* is a major pest. They noted that nanogel might be used as a crop protection agent because of its excellent performance, positive safety profiles, long-lasting residual activity, and effectiveness in rainy climates.

1.5.2 The Mode of Action of Nano-Formulations against Insect Pests

Recently, there has been a lot of scientific interest in nanoparticles created by diverse synthesis processes as novel pesticides. Although numerous research have been conducted to assess their toxicity against various insect pests, it is still unclear exactly how they work against insects (Athanassiou et al., 2018; Jasrotiaet al., 2022).

NPs method of action against insect pests is influenced by their size, shape, surface charge, composition, and functionalization. The following are some of the potential mechanisms:

Physical damage: By penetrating through the integument or digestive tract, NPs can physically damage the insect cuticle or internal organs. This may lead to membrane disruption, mechanical abrasion, pore formation, or organelle damage (Athanassiou et al., 2018; Lade & Gogle, 2019). For instance, by absorbing lipids and water, silica nanopesticides physically damage and desiccation the insect cuticle (Hayles et al., 2017).

Oxidative stress: Reactive oxygen species (ROS) generated by nanoparticles (NPs) or their interference with the antioxidant defense system can induce oxidative stress in insect cells. This oxidative stress can lead to various detrimental effects, such as cell death, protein oxidation, DNA damage, and lipid peroxidation (Mishra & Panda 2021; Shahzad & Manzoor, 2021). For instance, by interfering with antioxidants and detoxifying enzymes, silver nanopesticides reduce acetylcholinesterase function, resulting in oxidative stress and cell death (Zaheer et al., 2022).

Hormonal disruption: By attaching to or mimicking the insect hormones or their receptors, NPs can alter the hormonal regulation of insect development and reproduction. This may lead to fertility, molting, metamorphosis, or abnormal development (Shang et al., 2019; Eskin & Nurullahoglu, 2022).

1.6 MAJOR CHALLENGES IN NANOTECHNOLOGY ADVANCEMENT

Numerous uses of nanotechnology in disease and pest management have been shown, highlighting the fields potential for innovation. However, there are also some major risks and challenges related to the use of nanotechnology in agriculture, such as ethical, social, and environmental concerns. Some of these challenges include the lack of knowledge about the long-term effects of nanoparticles on the environment, the high cost of producing nanoparticles, the difficulty in controlling the size and shape of nanoparticles, and the potential for nanoparticles to accumulate in soil and water (Kole et al., 2013; Mishra et al., 2017).

The lack of proper regulation and standardization of NMs and their products is one of the major challenges with applying nanotechnology to agriculture. What makes a nanomaterial, how to quantify its impacts and qualities, and how to guarantee its safety and quality are all vague concepts. Additionally, there is no uniform framework for evaluating and managing risks associated with NMs in agriculture, which could result in uneven and inadequate regulation across different geographic regions and nations (Chen et al., 2011, Prasad et al., 2017; Fu et al., 2020;).

The potential environmental impact of nanoparticles is another challenge when using nanotechnology to agriculture. Due to their small size, large surface area, high reactivity, and mobility, NMs may behave and ultimately end up in the environment differently than their bulk counterparts. NMs may interact with soil, water, air, plants, animals, and microbes in complicated and unforeseen ways. This interaction may negatively impact the environment, including toxicity, bioaccumulation, bioavailability, biotransformation, and biodegradation. Additionally, NMs have the potential to modify the structure and function of food webs and biogeochemical cycles, which could have an impact on the ecological balance and biodiversity of natural ecosystems (Prasad et al., 2017; Usman et al., 2020).

The ethical and social implications of using nanoparticles in agriculture are another challenge with using nanotechnology. The ownership and patenting of NMs and their products, the potential misuse or abuse of nanotechnology for criminal or military purposes, the respect for human dignity and animal welfare, and public perception and acceptance of nanotechnology are some ethical issues that nanotechnology may bring up (Chowdhury et al., 2017; Mwaanga, 2018).

Therefore, nanotechnology in agriculture is a promising but challenging field requiring more research, innovation, regulation and communication to ensure sustainable and responsible development. So, the future prospective for nanotechnology applications in pest and disease management should address and resolve some challenges necessary for further expansion.

1.7 CONCLUSIONS AND FUTURE PROSPECTS

The use of NMs in agriculture has considerable promise for overcoming significant issues with modern farming methods. For effective and environmentally friendly agricultural production, it is essential to focus on reducing the use of plant protection products, minimizing nutrient losses during fertilization, and increasing yields through enhanced nutrient management. Nevertheless, despite the potential benefits that nanotechnology may provide, the use of NMs in agriculture is still only a small part of the industry. The agricultural industry has not yet fully embraced the use of NMs on a broader scale, in comparison to other applications of nanotechnology, such as electronics or healthcare. Agricultural systems are expected to function better thanks to engineered nanoparticles, a novel technical advancement. Employing innovative nanotechnology approaches can revolutionize agricultural systems and improve agricultural productivity by using innovative tools to monitor soil parameters, fertiliser use, pest management, chemicals, pollutants, water use, and climate measurements.

Furthermore, nanotechnological interventions could help address agricultural challenges like biotic stresses and abiotic stresses, enhancing productivity. Nanotechnology has been identified as an innovative approach for sustainable agricultural pests and disease management. Also, the development of new-generation pesticides and their carriers, increasing the effectiveness of nutrient use through nanoformulations of fertilizers, breaking yield barriers with bionanotechnology, monitoring and controlling pests and diseases, comprehending molecular mechanisms of host-parasite interactions, preserving and packaging food and food additives, strengthening natural fibers, and removing contaminants from soil and water are some of the future prospects. However, the application of nanotechnology for environmentally responsible management of plant diseases and pests necessitates the collaboration of researchers from multidisciplinary backgrounds.

More research, development, and education are imperative to overcome obstacles and promote widespread adoption of nanotechnology in agriculture. By doing so, we can fully maximize the advantages for farmers, the environment, and global food security, ensuring a sustainable and secure food supply across the globe. As scientists investigate the potential of nanotechnology, we can anticipate substantial advancement in the upcoming years. We may anticipate a future in which food is safer, agriculture is more robust, and societal well-being is improved as we wisely harness the potential of nanotechnology.

REFERENCES

Abdelkhalek, A., El-Gendi, H., Alotibi, F. O., Al-Askar, A. A., Elbeaino, T., Behiry, S. I., & Moawad, H. (2022). *Ocimum basilicum*-mediated synthesis of silver nanoparticles induces innate immune responses against cucumber mosaic virus in squash. *Plants, 11*(20), 2707.

Abrantes, D. C., Rogerio, C. B., de Oliveira, J. L., Campos, E. V., de Araujo, D. R., Pampana, L. C., & Fraceto, L. F. (2021). Development of a mosquito repellent formulation based on nanostructured lipid carriers. *Frontiers in Pharmacology*, 12, 760682.

Abubakar, Y., Tijjani, H., Egbuna, C., Adetunji, C. O., Kala, S., Kryeziu, T. L., ... & Patrick-Iwuanyanwu, K. C. (2020). Pesticides, history, and classification. In *Natural remedies for pest, disease and weed control* (pp. 29–42). Academic Press.

Ahmed, B., Rizvi, A., Ali, K., Lee, J., Zaidi, A., Khan, M. S., & Musarrat, J. (2021). Nanoparticles in the soil plant system: a review. *Environmental Chemistry Letters*, 19, 1545–1609.

Ahsan, T. (2020). Biofabrication of silver nanoparticles from *Pseudomonas fluorescens* to control tobacco mosaic virus. *Egyptian Journal of Biological Pest Control*, 30(1), 66.

Akpinar, I., Unal, M., & Sar, T. (2021). Potential antifungal effects of silver nanoparticles (AgNPs) of different sizes against phytopathogenic *Fusarium oxysporum* f. sp. radicis-lycopersici (FORL) strains. *SN Applied Sciences*, 3, 506.

Aktar, W., Sengupta, D., & Chowdhury, A. (2009). Impact of pesticides use in agriculture: their benefits and hazards. *Interdisciplinary Toxicology*, *2*(1), 1.

Al-Askar, A. A., Aseel, D. G., El-Gendi, H., Sobhy, S., Samy, M. A., Hamdy, E., ... & Abdelkhalek, A. (2023). Antiviral activity of biosynthesized silver nanoparticles from Pomegranate (*Punica granatum* L.) peel extract against tobacco mosaic virus. *Plants*, *12*(11), 2103.

Ali, S., Mehmood, A., & Khan, N. (2021). Uptake, translocation, and consequences of nanomaterials on plant growth and stress adaptation. *Journal of Nanomaterials*, 20 (21), 1–17.

Amin, A. R. H., Bayoumi, A. E. D., Dimetry, N. Z., & Youssef, D. A. (2019). Efficiency of Nano-formulations of neem and peppermint oils on the bionomics and enzymatic activities of *Agrotis ipsilon* larvae (Lepidoptera: Noctuidae). *International Journal of Natural Resource Ecology and Management*, *4*, 102–111.

An, C., Sun, C., Li, N., Huang, B., Jiang, J., Shen, Y., ... & Wang, Y. (2022). Nanomaterials and nanotechnology for the delivery of agrochemicals: strategies towards sustainable agriculture. *Journal of Nanobiotechnology*, *20*(1), 1–19.

Anjali, C. H., Sharma, Y., Mukherjee, A., & Chandrasekaran, N. (2012). Neem oil (*Azadirachta indica*) nanoemulsiona potent larvicidal agent against *Culex quinquefasciatus*. *Pest Management Science*, *68*(2), 158–163.

Anonymous. (2010). New approaches needed to gauge safety of nanotech-based pesticides. ScienceDaily. Available from: Accessed on 25/May/2023.

Athanassiou, C. G., Kavallieratos, N. G., Benelli, G., Losic, D., Usha Rani, P., & Desneux, N. (2018). Nanoparticles for pest control: current status and future perspectives. *Journal of Pest Science*, *91*, 1–15.

Barik, T. K., Sahu, B., & Swain, V. (2008). Nano-silica from medicine to pest control. *Parasitology Research*, *103*, 253–258.

Bhagat, D., Samanta, S. K., & Bhattacharya, S. (2013). Efficient management of fruit pests by pheromone nanogels. *Scientific Reports*, *3*(1), 1294.

Bihal, R., Al-Khayri, J. M., Banu, A. N., Kudesia, N., Ahmed, F. K., Sarkar, R., ... & Abd-Elsalam, K. A. (2023). Entomopathogenic fungi: an eco-friendly synthesis of sustainable nanoparticles and their nanopesticide properties. *Microorganisms*, *11*(6), 1617.

Cai, L., Liu, C., Fan, G., Liu, C., & Sun, X. (2019). Preventing viral disease by ZnONPs through directly deactivating TMV and activating plant immunity in Nicotiana benthamiana. *Environmental Science: Nano*, *6*(12), 3653–3669.

Campolo, O., Giunti, G., Laigle, M., Michel, T., & Palmeri, V. (2020). Essential oil-based nanoemulsions: effect of different surfactants, sonication and plant species on physicochemical characteristics. *Industrial Crops and Products*, *157*, 112935.

Carriger, J. F., Rand, G. M., Gardinali, P. R., Perry, W. B., Tompkins, M. S., & Fernandez, A. M. (2006). Pesticides of potential ecological concern in sediment from South Florida canals: an ecological risk prioritization for aquatic arthropods. *Soil & Sediment Contamination*, *15*(1), 21–45.

Chandrashekharaiah, M., Kandakoor, S. B., Basana Gowda, G., Kammar, V., & Chakravarthy, A. K. (2015). Nanomaterials: a review of their action and application in pest management and evaluation of DNA-tagged particles. In *New Horizons in Insect Science: Towards Sustainable Pest Management* (pp. 113–126).

Chen, H., & Yada, R. (2011). Nanotechnologies in agriculture: new tools for sustainable development. *Trends in Food Science & Technology*, *22*(11), 585–594.

Chen, J., Sun, L., Cheng, Y., Lu, Z., Shao, K., Li, T., ... & Han, H. (2016). Graphene oxide-silver nanocomposite: novel agricultural antifungal agent against *Fusarium graminearum* for crop disease prevention. *ACS Applied Materials & Interfaces*, *8*(36), 24057–24070.

Chowdhury, P., Gogoi, M., Borchetia, S., & Bandyopadhyay, T. (2017). Nanotechnology applications and intellectual property rights in agriculture. *Environmental Chemistry Letters*, *15*, 413–419.

Dantas, J., Motta, I., Vidal, L., Bilio, J., Pupe, J. M., Veiga, A., ... & Albuquerque, E. V. (2021). A comprehensive review of the Coffee Leaf Miner *Leucoptera Coffeella* (Lepidoptera: Lyonetiidae), with special regard to neotropical impacts, pest management and control. *Insects*, *12*(12), 1130.

Das, S., & Pattanayak, S. (2020). Nanotechnological approaches in sustainable agriculture and plant disease management. In Das, S. K. (Ed.), *Organic Agriculture* (pp. 13–30). IntechOpen: London, UK.

Delshadi, R., Bahrami, A., Tafti, G., Barba, J., & Williams, L. (2020). Micro and nano-encapsulation of vegetable and essential oils to develop functional food products with improved nutritional profiles. *Trends in Food Science & Technology, 104,* 72–83.

Derbalah, A. S. H., & Elsharkawy, M. M. (2019). A new strategy to control Cucumber mosaic virus using fabricated NiO-nanostructures. *Journal of Biotechnology*, *306*, 134–141.

Dimetry, N. Z., Amin, A. H., Bayoumi, A. E., & Abdel-Raheem, M. A. (2019). Comparative toxicity of neem and peppermint oils Nano formulations against *Agrotis ipsilon* (Hufn.) larvae (Lepidoptera: Noctuidae). *Journal of Botanical Research*, *1*(1), 13–19.

Djiwanti, S. R., & Kaushik, S. (2019). Nanopesticide: future application of nanomaterials in plant protection. In *Plant Nanobionics: Volume 2, Approaches in Nanoparticles, Biosynthesis, and Toxicity* (pp. 255–298).

Duhan, J. S., Kumar, R., Kumar, N., Kaur, P., Nehra, K., & Duhan, S. (2017). Nanotechnology: The new perspective in precision agriculture. *Biotechnology Reports*, *15*, 11–23.

Dutta, P., Kumari, A., & Mahanta, M. (2022). *Chitosan Nanoparticle: Synthesis, Characterization, and Use as Plant Health Materials*. *2*, ppt 386.

Dutta, P., Kumari, A., Mahanta, M., Upamanya, G. K., Heisnam, P., Borua, S., ... & Vijayreddy, D. (2023). Nanotechnological approaches for management of soil-borne plant pathogens. *Frontiers in Plant Science*, *14*, 1136233.

Elbeshehy, E. K., Elazzazy, A. M., & Aggelis, G. (2015). Silver nanoparticles synthesis mediated by new isolates of Bacillus spp., nanoparticle characterization and their activity against Bean Yellow Mosaic Virus and human pathogens. *Frontiers in Microbiology*, *6*, 453.

Elizabath, A., Babychan, M., Mathew, A. M., & Syriac, G. M. (2019). Application of nanotechnology in agriculture. *International Journal of Pure and Applied Bioscience*, *7*(2), 131–139.

Elmer, W. H., & White, J. C. (2016). The use of metallic oxide nanoparticles to enhance growth of tomatoes and eggplants in disease infested soil or soilless medium. *Environmental Science: Nano*, *3*(5), 1072–1079.

Elmer, W., Ma, C., & White, J. (2018). Nanoparticles for plant disease management. *Current Opinion in Environmental Science & Health*, *6*, 66–70.

El-Naggar, M. E., Abdelsalam, N. R., Fouda, M. M., Mackled, M. I., Al-Jaddadi, M. A., Ali, H. M., ... & Kandil, E. E. (2020). Soil application of nano silica on maize yield and its insecticidal activity against some stored insects after the post-harvest. *Nanomaterials*, *10*(4), 739.

El-Sawy, M. M., Elsharkawy, M. M., Abass, J. M., & Hagag, E. S. (2017). Inhibition of tomato yellow leaf curl virus by zingiber off cinaleand mentha longifolia. *International Journal of Antivirals & Anti retrovirology*, *1*(1), 1–6.

Eskin, A., & Nurullahoglu, Z. U. (2022). Effects of zinc oxide nanoparticles (ZnO NPs) on the biology of *Galleria mellonella* L.(Lepidoptera: Pyralidae). *The Journal of Basic and Applied Zoology*, *83*(1), 54.

Esquivel, D. M., Wajnberg, E., Do Nascimento, F. S., Pinho, M. B., de Barros, H. L., & Eizemberg, R. (2007). Do geomagnetic storms change the behaviour of the stingless bee guirucu (*Schwarziana quadripunctata*)?. *Naturwissenschaften*, *94*, 139–142.

FAO. (2019). *New Standards to Curb the Global Spread of Plant Pests and Diseases*. Food and Agriculture Organization of the United Nations.

FAO. (2020). International Year of Plant Health 2020. Protecting Plants, Protecting Life. Available from: accessed on 06/01/2023.

FDA. (2011). *Considering Whether an FDA-Regulated Product Involves the Application of Nanotechnology*. USA: Federal Drug Administration. Available from. Accessed on 06/01/2023.

Feynman, R. P. (1959). There is plenty of room at the bottom, an invitation to enter a new field of physics. *Report of Annual Meeting of the American Physical Society, California*. Available from: ppt 334

Fu, L., Wang, Z., Dhankher, O. P., & Xing, B. (2020). Nanotechnology as a new sustainable approach for controlling crop diseases and increasing agricultural production. *Journal of Experimental Botany*, *71*(2), 507–519.

Gajbhiye, M., Kesharwani, J., Ingle, A., Gade, A., Rai, M. (2009). Fungus mediated synthesis of silver nanoparticles and its activity against pathogenic fungi in combination of fluconazole. *Nanomedicine*, *5*(4), 282–286.

Galdiero, S., Falanga, A., Vitiello, M., Cantisani, M., Marra, V., & Galdiero, M. (2011). Silver nanoparticles as potential antiviral agents. *Molecules*, *16*(10), 8894–8918.

Gao, Y., Lei, Z., & Reitz, S. R. (2012). Western flower thrips resistance to insecticides: detection, mechanisms and management strategies. *Pest Management Science*, *68*(8), 1111–1121.

Gilbertson, L. M., Pourzahedi, L., Laughton, S., Gao, X., Zimmerman, J. B., Theis, T. L., ... & Lowry, G. V. (2020). Guiding the design space for nanotechnology to advance sustainable crop production. *Nature Nanotechnology*, *15*(9), 801–810.

Giraldo, J. P., Landry, M. P., Faltermeier, S. M., McNicholas, T. P., Iverson, N. M., Boghossian, A. A., ... & Strano, M. S. (2014). Plant nanobionics approach to augment photosynthesis and biochemical sensing. *Nature Materials*, *13*(4), 400–408.

Goswami, A., Roy, I., Sengupta, S., & Debnath, N. (2010). Novel applications of solid and liquid formulations of nanoparticles against insect pests and pathogens. *Thin Solid Films*, *519*(3), 1252–1257.

Graily-Moradi, F., & Lajayer, B. (2021). Nanoinsecticides: preparation, application, and mode of action. In N. Saglam, et al. (Eds.), *Nanotechnology Applications in Health and Environmental Sciences, Nanotechnology in the Life Sciences.* Switzerland: Springer Nature.

Gupta, D., & Chauhan, P. (2016, April). Fungicidal activity of silver nanoparticles against *Alternaria brassicicola. AIP Conference Proceedings*, *3*(1), 1724,

Gupta, A., Sahu, P. K., & Tiwari, R. K. (2021). Nanotechnology in insect pest management. *Molecular Approaches for Sustainable Insect Pest Management*, 377–394.

Guzman, M. G., Dille, J., & Godet, S. (2009). Synthesis of silver nanoparticles by chemical reduction method and their antibacterial activity. *International Journal of Chemical and Biomolecular Engineering*, *2*(3), 104–111.

Hamad, A., Khashan, K. S., & Hadi, A. (2020). Silver nanoparticles and silver ions as potential antibacterial agents. *Journal of Inorganic and Organometallic Polymers and Materials*, *30*(12), 4811–4828.

Hayles, J., Johnson, L., Worthley, C., & Losic, D. (2017). Nanopesticides: a review of current research and perspectives. *New Pesticides and Soil Sensors*, El Sevier, Vol (1) 193–225.

Hernandez-Giottonini, K. Y., Rodriguez-Cordova, R. J., Gutierrez-Valenzuela, C. A., Penunuri-Miranda, O., Zavala-Rivera, P., Guerrero-German, P., & Lucero-Acuna, A. (2020). PLGA nanoparticle preparations by emulsification and nanoprecipitation techniques: effects of formulation parameters. *RSC Advances*, *10*(8), 4218–4231.

Hernandez-Hernandez, H., Gonzalez-Morales, S., Benavides-Mendoza, A., Ortega-Ortiz, H., Cadenas-Pliego, G., & Juarez-Maldonado, A. (2018). Effects of chitosan PVA and Cu nanoparticles on the growth and antioxidant capacity of tomato under saline stress. *Molecules*, *23*(1), 178.

Holt, K. B., & Bard, A. J. (2005). Interaction of silver (I) ions with the respiratory chain of *Escherichia coli*: an electrochemical and scanning electrochemical microscopy study of the antimicrobial mechanism of micromolar Ag+. *Biochemistry*, *44*(39), 13214–13223.

Horrigan, L., Lawrence, R. S., & Walker, P. (2002). How sustainable agriculture can address the environmental and human health harms of industrial agriculture. *Environmental Health Perspectives*, *110*(5), 445–456.

Hulkoti, N. I., & Taranath, T. C. (2014). Biosynthesis of nanoparticles using microbes a review. *Colloids and surfaces B: Biointerfaces*, *121*, 474–483.

Imada, K., Sakai, S., Kajihara, H., Tanaka, S., & Ito, S. (2016). Magnesium oxide nanoparticles induce systemic resistance in tomato against bacterial wilt disease. *Plant Pathology*, *65*(4), 551–560.

Janmohammadi, M., & Sabaghnia, N. (2015). Effect of pre-sowing seed treatments with silicon nanoparticles on germinability of sunflower (*Helianthus annuus*). *Botanica Lithuanica*, *21*(1), 13–21.

Jasrotia, P., Nagpal, M., Mishra, C. N., Sharma, A. K., Kumar, S., Kamble, U., ... & Singh, G. P. (2022). Nanomaterials for postharvest management of insect pests: current state and future perspectives. *Frontiers in Nanotechnology*, 3.

Javaid, M., Haleem, A., Khan, I. H., & Suman, R. (2023). Understanding the potential applications of artificial intelligence in agriculture sector. *Advanced Agrochem*, *2*(1), 15–30.

Jeevanandam, J., Barhoum, A., Chan, Y. S., Dufresne, A., & Danquah, M. K. (2018). Review on nanoparticles and nanostructured materials: history, sources, toxicity and regulations. *Beilstein Journal of Nanotechnology*, *9*(1), 1050–1074.

Jo, Y. K., Kim, B. H., & Jung, G. (2009). Antifungal activity of silver ions and nanoparticles on phytopathogenic fungi. *Plant Disease*, *93*(10), 1037–1043.

Johnson, M. S., Sajeev, S., & Nair, R. S. (2021, March). Role of Nanosensors in agriculture. In *2021 International Conference on Computational Intelligence and Knowledge Economy (ICCIKE)* (pp. 58–63). IEEE.

Kah, M., Johnston, L. J., Kookana, R. S., Bruce, W., Haase, A., Ritz, V., ... & Gubala, V. (2021). Comprehensive framework for human health risk assessment of nanopesticides. *Nature Nanotechnology*, *16*(9), 955–964.

Kang, B. C., Yeam, I., & Jahn, M. M. (2005). Genetics of plant virus resistance. *Annual Review of Phytopathology*, *43*, 581–621.

Kashyap, P. L., Rai, P., Sharma, S., Chakdar, H., Kumar, S., Pandiyan, K., & Srivastava, A. K. (2016). Nanotechnology for the detection and diagnosis of plant pathogens. *Nanoscience in Food and Agriculture*, *2*, 253–276.

Khan, F., Pandey, P., & Upadhyay, T. K. (2022). Applications of nanotechnology-based agrochemicals in food security and sustainable agriculture: an overview. *Agriculture*, *12*(10), 1672.

Khooshe-Bast, Z., Najmeh, S., Ghaffari-Moghaddam, M., & Mirshekar, A. (2016). Insecticidal effects of zinc oxide nanoparticles and Beauveria bassiana TS11 on Trialeurodes vaporariorum (Westwood, 1856) (Hemiptera: Aleyrodidae). *Acta Agriculturae Slovenica*, 107(2), 299–309.

Kim, S. W., Jung, J. H., Lamsal, K., Kim, Y. S., Min, J. S., & Lee, Y. S. (2012). Antifungal effects of silver nanoparticles (AgNPs) against various plant pathogenic fungi. *Mycobiology*, *40*(1), 53–58.

Kim, S. W., Kim, K. S., Lamsal, K., Kim, Y. J., Kim, S. B., Jung, M., ... & Lee, Y. S. (2009). An in vitro study of the antifungal effect of silver nanoparticles on oak wilt pathogen Raffaelea sp. *Journal of Microbiology and Biotechnology*, *19*(8), 760–764.

Kole, C., Kole, P., Randunu, K. M., Choudhary, P., Podila, R., Ke, P. C., ... & Marcus, R. K. (2013). Nanobiotechnology can boost crop production and quality: first evidence from increased plant biomass, fruit yield and phytomedicine content in bitter melon (*Momordica charantia*). *BMC Biotechnology*, *13*(1), 1–10.

Kose, Y., Suvaci, E., & Atli, B. (2023). Improving properties of biodegradable chitosan/PVA composite polymers via novel designed ZnO particles. *Journal of the Australian Ceramic Society*, *59*(1), 245–257.

Kumar, A., Gupta, K., Dixit, S., Mishra, K., & Srivastava, S. (2019). A review on positive and negative impacts of nanotechnology in agriculture. *International Journal of Environmental Science and Technology*, *16*, 2175–2184.

Kumar, M. N. R. (2000). A review of chitin and chitosan applications. *Reactive and Functional Polymers*, *46*(1), 1–27.

Kumar, P., Pandhi, S., Mahato, D. K., Kamle, M., & Mishra, A. (2021). Bacillus-based nano-bioformulations for phytopathogens and insect pest management. *Egyptian Journal of Biological Pest Control*, *31*, 1–11.

Kumar, R. (2021). Nanotechnology: advancement for agricultural sustainability. In P. Singh et al. (Eds). *Plant-Microbes-Engineered Nanoparticles (PM-ENPs) Nexus in Agro-Ecosystems: Understanding the Interaction of Plant, Microbes and Engineered Nanoparticles (ENPS)* (pp. 19–27). Cham: Springer. Springer International Publishing.

Kumar, S., Chung-Jung, T., & Ruth, N., (2000). Factors enhancing protein thermostability. *Protein Engineering, Design and Selection, 13*(3), 179–191.

Lade, B. D., & Gogle, D. P. (2019). Nano-biopesticides: synthesis and applications in plant safety. *Nanobiotechnology Applications in Plant Protection, 2*, 169–189.

Li, L., Xu, Z., Kah, M., Lin, D., & Filser, J. (2019). Nanopesticides: a comprehensive assessment of environmental risk is needed before widespread agricultural application. *Environmental Science & Technology, 3*(14), 7923–7924.

Li, Y., Xu, X., Wu, Y., Zhuang, J., Zhang, X., Zhang, H., & Liu, Y. (2020). A review on the effects of carbon dots in plant systems. *Materials Chemistry Frontiers*, *4*(2), 437–448.

Lobato Rodrigues, A. B., Martins, R. L., Rabelo, E. D. M., Tomazi, R., Santos, L. L., Brandao, L. B., ... & de Almeida, S. S. M. D. S. (2021). Development of nanoemulsions based on *Ayapana triplinervis* essential oil for the control of Aedes aegypti larvae. *PLoS One*, *16*(7), e0254225.

Lowry, G. V., Avellan, A., & Gilbertson, L. M. (2019). Opportunities and challenges for nanotechnology in the agri-tech revolution. *Nature Nanotechnology*, *14*(6), 517 522.

Mahdy, A. M., Sherif, F. K., Elkhatib, E. A., Fathi, N. O., & Ahmed, M. H. (2020). Seed priming in nanoparticles of water treatment residual can increase the germination and growth of cucumber seedling under salinity stress. *Journal of Plant Nutrition*, *43*(12), 1862–1874.

Meyer, W. L., Gurman, P., Stelinski, L. L., & Elman, N. M. (2015). Functional nano-dispensers (FNDs) for delivery of insecticides against phytopathogen vectors. *Green Chemistry*, *17*(8), 4173–4177.

Min, J. S., Kim, K. S., Kim, S. W., Jung, J. H., Lamsal, K., Bin Kim, S., ... & Lee, Y. S. (2009). Effects of colloidal silver nanoparticles on sclerotium-forming phytopathogenic fungi. *Plant Pathology Journal*, *25*(4), 376–380.

Mishra, M., & Panda, M. (2021). Reactive oxygen species: the root cause of nanoparticle-induced toxicity in *Drosophila melanogaster*. *Free Radical Research*, *55*(8), 919–935.

Mishra, P., Kumari, S., Saini, P., & Meena, R. K. (2022). Role of nanotechnology in management of plant viral diseases. *Materials Today: Proceedings*, *69*, 1–10.

Mishra, S., Keswani, C., Abhilash, P. C., Fraceto, L. F., & Singh, H. B. (2017). Integrated approach of agri-nanotechnology: challenges and future trends. *Frontiers in Plant Science*, *8*, 471.

Mishra, S., Singh, B. R., Singh, A., Keswani, C., Naqvi, A. H., & Singh, H. B. (2014). Biofabricated silver nanoparticles act as a strong fungicide against *Bipolaris sorokiniana* causing spot blotch disease in wheat. *Plos One*, *9*(5), e97881.

Mittal, D., Kaur, G., Singh, P., Yadav, K., & Ali, S. A. (2020). Nanoparticle-based sustainable agriculture and food science: recent advances and future outlook. *Frontiers in Nanotechnology*, *2*, 579954.

Mondal, N. K., Chowdhury, A., Dey, U., Mukhopadhya, P., Chatterjee, S., Das, K., & Datta, J. K. (2014). Green synthesis of silver nanoparticles and its application for mosquito control. *Asian Pacific Journal of Tropical Disease*, *4*, 204 210.

Mukherjee, A., Majumdar, S., Servin, A. D., Pagano, L., Dhankher, O. P., & White, J. C. (2016). Carbon nanomaterials in agriculture: a critical review. *Frontiers in Plant Science*, *7*, 172.

Mwaanga, P. (2018). Risks, uncertainties, and ethics of nanotechnology in agriculture. *New Visions in Plant Science*, *22*(3). 225 pages.

Myers, S. S., Zanobetti, A., Kloog, I., Huybers, P., Leakey, A. D., Bloom, A. J., ... & Usui, Y. (2014). Increasing CO2 threatens human nutrition. *Nature*, *510*(7503), 139–142.

Ngwuluka, N. C., Abu-Thabit, N. Y., Uwaezuoke, O. J., Erebor, J. O., Ilomuanya, M. O., Mohamed, R. R., ... & Ebrahim, N. A. (2021). Natural polymers in micro-and nanoencapsulation for therapeutic and diagnostic applications: part II-polysaccharides and proteins. *Nano- and Microencapsulation Techniques and Applications*, *8*, 9.

Nikmaram, N., Roohinejad, S., Hashemi, S., Koubaa, M., Barba, J., Abbaspourrad, A., & Greiner, R., (2017). Emulsion-based systems for fabrication of electrospun nanofibers: food, pharmaceutical and biomedical applications. *RSC Advances, 7,* 28951–28964.

Nisar, P., Ali, N., Rahman, L., Ali, M., & Shinwari, Z. K. (2019). Antimicrobial activities of biologically synthesized metal nanoparticles: an insight into the mechanism of action. *Journal of Biological Inorganic Chemistry*, *24*, 929–941.

Nithya, V. (2011). Nanotechnology applications in drug development. *Inventi Rapid: Pharm Tech*, 2(1). 16

Ocsoy, I., Paret, M. L., Ocsoy, M. A., Kunwar, S., Chen, T., You, M., & Tan, W. (2013). Nanotechnology in plant disease management: DNA-directed silver nanoparticles on graphene oxide as an antibacterial against *Xanthomonas perforans*. *ACS Nano*, *7*(10), 8972–8980.

Oerke, E. (2006). Crop losses to pests. *The Journal of Agricultural Science*, *144*(1), 31–43. 10.1017/S0021859605005708

Pan, X., Guo, X., Zhai, T., Zhang, D., Rao, W., Cao, F., & Guan, X. (2023). Nanobiopesticides in sustainable agriculture: developments, challenges, and perspectives. *Environmental Science: Nano, 10,* 41–61

Pandey, G. (2018). Challenges and future prospects of agri-nanotechnology for sustainable agriculture in India. *Environmental Technology & Innovation*, *11*, 299–307.

Pascoli, M., Lopes-Oliveira, P. J., Fraceto, L. F., Seabra, A. B., & Oliveira, H. C. (2018). State of the art of polymeric nanoparticles as carrier systems with agricultural applications: a minireview. *Energy, Ecology and Environment*, *3*, 137–148.

Patil, C. D., Borase, H. P., Suryawanshi, R. K., & Patil, S. V. (2016). Trypsin inactivation by latex fabricated gold nanoparticles: a new strategy towards insect control. *Enzyme and Microbial Technology*, *92*, 18–25.

Pena, G. A., da Costa Lopes, A. S., de Morais, S. H. S., do Nascimento, L. D., Dos Santos, F. R. R., da Costa, K. S., ... & Lameira, J. (2022). Host-guest inclusion complexes of natural products and nanosystems: applications in the development of repellents. *Molecules*, *27*(8), 2519.

Pradeep. (2007). *NANO: The Essentials: Understanding Nanoscience and Nanotechnology*. 1st ed. New Delhi McGraw-Hill Offices: Tata McGraw-Hill Publishing Company Limited, 435.

Prasad, R., Bhattacharyya, A., & Nguyen, Q. D. (2017). Nanotechnology in sustainable agriculture: recent developments, challenges, and perspectives. *Frontiers in Microbiology*, *8*, 1014.

Prasad, R., Kumar, V., & Prasad, K. S. (2014). Nanotechnology in sustainable agriculture: present concerns and future aspects. *African Journal of Biotechnology*, *13*(6), 705–713.

Prasad, T. N. V. K. V., Sudhakar, P., Sreenivasulu, Y., Latha, P., Munaswamy, V., Reddy, K. R., ... & Pradeep, T. (2012). Effect of nanoscale zinc oxide particles on the germination, growth and yield of peanut. *Journal of Plant Nutrition*, *35*(6), 905–927.

Priyadarshini, E., Priyadarshini, S. S., Cousins, B. G., & Pradhan, N. (2021). Metal-Fungus interaction: review on cellular processes underlying heavy metal detoxification and synthesis of metal nanoparticles. *Chemosphere*, *274*, 129976.

Radzig, M. A., Nadtochenko, V. A., Koksharova, O. A., Kiwi, J., Lipasova, V. A., & Khmel, I. A. (2013). Antibacterial effects of silver nanoparticles on gram-negative bacteria: influence on the growth and biofilms formation, mechanisms of action. *Colloids and Surfaces B: Biointerfaces*, *102*, 300–306.

Rai, M., Yadav, A., & Gade, A. (2009). Silver nanoparticles as a new generation of antimicrobials. *Biotechnology Advances*, *27*(1), 76–83.

Rajani, P. M., Kumari, S., Saini, P., & Meena, R. K. (2022). Role of nanotechnology in management of plant viral diseases. *Materials Today Proceeding*. pp. 1–10

Rajput, V. D., Singh, A., Singh, V. K., Minkina, T. M., & Sushkova, S. (2021). Impact of nanoparticles on soil resource. In *Nanomaterials for Soil Remediation,13*(2), (pp. 65–85.

Ramalingam, B., Parandhaman, T., & Das, S. K. (2016). Antibacterial effects of biosynthesized silver nanoparticles on surface ultrastructure and nanomechanical properties of gram-negative bacteria viz. Escherichia coli and Pseudomonas aeruginosa. *ACS Applied Materials & Interfaces*, *8*(7), 4963–4976.

Rasheed, A., Li, H., Tahir, M. M., Mahmood, A., Nawaz, M., Shah, A. N., ... & Wu, Z. (2022). The role of nanoparticles in plant biochemical, physiological, and molecular responses under drought stress: a review. *Frontiers in Plant Science*, *13*, 976179.

Romero, A., Munevar, F., & Cayon, G. (2011). Silicon and plant diseases. A review. *Agronomia Colombiana*, *29*(3), 473–480.

Rouhani, M., Samih, A., & Kalantri, S. (2013). Insecticidal effects of silica and silver nanoparticles on the cowpea seed beetles *Callosobruchus maculatus*. *Journal of Entomological Research, 4(4):*297–305.

Sabbour, M. M. (2016). Observations of the effect of Nano chitosan against the locust Schistocerca gregaria (Orthoptera: Acrididae). *Journal of Nanoscience and Nanoengineering*, *2*(4), 28–33.

Saleem, H., & Zaidi, S. J. (2020). Recent developments in the application of nanomaterials in agroecosystems. *Nanomaterials*, *10*(12), 2411.

Servin, A., Elmer, W., Mukherjee, A., De la Torre-Roche, R., Hamdi, H., White, J. C., ... & Dimkpa, C. (2015). A review of the use of engineered nanomaterials to suppress plant disease and enhance crop yield. *Journal of Nanoparticle Research*, *17*, 1–21.

Shahzad, K., & Manzoor, F. (2021). Nanoformulations and their mode of action in insects: a review of biological interactions. *Drug and Chemical Toxicology*, *44*(1), 1–11.

Shang, Y., Hasan, M. K., Ahammed, G. J., Li, M., Yin, H., & Zhou, J. (2019). Applications of nanotechnology in plant growth and crop protection: a review. *Molecules*, *24*(14), 2558.

Sharma, J., Singh, V. K., Kumar, A., Shankarayan, R., & Mallubhotla, S. (2018). Role of silver nanoparticles in treatment of plant diseases. In *Microbial Biotechnology: Volume 2. Application in Food and Pharmacology* (pp. 435–454).

Shekhar, S., Sharma, S., Kumar, A., Taneja, A., & Sharma, B. (2021). The framework of nanopesticides: a paradigm in biodiversity. *Materials Advances*, *2*(20), 6569–6588.

Singh, B. R., Singh, B. N., Singh, A., Khan, W., Naqvi, A. H., & Singh, H. B. (2015). Mycofabricated biosilver nanoparticles interrupt Pseudomonas aeruginosa quorum sensing systems. *Scientific Reports*, *5*(1), 13719.

Singh, P. M., Tiwari, A., Maity, D., & Saha, S. (2022). Recent progress of nanomaterials in sustainable agricultural applications. *Journal of Materials Science*, *57*(24), 10836–10862.

Srujana, S., & Bhagat, D. (2022). Sensors for agricultural pest management using nanotechnology/nanomaterials. *Nanotechnology for Environmental Engineering*, *7*(1), 319–323.

Su, Y., Ashworth, V., Kim, C., Adeleye, A. S., Rolshausen, P., Roper, C., ... & Jassby, D. (2019). Delivery, uptake, fate, and transport of engineered nanoparticles in plants: a critical review and data analysis. *Environmental Science: Nano*, *6*(8), 2311–2331.

Tripathi, D. K., Singh, S., Singh, S., Pandey, R., Singh, V. P., Sharma, N. C., ... & Chauhan, D. K. (2017). An overview on manufactured nanoparticles in plants: uptake, translocation, accumulation and phytotoxicity. *Plant Physiology and Biochemistry*, *110*, 2–12.

Usman, M., Farooq, M., Wakeel, A., Nawaz, A., Cheema, S. A., ur Rehman, H., ... & Sanaullah, M. (2020). Nanotechnology in agriculture: current status, challenges and future opportunities. *Science of the Total Environment*, *721*, 137778.

Vargas-Hernandez, M., Macias-Bobadilla, I., Guevara-Gonzalez, R. G., Rico-Garcia, E., Ocampo-Velazquez, R. V., Avila-Juarez, L., & Torres-Pacheco, I. (2020). Nanoparticles as potential antivirals in agriculture. *Agriculture*, *10*(10), 444.

Vemula, M., & Reddy, A. (2023). Polymeric nanoparticles as effective delivery systems in agriculture sustainability. *Nanotechnology for Environmental Engineering*, *8*(3) , 805–814.

Wang, L., Li, X., Zhang, G., Dong, J., & Eastoe, J. (2007). Oil-in-water nanoemulsions for pesticide formulations. *Journal of Colloid and Interface Science*, *314*(1), 230–235.

Wani, T. A., Masoodi, F. A., Baba, W. N., Ahmad, M., Rahmanian, N., & Jafari, S. M. (2019). Nanoencapsulation of agrochemicals, fertilizers, and pesticides for improved plant production. In *Advances in Phytonanotechnology* (pp. 279–298). Academic Press.

Wu, L., & Liu, M. (2008). Preparation and properties of chitosan-coated NPK compound fertilizer with controlled-release and water-retention. *Carbohydrate polymers*, *72*(2), 240–247.

Xin, X., Li, Y., Yu, L., Li, W., Li, J., & Lu, R. (2021). Facile synthesis of Ag@ C@ Ag hybrid nanoparticles as SERS substrate. *Analytical and Bioanalytical Chemistry*, *413*, 5767–5777.

Xu, Z., Tang, T., Lin, Q., Yu, J., Zhang, C., Zhao, X., ... & Li, L. (2022). Environmental risks and the potential benefits of nanopesticides: a review. *Environmental Chemistry Letters*, *20*(3), 2097–2108.

Xue, J., & Luo, Y. (2021). Sustainable food and agriculture system: a nanotechnology perspective. *ES Food Agrofor*, *5*, 1–3.

Yadav, A., & Yadav, K. (2018). Nanoparticle-based plant disease management: tools for sustainable agriculture. *Nanobiotechnology Applications in Plant Protection*, *2*(2), 29–61.

Yadav, N., Garg, V. K., Chhillar, A. K., & Rana, J. S. (2023). Recent advances in nanotechnology for the improvement of conventional agricultural systems: a review. *Plant Nano Biology*, *4*(3), 100032.

Younas, A., Yousaf, Z., Rashid, M., Riaz, N., Fiaz, S., Aftab, A., & Haung, S. (2020). Nanotechnology and plant disease diagnosis and management. *Nanoagronomy*, *2*(4), 101–23.

Zacharia, J. T. (2011). Identity, physical and chemical properties of pesticides. In *Pesticides in the Modern World-Trends in Pesticides Analysis,* ISBN: 978-953-307-437-5, InTech, (pp. 1–18).

Zaheer, T., Ali, M. M., Abbas, R. Z., Atta, K., Amjad, I., Suleman, A., ... & Aqib, A. I. (2022). Insights into nanopesticides for ticks: the superbugs of livestock. *Oxidative Medicine and Cellular Longevity*, *22*(1), 1–18.

Zannat, R., Rahman, M. M., & Afroz, M. (2021). Application of nanotechnology in insect pest management: a review. *SAARC Journal of Agriculture*, *19*(2), 1–11.

Zhou, P., Adeel, M., Shakoor, N., Guo, M., Hao, Y., Azeem, I., ... & Rui, Y. (2020). Application of nanoparticles alleviates heavy metals stress and promotes plant growth: An overview. *Nanomaterials*, *11*(1), 26.

Zhu, L., Chen, L., Gu, J., Ma, H., & Wu, H. (2022). Carbon-based nanomaterials for sustainable agriculture: their application as light converters, nanosensors, and delivery tools. *Plants*, *11*(4), 511.

2 Revolutionizing Nematode Management – Nanomaterials as a Promising Approach for Managing Economically Important Plant-Parasitic Nematodes

Current Knowledge and Future Challenges

Mahfouz M. M. Abd-Elgawad

2.1 INTRODUCTION

In the last few decades, globally agricultural crops have faced the problem of abundant food production that fills the needs of the increasing numbers of people, and their domestic animals, as a result of the improvement in health and social care as well as the general rise in the relative standard of living of the world's population. This situation has led to heightened levels in human population numbers at rates faster than any actual food production without authentic scientific intervention (FAO 2017). Consequently, the researchers have come to know about the vertical increase in production that boosts the actual yield of the same land area as well as the horizontal increase that raises the cultivated area by adding new arable lands especially after reclamation. In both, vertical and horizontal, types of expansion, many categories of chemical pesticides are used as an effective approach in controlling plant pests and pathogens in order to increase crop production in size and quality. Therefore, many dangerous and serious side-effects of such chemicals to mankind and the relevant setting have been documented by constant research (Abd-Elgawad and Askary 2015). It goes without saying that these harmful effects had extended to various components, including agricultural and animal products, water, soil and air (Coupland et al. 2017). Their unfavourable impacts prompted scientists to confront new dangers. They should face the related cancerous diseases as well as kidney/liver failure, and the likes at unfamiliar rates; by searching for clean, safe, and abundant crop yields. This is especially needful for countries with constantly increasing populations, given the desperate needs to provide food, feed, and clothing. It is not imagined that the solution is to return to primitive farming methods or even traditional chemical pesticides. This is simply because plant pests and pathogens have become fiercer and more damaging over time. Control of such biotic stressors in effective and safe ways has been ongoing. Such pressures are diverse and must be confronted, for example, to improve the percentage of peanut germination inflicted by insect pests (Shehata et al.

DOI: 10.1201/9781003256762-2

2021) with superior biocontrol agents (Shehata et al. 2019), avoid root decay of strawberry plant, followed by lack of its productivity when infected with several fungal species (Abd-El-Kareem et al. 2021), protect citrus trees from nematode and fungal pathogens (Abd-Elgawad et al. 2010), and face quality problems that impede the export of the Egyptian potato crop due to the brown root rot disease caused by the bacterium, *Ralstonia solanacearum* (Kabeil et al. 2016).

Plant-parasitic nematodes (PPNs) are pathogens of rank and reproduction. They can be the cause of substantial crop losses worldwide. They represent big groups of mostly subterranean pests and pathogens with various population levels, vertical and horizontal distributions, and importance (Jones et al. 2013). Growers and stakeholders are often unaware of damages brought about by PPNs, because most of them are underground-living organisms. So, much of the losses caused by them are ascribed to other sources (Abd-Elgawad and Askary 2015). Hence, awareness raising, and adequate control measures of these nematodes should tackle their induced losses. Nematodes parasitizing plant roots may be ectoparasites, semi-endoparasites, sedentary endoparasites, and migratory endoparasites (Jones et al. 2013). Above-ground PPNs are less copious and include seed gall nematodes, stem and bulb nematodes, and foliar nematodes. Comprehensive reviews of PPNs, their damage, and management in specific locations and on definite crops such as tomato (Abd-Elgawad 2020a), pepper (Abd-Elgawad 2020b), and potato (Abd-Elgawad 2020c) are recently published.

This chapter highlights yield losses in different crops caused by wide array of PPNs. It also discusses the current challenges in nematode management along with options for their control. Nanotechnology is one of the promising options in now-a-days, and the exploitation of nematocidal nanomaterials (NNM) for managing the PPNs can help us find sustainable solutions in different cropping systems. The chapter focuses on the ongoing applications, different mode of actions of these NNM, and their integration with other agricultural inputs to optimize the nematode control and crop yields. Merits and demerits of using the NNM for PPN control pertaining to other commonly used tactics are highlighted.

2.2 DAMAGE AND CONSEQUENCE OF PPNS

These nematodes, generally known as phytonematodes, rank high among plant pests and pathogens (Jones et al. 2013; Ansari and Mahmood 2017a; Ansari and Mahmood 2019a; Ansari and Mahmood 2019b; Ansari et al., 2020). They can pose a big threat to qualitative and quantitative yields of many economically important food and commercial crops. Staggering figures of such losses on numerous crops have been globally reported over the years. A recent and brief record has mentioned crop yield loss caused by PPNs annually to be over \$100 billion worldwide (Forghani and Hajihassani 2020). Abd-Elgawad and Askary (2015) adopted several questionnaires so that they may reflect the real losses caused by PPNs worldwide and may be considered for locally-oriented objectives for pest management as well. Their questionnaires were backed by 371 replies from the previous International *Meloidogyne* Project (IMP) for twenty life sustaining plant species as well as 20 other economically significant crops. Thus, for the twenty plant species that satisfy the human need from hunger, 12.6% is assessed as an average global loss in these crops due to PPNs, corresponded to \$ 215.77 plant damage annually. The other 20 plant species that have considerable food, clothing, export, or industrial value were inflicted by 14.45% loss annually that amounted to \$ 142.47. These 40 crops showed an average of 13.5% losses (Abd-Elgawad and Askary 2015). Admittedly, such estimates will probably go up as other PPNs-infected plants throughout the world may be added to the list due to significant number of substantial reasons including environmental fluctuations.

The economic ramifications of crop damages caused by PPNs may be distributed to several levels of losses; the grower/farmer, purchaser, and society. The farmer experiences the lack of revenue and/or the raise in production costs due to the low yield, the degradation in the marketable product quality, and/or costs of nematicides and their application (Abd-Elgawad and Askary 2015). Many farmers suffer from the factor of uncertainty; they doubt the importance of nematicides to achieve a profitable return. So, if the extra costs of such control measures cannot save significant return/value, these measures cannot be considered as profitable. Another suffering level is the consumer in terms of individuals, households, companies and industry in general. Those consumers bear the raise in price, the drop in quality of the used output, or both. Some human diseases as some types of cancer and reproductive diseases might underlie the consumption of produce/food treated with banned pesticides especially in developing countries (Taskinen et al., 2011). Moreover, the upper suffering level is the society as both the grower and consumer are a part of it.

2.3 CHALLENGES IN NEMATODE MANAGEMENT

Even in developed countries, production of many PPN-sensitive crops such as vegetables had been dependent on fumigation with methyl bromide (MB). So, farmers just disregarded PPNs, relying on MB as efficient control measure against a variety of soil-borne pathogens and pests (e.g., nematodes, weeds, fungi, bacteria, and insects) even at severe field infestations (Abd-Elgawad and Askary 2015). They must raise PPN-susceptible crops in these sites but without the currently banned MB. Admittedly, banning of MB has created a considerable void in chemical arsenal utilized against soil-borne pests and plant diseases worldwide. This is quite intelligible from the current voluminous research and review of global field research trials seeking for benign alternatives of MB. For instance, in Florida (USA), various soil fumigants are assessed in fields for pest and pathogen suppression and subsequent response in yields. Noling (2019) summarized the general efficacy of some current fumigants in such trials. For instance, chloropicrin is reliable for fungal diseases but not so effective to control weeds/nematodes. Telone is very effective nematicide but not so reliable to control weeds/diseases. Bacterial diseases are still being searched for effective fumigants. Metam potassium and metam sodium can offer effective weed control. Yet, ongoing research is still necessary to optimize rate, application technology, and placement of such fumigants, and resolve issues of inconsistent control against some pathogens/pests (Noling 2019; Ansari et al., 2017; Ansari et al., 2023; Boyno et al., 2024; Fabiyi et al., 2024).

Also, non-fumigant liquid and/or granular nematicides are frequently not as efficient as fumigants in boosting crop yields as they do not possess wide-spectrum activity (Sikora et al. 2005). Nevertheless, the two main categories of these nematicides are facing possible restrictions or even banning too. These are organophosphates (cadusaphos, fenamiphos, and ethoprophos), and oxime-carbamates (oxamyl, aldicarb, and carbofuran) (Abd-Elgawad 2008). The outcomes of seeking safe replacements via many research trials have offered foundation at least for overall generalization of effective nematicides. This does not negate that we are challenging substantial re-directions in the choice and categories of usable and effective nematicides.

The current intensive agro ecosystems usually apply high or even extra amounts of backing pesticides and fertilizers to attain maximum yields. Thus, the negative impacts of intensive farming seem to be unfavourable for sustainable agriculture. This is being reflected on pollution of air and groundwater, soil deterioration, salinization of watered areas, erosion of biodiversity, and pesticidal resistance (Abd-Elgawad and Askary 2015). For example, recent intensified citriculture regime suppressed beneficial biocontrol agents and consequently led to worsen herbivory especially via certain insect species (Campos-Herrera et al. 2013). On the other hand, the recent climate change is probably shifting the geographical allocation of both plant hosts and their pathogens. As this change is associated with rise in temperature, it coincides with favouring some PPN species reproduction and activities (Dutta and Phani 2023). The unfavourable influences of climate change including universal warming, increased CO_2, precipitation shift, and other extremes of heat waves, floods, droughts, storms, and wildfires considerably impact the biogeographic range, abundance, distribution, fitness, survival, parasitic potential, and reproduction of the PPNs. Changes in such PPN parameters rely on the relative adaptability the nematode species. In this context, it is expected that important PPN genera such as the root-knot nematodes (RKNs) will have more spread and increase. This is simply because more pathogen generations/ year are expected to be increased in warmer climate (Abd-Elgawad and Askary 2015). Advances in molecular biology to investigate various aspects of plant-nematode interactions and how plants respond to different invading pathogens may lead to an optimistic and efficient PPN control. Genetic basis of plant resistance to nematodes accompanied with such interactions driving to incompatible plant responses should be grasped and exploited (Ibrahim et al. 2019; Zheng et al. 2021). Furthermore, complexities associated with nematode-genetic manipulation should be unravelled and developed.

Adequate quarantine plans and certification programs should be applied in earnest to limit the diffusion of PPNs. As quarantine programs cannot be done to entire units of transferable soil, plants, cuttings, and pots, thorough measures should be applied for perfect protection of any PPN species that has not yet broken into a country (Salama and Abd-Elgawad 2003). Alert and accurate protocols should be continuously updated to block the introduction and diffusion of new/quarantine species (Handoo et al. 2013). Current and new molecular tools should be dedicated to accurately identify PPNs species and races. This would avoid the confusion in identification of some species/races. While these approaches will offer a precise and reliable description of quarantine PPNs, they will authorize easier and clearer materials to study them. Moreover, genome-constructed phylogenetic trees based on accurate PPN identification will be perfected. This will lead to exact comparative research results that can rightly define their diversity and construct sound relationship.

Additional challenge is to face inconsistency in PPN technological advances between the South and the North countries (Luc et al. 2005). While managing PPNs in tropical and subtropical countries are lagging behind those in developed countries, boosted application of molecular diagnostics and taxonomy may broaden these gaps between the two categories. Thus, difficulties to detect emerging/new PPN threats in the former countries block the timely application of management plans and aggravate crop losses. Centres and laboratories with frontiers in nematode diagnostics and systematics like those in developed countries should be expanded in developing countries. Moreover, tackling nematode problems should be continuous actions especially in heavily infested areas. Neglected nematode issues can reoccur when crop rotations are changed, or new cultivars are planted. Forgotten PPN problems led to new outbreaks of sugar beet stem nematode and potato cyst nematode (Luc et al. 2005). Finally, all such challenges should be adequately and financially solved by economically-directed PPN research. This is especially significant for several plant species that usually suffer severe PPN infection and yield losses; e.g. maize, rice, and vegetables (Abd-Elgawad and Askary 2015).

2.4 OPTIONS FOR NEMATODE CONTROL

Currently, PPN are ordinarily managed through distinct production exercises (bionematicides, chemical nematicides, fallowing, flooding, soil amendments, tillage, use of certified seeds/transplants (Fabiyi et al., 2020; Patil et al., 2024; Solanki et al., 2021; Sumbul et al., 2020). PPNs spend their lives in plant roots and within the soil, delivery of chemicals to the sites occupied by PPN is a real challenge ahead the researchers which needs immediate attention (Chitwood 2003). Unfavourable impact of non-judicious introduction of chemical fertilizers and nematicides on human health and environmental attributes have compelled the regulatory authority to take strict actions against the non-judicious users. Additionally, the excessive pesticidal applications have led to the appearance of resistance-breaking nematode pathotypes/strains (Fullana et al. 2023; Ansari et al., 2020; Ansari et al., 2019; Mahmood et al., 2019; Ansari and Khan 2012a; Ansari and Khan 2012b). The newly emerged nematode pathotypes, resistant to these chemicals, have entailed the increased search of viable and safe alternatives. Using bionematicides, in forms of biocontrol agents (BCAs) and/or their bioactive compounds as well as naturally occurring materials such as antagonistic phytochemicals, are mostly benign alternatives for PPN management methods. However, they are many times less effective, more inconsistent, and slower influential than traditional PPN control normally achieved through chemical application (Abd-Elgawad 2021; de Freitas Bueno et al., 2019; Ibrahim et al., 2020; Devi et al., 2020; Ibrahim et al., 2022; Keshari et al., 2024; Prasad et al., 2020; Solanki et al., 2020). Robust investigations of the ecology, biology, and mode of action of these BCAs is desperately needed to back, conserve and realize their use. Botanical additives and secondary metabolites of many BCAs have been proven to be economical and cheaper but time-consuming that has slowed down their execution in different agroecosystem (Ntalli et al. 2020). Basic requirements for these materials and any new nematicides are their reliable nematocidal effect, safety, and favourable economics (Abd-Elgawad 2021a). In this context, volatile and toxic isothiocyanates (ITCs) are active components of several synthetic nematicides (Ntalli and Caboni 2017). Nonetheless, natural ITCs that possess the same biochemical mode of action against nematodes are utilized as biofumigants to manage the PPNs. The adverse impacts of ITCs in biofumigants like mustard have brought about instability and vulnerability of soil food webs with consequent harming of useful fauna/flora (Dutta et al. 2019). Ntalli and Caboni (2017) concluded that both natural and synthetic components of ITCs react in an irreversible and nonspecific trend with proteins and amino acids so that non-target and beneficial organisms are also negatively impacted. Comparatively, chemical fumigants mostly have a wide-spectrum and are more effective against PPN populations than organic and inorganic soil amendments (Dutta et al. 2019). A variety of organic soil amendments can control PPNs, but specific techniques are needed to develop their efficient utilization. Neither fallowing nor flooding is an economic measure in the management of PPN. Fallowing may misuse the perfect investment of arable lands (Abd-Elgawad 2022). Flooding may only succeed in countries with water abundance and warm climates as it requires excess water and a prolonged time that may harm the plant (Abd-Elgawad 2022). Moreover, field solarization needs long periods during which no cultivation is practiced. Crop rotation and utilizing available resistant/tolerant crop cultivars are quite safe approaches for PPN management (Fullana et al. 2023). However, main obstacles for growing resistant/tolerant cultivars are the presence of nematode-physiological races that can break plant resistance. Cultivars bred for resistance to one race, when cultivated in soil infested with different race of the same nematode species, may become severely infected which may offer secondary pathogens to colonize the plants and harm together (Timper 2014). Furthermore, the level of PPN control through crop rotation is relied on the capability of the planted resistant/tolerant crops. Additionally, heat instability or *Mi* gene sensitivity to temperature may weaken or mask resistance in resistant tomato plants (Noling 2019). Agricultural operations as tillage, crop rotation, chemical applications, and fallow can directly/indirectly damage populations of antagonistic or useful organisms and consequently increase damage from pests (Timper 2014). Eventually, these PPN control methods are not partially successful because of the nematodes' diverse and uneven distribution, broad host range, and subterranean and within plant life stages (Abd-Elgawad 2021). Thus, successful exercises and continuation of production practices to control PPNs infecting important plant species require economic and extensive programs with considerable investments.

2.5 RATIONALE IN USING NANOMATERIALS FOR PPN MANAGEMENT

It is easy to conclude the necessity of emerging/novel tactics which are economically sound, safe, and effectively managing PPNs. Nanomaterials seem to be fit as prerequisites for PPN management. Nanotechnology focuses on generating very minute 'nano' structures, e.g. atomic and molecular molecules, with augmented functionalities. Nanoparticles (NPs) address particulate matter in the form of 1–100 nm. This size can provide the merit of possessing a big surface area to volume ratio (Christian et al. 2008). This merit combined with both nanoparticle forms and distributions in solution give a share in their boosted chemical and physical properties that are useful in numerous domains (Moodley et al. 2018; Mittal et al. 2020). Moreover, with the existence of such ingredients in so diminished but functional concentration of formulations, these nanoparticles at the intended sites

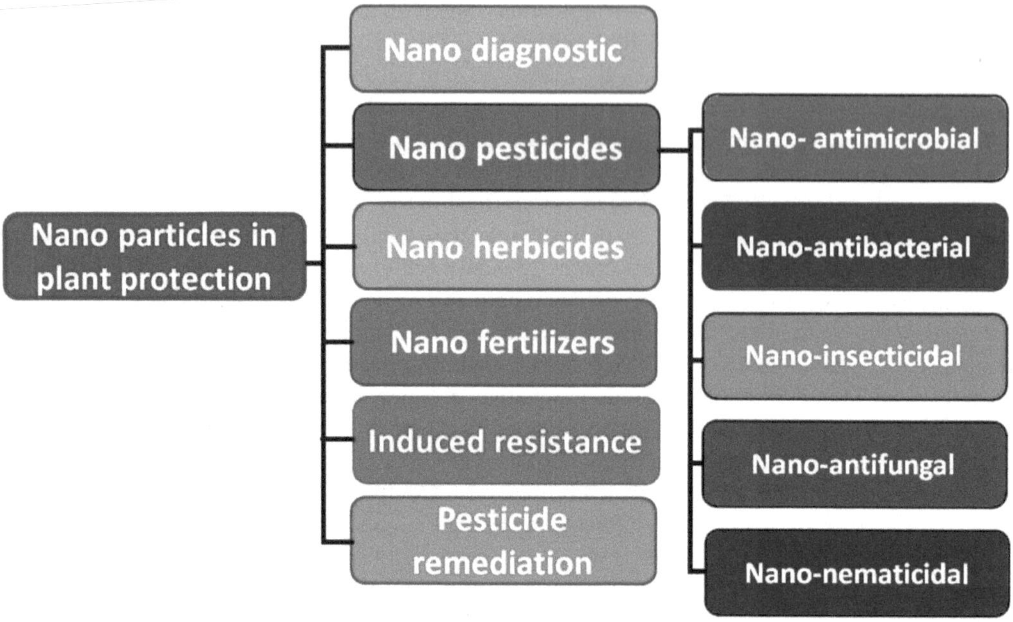

FIGURE 2.1 Nanoparticles are used for various approaches in plant protection including application of diagnostics, herbicides, fertilizers, inducing plant resistance, pesticides remediation, and pesticides. The latter approaches of pesticides comprise antimicrobial, antibacterial, anti-insecticidal, antifungal, and anti-nematocidal nanoparticles (Modified from Shang et al. 2019.)

can assure better plant protection from pest/nematode invasion and subsequent crop damage (Sabry 2019; Baronia et al. 2020; Shekoohi et al. 2021). Therefore, the development of novel formulations for plant protection and pest control has long been an attractive field of agricultural experimentation. Strictly speaking, nanoformulation technology or encapsulation of nematicides can revolutionize the plant protection measures and approaches (Abd-Elgawad 2024). This emerging technology in the pesticidal domain provides very tiny sizes/types of devices that operate as effective components of pesticides. Nonetheless, other engineered nano-structures can also possess useful pesticidal attributes (Zhao et al. 2013; Haq and Ijaz 2019; Safeena and Zakeel 2020). The term 'nanoencapsulation' of pesticides usually refers to the coating of active pesticidal ingredients with additional material of different nano-range. The encapsulated materials are known as inside phase of the essence material (e.g., nematicides). The coating nanomaterials are the capsulation materials or the external matter (Nuruzzaman et al. 2016). Also, the so called 'Green synthesis of nanoparticles' offered additional safety factor over classical chemical and physical methods for preparing metal nanoparticles via adding safety chemicals (Kalaiselvi et al. 2017). Contrary to classical synthetic strategies of nanoparticles, green synthesis of these very minute materials always utilizes environmentally compatible items as plants, fungi, and bacteria (Heflish et al. 2021). In addition to the inexpensive costs needed to manufacturing the nanoparticles via downstream processing, the green synthesis has the merits of providing high yields within short time (Moodley et al. 2018; Figure 2.1).

Hence, many researchers believe that nanomaterials will assist to alleviate the above-mentioned safety concerns of many chemical nematicides via global, ecologically friendly, effective nanomaterials at a cost comparable to these harmful chemicals. In addition to offering excellent alternatives to classical nematicides, these nanomaterials proved to possess multi-site modes of action to effectively kill the nematodes (Khan et al. 2023). The latter authors recently reviewed such modes of action against the nematodes. Basically, the concrete structure of NNM in terms of shape, size, and homogeneity is usually joined to their inhibitory impact, which significantly contribute to penetration of PPN egg cell walls causing their malfunction. Interestingly, NNMs as Ag-NPs are not species-specific in inducing toxicity. They can be used not only against PPNs but also to control fungal pathogens of many plants. Their mechanism is linked to the disturbance and unfit of distinct cellular systems of eukaryotic/prokaryotic cells (Lim et al. 2012). Algae-synthesized ZnO NPs showed their favourable influence on RKN control in both roots and soil with a significant impact on RKN-infected banana (El-Ansary et al. 2022). Although these NNM displayed a major biostimulant potential in banana growth, they had no phytotoxicity. Additionally, the rapid progress in nanotechnology enabled the possible utilization of numerous nanomaterials/nano-products to control a broad variety of plant-parasitic nematode species whether individually or as polyspecific nematode populations (Safeena and Zakeel 2020). The used NPs have several merits over the general performance of most bionematocidal formulations as it is not dependent on living organisms as the active ingredient(s). Eventually, the synthesis of efficient and benign nematicides is generally difficult and costly, but nanomaterials may offer a new and optimized solution especially for secondary agriculture which is the need of hour (Ur Rahim et al. 2021).

Such PPNs are soil-dwelling pathogens and the use of such a profitable management option is needed to maintain their population levels below the damaging density or mainly the economic threshold level. Hence, NPs synthesis based on such materials as plant extracts displays new avenues for favourable industrial products. It can offer economical and commercial production of effective nematicides in the form of nanoparticles.

2.6 CURRENT USE OF NANOMATERIALS IN THE MANAGEMENT OF ECONOMICALLY IMPORTANT PPNS

2.6.1 NPs Synthesis

Three major classes, i.e. chemical, physical, and biological operations have been generally forwarded as broad classes for synthesis of nanomaterials (Safeena and Zakeel 2020). Nonetheless, the first two classes are encumbered with different issues comprising utilizing unfriendly chemicals and production of hurtful commodities in addition to being costly and technically arduous (Abd-Elsalam 2021a). Biological synthesis is considered as healthier and less expensive method than chemical or physical techniques (Safeena and Zakeel 2020). Synthesis of nanomaterials via biological techniques can offer better opportunities to grasp the molecular/biochemical mechanisms of nanoparticles. Also, formation as functional and catalytic data can be easily obtained. Thus, various biological systems like viruses, plants, fungi, algae, and bacteria have been utilized to attain nanomaterials in eco-friendly, inexpensive, energy-efficient, and metallic forms (Abdellatif et al. 2016; Moodley et al. 2018; Safeena and Zakeel 2020).

2.6.1.1 Physical Approach

In evaporation-condensation, starting and pure metals are used. This procedure avoids using potentially poisonous and costly precursors. For instance, 4–50 nm of copper nanoparticles could be synthesized in an arc furnace technique (Forster et al. 2012). A great challenge is the large scale and reliable production of nanoparticles over a domain of constant size, composition, and high mono-dispersity. Yet, it has the privilege of generating pure and uncontaminated metal colloids with even nanoparticles distribution for use. Also, laser ablation method acts on ablation of an intended substance by a heavy radiation on a firm substance in a liquid. This leads to fusing and then ablation of the substance from object ending with the release of NPs. As a good and effective technique, Bhau et al. (2016) reported the following advantages of laser ablation in liquids: i) unanimous application with numerous materials and solvents to synthesize nanoparticles, ii) contrary to fabrication in gas, it has no problems in the particles collection, iii) it generates full pure particles,

iv) it offers occupational safety as the generated nanoparticle colloids are not inhalable, and v) the nanomaterials shape and size may be controlled via available ablation parameters. Zamiri et al. (2012) found that silver nanoparticles generated via laser ablation in natural polymer technique have more stability and efficacy. Also, utilizing a silver NP suspension in deionized water via the arc discharged technique did not require addition of surfactants (Tien et al. 2008). In this technique, silver wires functioned as electrodes in the deionized water at 100 mg/min consumption rate of the silver rod to produce 10 nm silver NPs.

2.6.1.2 Chemical Approach

This technique is often commercially used for the nanoparticles synthesis. It includes various methods such as sonochemical method, colloidal method, and reduction method. Generally, any used method should be compatible with the employed material. The commonly used chemical reaction via inorganic and organic reducing agents can generate nanoparticles after reducing the metallic ions. This method provides protective agents, stabilizers, to stabilize dispersive nanoparticles away from agglomeration (Dang et al. 2011; Usman et al. 2013). Otherwise, inorganic nanoparticles can be ideally prepared by colloidal/micro-emulsion technique (Yu et al. 2010). This technique qualifies for controlling the obtained properties, e.g. morphology, surface area, homogeneity, size control, geometry, and mechanisms of particle (Malik et al. 2012). Yet, the sonochemical technique is used extensively to synthesize nanoparticles with higher surface area, more controllable morphologies, and smaller size than those of other techniques (Bhau et al. 2016).

2.6.1.3 Biological Approach

As indicated above, this relatively recent approach for generating nanoparticles has distinct merits. It has more usable control of its ingredients, less wastage of insertions, and less need of energy. Bhau et al. (2016) recorded biologicals; e.g. fungi, yeasts, plants, bacterial, and viruses; that are utilized in the extracellular or intracellular synthesis of various nanoparticles (Table 2.1). It can utilize metabolites gained from these biologicals; e.g. biocontrol agents. Ultimately, the obtained particle sizes of the nanoparticles have wide distributions according to the species used (Abbassy et al. 2017; Abd-Elsalam 2021a).

2.6.2 Nanotechnology for PPN Management

Encouraging and increased references on nanotechnology for PPN management to acquire better crop yield/production have been published in the last two decades. For instance, Cromwell et al. (2014) stated that the bio-efficacy of silver nanoparticles (AgNPs) on *Meloidogyne incognita* J2 was demonstrated in suppressing the nematodes in the laboratory assays but field results were not decisive for controlling *Meloidogyne graminis* on bermudagrass. While their applied AgNPs could decrease gall formation on the plant roots in the two tested seasons and enhance turfgrass quality in one year unaccompanied by phytotoxicity, AgNPs did not considerably lower the number of J2. Generally, their field trials showed AgNPs merits mitigating damage imposed by RKN in bermudagrass. Meanwhile, Jo et al. (2014) emphasized that AgNPs, used to manage nematode infecting bermudagrass, did not bring about any plant toxicity.

More recently, AgNPs demonstrated favourable impacts for suppressing *M. incognita* infecting tomato seedlings (Hassan et al. 2016) and eggplants (Abdellatif et al. 2016). The impact of silicon oxide (SiO_2 NP, 11–14 nm), silver (AgNPs, 20 nm), and titanium oxide (TiO_2 NP, 20 nm) at different concentrations against *Meloidogyne incognita* on tomato roots was assessed (Ardakani 2013). The bioassays indicated that full second stage juveniles (J2) mortality/immobility occurred at 200–800 mg/ml Ag NP; the lethal concentration required for 50% mortality (LC_{50}) value equalled 100 mg/ml. All tested concentrations of AgNPs and 0.02% TiO_2 NP could fully control *M. incognita* in pot trials. As silver ion (Ag^+) and its mixtures have biocidal efficacies on

TABLE 2.1
Exploring biological resources for nanoparticle synthesis and their potential nematocidal applications in agriculture

Biological resources	Nanoparticles	Nematocidal applications	References
		Bacteria	
Bacillus licheniformis	Gold or Silver	Biweekly applications with high doses can achieve effective field efficacy to control the root-knot nematodes	(Thakur and Shirkot 2017)
Nostoc sp.	Silver	cyanobacterium-based Ag-NPs significantly suppressed *Meloidogyne javanica* egg hatching and induce J2 mortality besides their good effect on growth of faba bean	(Hamed et al. 2019)
		Fungi	
Duddingtonia flagrans	Silver	The fungal extract can synthesize AgNP, and they also penetrate the cuticle of the nematode larvae, inducing alterations in the tegmentum, and kill the nematodes	(Barbosa et al. 2019)
Fusarium oxysporum	Various NPs	It is involved to produce various NPs, e.g. silver, lead sulfide, cadmium sulfide, molybdenum sulfide, and zinc sulfide; proper salts may be added to the growth medium. NPs function as nematicides, capped via fungal proteinaceous matter	(Fabiyi et al. 2020)
		Viruses	
tobacco mild green mosaic virus	Pesticidal nanoparticles	Corn roots infected by endoparasitic nematodes were subjected to NPs encapsulated in tobacco mosaic virus. The virus boosts the diffusion of the nematicide in the plant-root level, where the nematodes are found and killed	(Chariou and Steinmetz 2017)
Red clover necrotic mosaic virus	Biological pesticide (Abamectin or Abm)	Encapsulated Abm within the virus produced plant virus nanoparticles with more effective delivery system for Abm, enlarged the protection area from *Meloidogyne hapla*, backed healthier tomato root growth, and reduced nematode galls	(Cao et al. 2015)
		Algae	
Green macroalga *Cladophora glomerata*	Silver	NPs showed better nematocidal effects on *Meloidogyne javanica* egg hatchability & juvenile's mortality than the chemical nematicide Rugby with significant reduction in *M. javanica* galls number, egg masses, and females per tomato root system	(Ghareeb et al. 2020)
Ulva lactuca	Silver	NPs at 17 $mg.mL^{-1}$ gotten by *U. lactuca* showed more impact in decreasing *M.javanica* juveniles (69.44%) in soil than the other alga *Turbinaria turbinate* but NPs of both algae enhanced growth parameters of *M. javanica*-infected eggplants	(Abdellatif et al. 2016)
		Plants	
Rosemary (*Rosmarinus officinalis*)	Sulfur nanoparticles	The significant cytotoxic effects of the Sulfur-NPs on suppressing egg hatching and inducing larval mortality of *Meloidogyne javanica* were apparent.	(Al Banna et al. 2020)
Copperleaf (*Acalypha wilkesiana*)	Silver	The capacity of bio-Ag NPs considerably decreased *Meloidogyne incognita* vitality and egg hatching while increased nematode mortality (53.3%)	(Heflish et al. 2021)

numerous microorganisms, nanoparticles (AgNPs) related to various compounds and preparing methods were effectively tried against PPNs. These Ag NP as NNM have been proving efficacy against various PPN species (Cromwell et al. 2014; Abdellatif et al. 2016; Hassan et al. 2016; Nassar 2016; Taha 2016; Nour El-Deen and El-Deeb 2018; Sabry 2019; Baronia et al. 2020;

Shekoohi et al. 2021). The suppression was so advantageous that it was comparable to the chemical nematicides; oxamyl and fenamiphos at 0.003 ml/kg soil (Hassan et al. 2016). Likewise, FeNP proved effective against *M. incognita* on okra (Sharma et al. 2017). On the other hand, gold nanoparticles, though costly, were so effective in controlling PPNs in India (Thakur and Shirkot 2017). Nour El-Deen and El-Deeb (2018) prepared AgNPs at concentrations of 1, 0.5, and 0.25 mM via biological and chemical syntheses; $AgNO_3$ reacted with *Zingiber officinale* (ginger) rhizomes aqueous extract and $NaBH_4$, respectively. The prepared AgNPs was superior to silver nitrate and non-treated control in both enhancing plant growth and suppressing RKN (*M. incognita*) populations on tomato roots. The highest concentration of ginger extract (1 mM) gave the best nematode control.

Gkanatsiou et al. (2019) tested copper/iron based nanoparticles for controlling RKNs (*M. incognita* and *M. javanica*) on tomato plants. They used same sized Pegylated Cu, CuFe and $CuFeO_2$ NP for assessing the two nematode species in relation to RKN paralysis and biological cycle seizure within plant roots; fosthiazate, commercial nematicide, was also used as a check. Although all compounds showed nematocidal activity, CuFe NP demonstrated the best results; with the lowest half maximal effective concentration (EC_{50}) value (0.03 μg ai g/soil). Moreover, growth parameters of tomato plants with regard to fresh root and shoot weight was the best in plants cured with CuFe NPs. El-Sherif et al. (2019) recorded that each of three nano-fertilizers, i.e. Fe-Zn oxide, Nano Fe oxide, and Nano Nano-Zn oxide each applied as 5 gm/pot, was superior to its mineral one used at the same rate. The superiority was reflected regarding not only lowering *M. incognita* egg masses and galls but also boosting growth criteria of tomato plants. The ZnO NP showed the most superiority over all other treatments including the mineral forms and the chemical nematicide Oxamyl. Sabry (2019) reviewed other trials documenting the promising impacts of nanomaterials on controlling PPNs.

Thereafter, Baronia et al. (2020) further demonstrated that as low as 0.1 μg/ml of AgNPs is considered the minimum concentration for full irreversible mortality of *Meloidogyne graminicola* in laboratory assays. However, this concentration was raised to 2 μg/ml in the sand screening tests in order to achieve the full irreversible mortality. In a hydroponic system, 1 μg/ml concentration of AgNPs attained considerable suppression of *M. graminicola* galls on rice roots. In the field, 3 μg/ml was so effective dosage in killing *M. graminicola*; significantly fewer than the 150 μg/ml as a previously recorded value (Baronia et al. 2020). The latter authors emphasized the robust nematocidal efficacy of AgNPs against *M. graminicola* on rice without any visible negative impact on plant growth or seed germination. Very recently, Shekoohi et al. (2021) synthesized AgNPs using plant (*Malva sylvestris*) leaves as aqueous extract. They noticed that raising the concentration of AgNPs decreased nematode egg hatching and enhanced J_2 mortality of *Meloidogyne javanica* infecting eggplant. The number of egg masses, galls, eggs, and reproduction factor (RF) in eggplant roots infected with 4 eggs per cm^3 soil and treated with AgNPs decreased by 23.5, 24.7, 43.4, and 43.2%, but decreased by 63.1, 25.5, 64.5, and 64.5%, respectively, relative to control plants in roots infected with 8 eggs per cm^3 soil. When AgNPs was combined with *β*-aminobutyric acid (1 mM), the RF decreased by 55.9 and 51.6% on eggplants infected with 8 and 4 eggs per cm^3 soil, respectively, relative to check plants (Shekoohi et al. 2021).

Eventually, it is clear that the swift development in syntheses, formulation and application of nanotechnology is progressing rapidly (Fig. 2.1) which can open new avenues for PPN management. The expansion in the use of nanomaterials via fitting them into current or emerging pest management strategies can enhance the performance of classical nematicide formulations as it introduces novel construction of very minute materials for nanotechnology-based PPN control plans.

2.6.3 Mechanisms Implicated in Nematode Management

Basically, nanoparticles have superior efficacy over microparticles (having identical elemental sequence) against many PPN species and the likes of various plant pathogens due to two major elements. These are the quantum effect and the surface effect. Such factors enable altering in chemical reactivity of the material used as well as its magnetic, visual, and mechanical properties (Amelia et al. 2012; Abd-Elsalam 2021a). The two factors are important against various pressures exerted on plants. This is especially influential as plants are subjected during their growth not only to PPN attack and damage but also to other general stresses. So, plants respond to these general stresses in various ways; e.g. stress-responsive gene expression, modifications in molecular mechanisms, and increasing the activities of antioxidative enzymes. Such responses can substantially assist in protecting the plants under these stressed conditions; e.g. invasion by pests and pathogens (Rejeb et al. 2014). For instance, plants resistant to nematodes (Abd-Elgawad 2021) and other soil borne pathogens (Abd-El-Kareem et al. 2019; Abd-Elsalam 2021a) usually show elevated activities in enzymes as peroxidase, endochitinase, and glucanase to protect them from nematode infection. Thus, nanoparticles offer an effective way or give support to plants by raising such enzymatic activities to back this protection mechanism (Mittal et al. 2020). Safeena and Zakeel (2020) reported two main processes that enable successful application of nanomaterial for PPM management: i) Impact on nematodes when NPs are set solely and applied as protectants or used in the nanopesticidal forms by generating nanomaterials; i.e. utilizing nanocarriers for nematicides, ii) Effect of nanomaterials on the biochemical and physiological activities of PPNs as NPs are very reactive materials due to their big surface-to-volume ratio and quantum effect.

The modes of action of specific minerals used in the form of nanoparticles are also investigated. For example, silver nanoparticles is generally linked to response to oxidative stress in the targeted organisms (e.g., PPNs), synthesizing the energy storage molecule adenosine triphosphate (ATP), and deactivating the cellular function at manifold level and greatly affect membrane permeability (Roh et al. 2009; Ahamed et al. 2010; Lim et al. 2012). Hence, Thiruvenkataswamy et al. (2021) concluded that the nematocidal effect of AgNPs against root-knot nematodes likely applies to other PPN genera and to fungal pathogens of plants, since its mechanism is not specific but related to damaging the aforementioned multiple cellular mechanisms, e.g. adenosine triphosphate (ATP) synthesis, response to oxidative stresses, and membrane permeability.

Moreover, green silver nanoparticles (GSN) combined with algae such as *Turbinaria turbinata* or *Ulva lactuca* could enhance GSN efficacy and applicability as nematicides. Therefore, grasping modes of action of these nematocidal combinations allows enhancing of GSN efficacy via appending normal materials, i.e. algae ingredients (Abdellatif et al. 2016). Such effective mechanisms should boost silver nanoparticles as broad spectrum nematocidal, antifungal, and antimicrobial agents. Ultimately, these novel biotechnologies may offer AgNPs very potential nematocidal activities as effective alternatives to unhealthy chemical forms of nematicides and costly or unreliable biologicals used against the nematodes (Bhau et al. 2016; Malandrakis et al. 2019; Sabry 2019; Shang et al. 2019; Baronia et al. 2020; Abd-Elsalam 2021a; Shekoohi et al. 2021).

2.6.4 Nanotechnology for Integrated Pest Management

Nanotechnology can assist to raise agricultural outputs by enhancing the efficiency of various inputs and lessening relevant yield losses. This should be sought via effective and wider reactions of NP-specific surface areas of several categories (Fig. 2.1) that should be combined with other inputs to maximize their benefits in integrated nematode management (INM) plans (Abbassy et al. 2017; Abd-Elgawad and Askary 2018; Shang et al. 2019; Abd-Elgawad 2021). Moreover, nanomaterials may be utilized as excellent carriers of agro-inputs to effectively enable the site-targeted controlled application/delivery of such inputs. Hence, nanoencapsulated marketable nematicides/pesticides with optimized attributes have recently been released (Nuruzzaman et al. 2016; Shang et al. 2019). In this regard, Polymer(s), e.g. polyethylene glycol (PEG) and polyester triblock copolymer polyethylene glycol-polybutyleneadipate-polyethylene glycol (PEG–PBA–PEG) have been utilized as cover substance for the synthesis/form of nanocapsules (Nuruzzaman et al. 2016).

Nanoencapsulated lansiumamide B (NCLB) indicated more nematocidal effect than lansiumamide B in pot experiment. The LC_{50} of NCLB equalled 19.3608 and 2.1407 mg/l to kill *M. incognita* J2 and *Bursaphelenehus xylophilus*, respectively after 24 h of treatment. Meanwhile, both the chemical nematicide (ethoprophos) and the chemical inducing the acquired resistance of plants to RKNs (lansiumamide B or the same chemical sequence in non-nanoform) displayed quite high LC_{50} relative to the NCLB (Yin et al. 2012). They encapsulated lanssiumamide B as nanocapsules also to boost environmental stability, bioavailability, and efficient maintenance. Thus, the suspensions of the nanocapsules could be maintained at 54 and 0 °C with reasonable stability. The synthesized NCLB was developed by the microemulsion polymerization technique. In this approach, they used the effective component lansiumamide B extracted from *Clausena lansium* (known as Wampee) kernels for encapsulation with methyl methacrylate and styrene, the emulsifier (sodium dodecyl sulfate), the auxiliary emulsifier (N-amyl alcohol), and the initiator (azodiisobutyronitrile). In such controlled-release formulations, the obtained data reflected the dependence of efficient pest control at the required period and level on the proper application rate and the polymer matrix used since the hydrophilic part impacts the active substance's release. Therefore, Loha et al. (2011) developed formulations that could retain their nematocidal activity for more times than the marketable β-cyfluthrin. Their half-release time ($t_{1/2}$) and period of optimum availability (POA) were more favourable than that of other pesticides such as imidacloprid. The $t_{1/2}$ and POA ranged 5.09–7.9 days, and 14–20.5 days, respectively for distinct formulations. Also, the bioefficacy examination of controlled-release formulations (CRFs) of the chemical nematicide carbofuran to control *M. incognita* infecting tomato (cv. PusaRuby) roots showed promising results in pot/field conditions regarding two different classes of polymerized polyethylene glycol (PEG), PEG-900 and PEG-600 . Both formulations at several concentrations displayed better efficacy in managing the J2-root penetration and development than carbofuran (Pankaj et al. 2012). Likewise, the rate of release decreased due to introducing clay (kaolinite, Fuller's earth, and bentonite) substances to the biodegradable clay compounds. When $t_{1/2}$ and POA of various controlled-release formulations was compared with their corresponding marketable formulations of carbofuran, the most promising values were noticed in CRFs (Choudhary et al. 2006). Thus, CRFs of carbofuran are more persistent in suppressing nematode populations than commonly marketable formulations. Such optimized applications in the integrated pest management (IPM) of these and other inputs (e.g., nematicides, fertilizers, insecticides, herbicides, fungicides) should assist in increasing crop protection and yield (Nuruzzaman et al. 2016; Priyanka et al. 2019; Shang et al. 2019; Safeena and Zakeel 2020; Shekoohi et al. 2021).

In order to boost their gains, these nanoparticles can also be comprised in IPM schemes via emerging and current approaches that fashion them superior or complimentary to current PPN management tactics (Cao et al. 2015; Abd-Elgawad and Askary 2018; Fu et al. 2018; Ansari et al., 2024; Ansari and Mahmood 2017b). This is especially timely as a few nematicides/acting nematicides can operate additively/synergistically with other additions. Acting nematicides may include chemical compounds that induce the acquired resistance of plants to PPNs. For instance, β-aminobutyric acid is an inducer of plant resistance against RKNs, *Meloidogyne* spp. Hence, a synergistically operating trend was apparent in combining AgNPs and β-aminobutyric acid for suppressing *M. javanica* populations but each of them individually had a lower effect (Shekoohi et al. 2021). In this respect, the reduction in RF of *M. javanica* at 4 eggs per cm^3 soil planted with eggplant recorded 43.2% due to AgNPs application but increased to 51.6% as combined with β-aminobutyric acid. Moreover, this RF on the eggplant roots infected with 8 and 4 eggs per cm^3 soil reduced by 7 and 6% after an application of AgNPss, 37 and 22% after an application of β-aminobutyric acid (1 mM) and 56 and 52% after the joint application of the β-aminobutyric acid (1 mM) and AgNPss, respectively, relative to the untreated plants (Shekoohi et al. 2021).

Other proper NP combinations to circumvent the lack of adequate use of various nematicides/pesticides, concerning their rates and efficacies, should be sought and practiced on wider scales/

bases. Within such IPM programs, nanomaterials can effectively boost pesticidal roles in suppressing pathogens and pests, enhance crop productivity, and decrease pollutions caused by microparticles of common chemical pesticides. Therefore, identification priorities in experimentation for harnessing the related nanotechnology in secondary agriculture and understanding any relevant biology/ecology/mode of action/ interaction with other production practices is direly required. In this context, the evolution of nematocidal nanohybrid agents that comprise conjugates of inorganic/organic materials, biopolymers, and biological components should be further researched to produce inexpensive, reliable and safe compounds for effective integration in plant pest control. On the contrary, crop production practices are currently still utilizing frequently over-dose and wide-scale of nematicides, insecticides, herbicides, and fungicides. Nuruzzaman et al. (2016) stated that > 90% of the classical pesticides are either environmentally lost or wasted away of its target sites. This does not only aggravate the costs of crop production, but also brings about pollution of the agro-ecosystems. Eventually, nanomaterials have a significant potential for boosting various aspects of modern agriculture (Fig. 2.1). For example, they can offer accurate diagnostics, proper biosensors, and targeted smart delivery systems (Abd-Elsalam 2021a, b). Therefore, these nanotools can back the development of many organic and inorganic farms. Furthermore, the integration of other disciplines and nanotechnology into nanosensors has considerably elevated their potential to sense and characterize the ecological settings or fragilities (Shang et al. 2019). Ultimately, commercial production and application of NNM should be backed by further ecological, intrinsic, technological, regulatory and marketing factors similar to those recorded for biological nematicides (Moosavi and Zare, 2015; Abd-Elgawad 2020).

2.7 SIDE EFFECTS AND REGULATIONS FOR NANOPARTICLES APPLICATIONS

As aforementioned, most NP utilized for large-scale applications are engineered NP, produced via biological, physical, and chemical techniques. Thus, the synthesized NP are linked to a rigid surface, and therefore, the risk of splitting and bringing about health problems are generally minimized (Zhao et al. 2013; Hardman 2006; Christian et al. 2008; Jeevanandam et al. 2018, Moon et al. 2019). Yet, specific toxicity on specific organisms should be considered. In this connection, silver ions, commonly used as NP against PPNs, are so reactive. They can link to proteins of cell membrane that lead to structural alterations in the cell walls and nuclear membranes, of nematodes or other pathogens, and finally results in cell death (Hachicho et al. 2014; Li et al. 2017). Hence, AgNPs have engaged many stakeholders because of their stability, unique morphology, and controlled geometry (Tran et al. 2013); these AgNPs are well recognized for their biocidal impact against many pathogens/pests (Khan et al. 2014) including the nematodes (Bhau et al. 2016; Sabry et al. 2019; Shekoohi et al. 2021). Hence, they can be utilized to control PPNs without excessive in order not to harm the beneficial nematodes (Taha and Abo-Shady 2016).

Moodley et al. (2018) invented safe approaches to optimize biosynthesis of AgNPs from moringa plant, *Moringa oleifera*. They used sun's rays as the major origin to generate energy necessary to form nanoparticles from fresh and freeze-dried leaf material. In such techniques, *Moringa* leaves offered a favourable and environmentally friendly candidate for green synthesis of antimicrobial agents with potential bioactive AgNPs applications. Also, size control and production of safe and effective nanopesticides to boost target efficiency and decrease undesirable side effects are ongoing (Wang et al. 2021). Those authors developed buprofezin nanoparticles with tunable size that have more homogeneity and reproducibility than those of the traditional method.

Nevertheless, precautions should be exercised to meet ISO international standards. They require safer compounds but fewer toxic solvents/reagents in order to synthesize metal nanoparticles (Carrillo-López et al. 2014). On the other hand, the Environmental Protection Agency allowed using herbicides in the formulation of, as carrier to deliver, PPN control nanomaterials or nematicides (Sabry 2019). Moreover, utilizing green nanoparticles engineering processes that have bio-compatibility for PPN control is encouraged (Zhao et al. 2013). Hence, using plant extracts, proteins, and microorganisms as biogenic approaches are receiving increased attention in forming metal nanoparticles (Adil et al. 2015; Shang et al. 2019; Safeena and Zakeel 2020). As molecules of plant extracts= and other biological materials are gained from the nature, synthesizing silver nanoparticles via the green forming approach is the best technique (Mie et al. 2014). In this respect, employment of not only plants (Yin et al. 2012; Kalaiselvi et al. 2017) but other biologicals such as algal species are also in use to synthesize nanoparticles (Abdellatif et al. 2016) due to their low cost of production, potential effects against PPNs, and/or stability to other abiotic parameters; e.g. tolerating high pressure and/or temperature (Sherly et al. 2011; Shekoohi et al. 2021). Although using such green silver nanoparticles did not cause any phytotoxicity in the treated eggplants, change in DNA profile was detected via using two molecular markers, random amplified polymorphic DNA and expressed sequence tag (Abdellatif et al. 2016). On the other hand, rice germination and growth was not adversely affected by AgNPs treatment that was effective in suppressing *Meloidogyne graminicola* populations on rice roots (Baronia et al. 2020). Safe and efficient tools such as the microwave device are suggested as an environmentally adaptive option to obtain the organic compounds from plants to synthesize NPs (Annu et al. 2017). Eventually, the toxicity and/or side impacts of nanoparticles are closely linked to their shape, chemical components and concentrations used, surface charge and properties, size, and fluid attributes (Rai et al. 2021).

2.8 CONCLUSIONS AND FUTURE PROSPECTS

As several traditional but unhealthy chemical nematicides are banned or restricted in use, efficient and safe alternatives should be established and developed for large scale applications. Using NNM are favourable alternatives that can be further optimized for sustainable farming approaches especially in integrated management programs of plant pathogens and pests. Nanomaterials have supreme merits in providing wide specific surface areas in the form of many fertilizers and pesticides such as nematicides. Furthermore, emerging strategies should employ NNM as distinctive carriers to ease the site-aimed controlled allocation of their related active ingredients to secure increased crop protection against the nematodes. Progress made in other nanotools, and disciplines should also back the new trend of using NNM. For example, sunlight could be used as environmentally safe source of energy in the syntheses of nanoparticles using favourable physical properties from dry and fresh plant leaf extracts. Nanosensors may be integrated with relevant biological and ecological factors to identify and subsequently optimize the ecological settings necessary for effective nematocidal applications. Furthermore, additional studies regarding the behaviour of NNM, their economic feasibility, and fate in the environment will assist to establish a regulatory framework for their sound commercialization. Boosting commercial demand of NNM should also be supported by further technological, ecological, intrinsic, regulatory and market-driving factors in parallel to those reported for other emerging nematicides.

ACKNOWLEDGEMENT

This study was supported in part by the US-Egypt project cycle 22 (no. 47725) entitled '**Setting and assessing integrated pest management tactics to optimize livestock and honeybee production**' related to STDF. The study was also supported in part by the nrc in house project no. 13050112 entitled 'pesticide alternatives against soil-borne pathogens and pests attacking economically significant export crops' headed by Prof. Dr. Mahfouz Abd-Elgawad.

REFERENCES

Abbassy, M. A., Abdel-Rasoul, M. A., Nassar, A. M. K. & Soliman, B. S. M. (2017). Nematocidal activity of silver nanoparticles of botanical products against root-knot nematode, *Meloidogyne incognita. Archives of Phytopathology and Plant Protection, 50*(17–18).

Abd-Elgawad, M. M. M. (2024). Nanonematicides: production, mechanisms, efficacy, opportunities and challenges. *Nematology, 26*(5), 479–489. https://doi.org/10.1163/15685411-bja10324

Abd-Elgawad, M. M. M. & Askary, T. H. (2015). Impact of phytonematodes on agriculture economy. In: Askary, T. H. & Martinelli, P. R. P. (eds) *Biocontrol Agents of Phytonematodes*. CAB International, Wallingford, UK, pp. 3–49.

Abd-Elgawad, M. M. M. & Askary, T. H. (2018). Fungal and bacterial nematicides in integrated nematode management strategies. *Egyptian Journal of Biological Pest Control, 28*, 74.

Abd-Elgawad, M. M. M. (2008). The current status of phytonematode management in Egypt with special reference to applicable nematicides. *Egyptian Journal of Agronematology, 6*, 33–46.

Abd-Elgawad, M. M. M. (2020). Plant-parasitic nematodes and their biocontrol agents: Current status and future vistas. In: Ansari, R. A., Rizvi, R. & Mahmood, I. (eds) *Management of Phytonematodes: Recent Advances and Future Challenges*. Springer Nature Singapore Pte Ltd., Singapore, pp. 171–204. www.springer.com/gp/book/9789811540868.

Abd-Elgawad, M. M. M. (2020a). Optimizing biological control agents for controlling nematodes of tomato in Egypt. *Egyptian Journal of Biological Pest Control, 30*, 58.

Abd-Elgawad, M. M. M. (2020b). Biological control agents in the integrated nematode management of pepper in Egypt. *Egyptian Journal of Biological Pest Control, 30*, 70.

Abd-Elgawad, M. M. M. (2020c). Biological control agents in the integrated nematode management of potato in Egypt. *Egyptian Journal of Biological Pest Control, 30*, 121.

Abd-Elgawad, M. M. M. (2021). Optimizing safe approaches to manage plant-parasitic nematodes. *Plants, 10*, 1911.

Abd-Elgawad, M. M. M. (2022). Exploiting plant-phytonematode interactions to upgrade safe and effective nematode control. *Life, 12*, 1916.

Abd-Elgawad, M. M., El-Mougy, N. S., El-Gamal, N. S., Abdel-Kader, M. M. & Mohamed, M. (2010). Protective treatments against soilborne pathogens in citrus orchards. *Journal of Plant Protection Research*, 50(4), 512–519.

Abd-El-Kareem, F., Elshahawy, I. E. & Abd-Elgawad, M. M. M. (2019). Local *Trichoderma* strains as a control strategy of complex black root rot disease of strawberry in Egypt. *Bulletin of the National Research Centre, 43*, 160.

Abd-El-Kareem, F., Elshahawy, I. E. & Abd-Elgawad, M. M. M. (2021). Application of *Bacillus pumilus* isolates for management of black rot disease in strawberry. *Egyptian Journal of Biological Pest Control, 31*, 25.

Abdellatif, K. F., Abdelfattah, R. H. & El-Ansary, M. S. M. (2016). Green nanoparticles engineering on root-knot nematode infecting eggplants and their effect on plant DNA modification. *Iran Journal of Biotechnology, 14*(4), 250–259.

Abd-Elsalam, K. A. (2021a). *Nanotechnology for Plant Protection: Silver Nanomaterials for Agri-Food Applications*. Elsevier Science Publishing Co Inc, New York, USA, p. 764.

Abd-Elsalam, K. A. (2021b). Special issue: Fungal nanotechnology. *Journal of Fungi, 7*, 583.

Adil, S. F., Assal, M. E., Khan, M., Al-Warthan, A., Siddiqui, M. R. H. & Liz-Marzán, L. M. (2015). Biogenic synthesis of metallic nanoparticles and prospects toward green chemistry. *Dalton Transactions, 44*, 9709–9717.

Ahamed, M., Posgai, R., Gorey, T. J., Nielsen, M., Hussain, S. M. & Rowe, J. J. (2010). Silver nanoparticles induced heat shock protein 70, oxidative stress and apoptosis in *Drosophila melanogaster. Toxicology* and *Applied Pharmacology, 242*(3), 263–269.

Ahmad, A., Mukherjee, P., Senapati, S., Mandal, D., Khan, M., Kumar, I. R. & Sastry, M. (2003). Extracellular biosynthesis of silver nanoparticles using the fungus *Fusarium oxysporum. Colloids Surfaces B: Biointerfaces, 27*, 313–318.

Al Banna, L. S., Salem, N. M. & Awwad, A. M. (2020). Green synthesis of sulfur nanoparticles using *Rosmarinus officinalis* leaves aqueous extract and nematocidal activity against *Meloidogyne javanica. Chemistry International, 6*(3), 137–143.

Amelia, M., Lincheneau, C., Silvi, S. & Credi, A. (2012). Electrochemical properties of CdSe and CdTe quantum dots. *Chemical Society Reviews*, *41*, 5728–5774.

Annu, S. A., Chaudhry, S. A. & Ikram, S. (2017). A review on biogenic synthesis of ZnO nanoparticles using plant extracts and microbes a prospect towards green chemistry. *Journal of Photochemistry and Photobiology B: Biology*, *166*, 272–284.

Ansari, R. A. & Khan, T. A. (2012a). Parasitic association of root-knot nematode, *Meloidogyne incognita* on guava. *e-Journal of Science Technology*, *5*(12), 65–67.

Ansari, R.A. & Khan, T.A. (2012b). Diversity and community structure of phytonematodes associated with guava in and around Aligarh, Uttar Pradesh, India. *Trends in Biosciences*, *5*(3), 202–204.

Ansari, R. A. & Mahmood, I. (2017a). Determination of disease incidence caused by *Meloidogyne* spp. and or *Fusarium udum* on pigeonpea in Aligarh district: A survey. *Trends in Biosciences*, *10*(24), 5239–5243.

Ansari, R. A. & Mahmood, I. (2017b). Optimization of organic and bio-organic fertilizers on soil properties and growth of pigeon pea. *Scientia Horticulturae*, *226*, 1–9.

Ansari, R. A. & Mahmood, I. (2019a). *Plant Health Under Biotic Stress – Volume II, Microbial Interactions*, Springer Nature, Singapore. ISBN: 978-981-13-6040-4.

Ansari, R. A. & Mahmood, I. (2019b). *Plant Health Under Biotic Stress – Volume I, Organic Strategies*, Springer Nature, Singapore. ISBN: 978-981-13-6043-5.

Ansari, R. A., Mahmood, I., Rizvi, R., Sumbul, A. & Safiuddin. (2017). Siderophores: augmentation of soil health and crop productivity. In: Kumar, V. (ed) *Probiotics in Agroecosystem*, Springer Nature, Singapore, pp. 291–312.

Ansari, R. A., Rizvi, R. & Mahmood, I. (2020). *Management of Phytonematodes: Management of Phytonematodes: Recent Advances and Future Challenges*, Springer Nature, Singapore. ISBN: 978-981-15-4087-5.

Ansari, R. A., Rizvi, R. & Mahmood, I. (2024). *Mycorrhizal Symbiosis and Agroecosystem Restoration*, Springer Nature, Singapore.

Ansari, R. A., Rizvi, R., Sumbul, A. & Mahmood, I. (2020). Plant-growth-promoting rhizobacteria (PGPR)-based sustainable management of phytoparasitic nematodes: Current understandings and future challenges. In: *Management of Phytonematodes: Recent Advances and Future Challenges*, pp. 51–71.

Ansari, R. A., Sumbul, A., Rizvi, R. & Mahmood, I. (2019). Organic soil amendments: Potential tool for soil and plant health management. In: *Plant Health Under Biotic Stress: Volume 1: Organic Strategies*, Springer Nature, Singapore, pp. 1–35.

Ansari, R. A., Sumbul, A., Rizvi, R. & Mahmood, I. (2024). Impact of environmental gases on mycorrhizal symbiosis and its influence on ecosystem functioning under the current climate change scenario. In: Ansari, R. A., Rizvi, R. & Mahmood, I. (eds) *Mycorrhizal Symbiosis and Agroecosystem Restoration*. Springer Nature Singapore, Singapore, pp. 51–76.

Ardakani, A. S. (2013). Toxicity of silver, titanium and silicon nanoparticles on the root-knot nematode, *Meloidogyne incognita* and growth parameters of tomato. *Nematology*, *15*(6), 671–677.

Barbosa, A., Silva, L. P. C., Ferraz, C. M., Tobias, F. L., de Araujo, J. V., Loureiro, B., Braga, G., Veloso, F. B. R., Soares, F. E. F., Fronza, M. & Braga, F. R. (2019). Nematocidal activity of silver nanoparticles from the fungus *Duddingtonia flagrans*. *International Journal of Nanomedicine*, *14*, 2341–2348.

Baronia, R., Kumar, P., Singh, S. P. & Walia, R. K. (2020). Silver nanoparticles as a potential nematicide against *Meloidogyne graminicola*. *Journal of Nematology*, *52*, 1–9.

Bhau, B., Phukon, P., Ahmed, R., Gogoi, B., Borah, B., Baruah, J., Sharma, D. & Wann, S. (2016). A novel tool of nanotechnology: Nanoparticle mediated control of nematode infection in plants. In: Singh, D., Singh, H. & Prabha, R. (eds) *Microbial Inoculants in Sustainable Agricultural Productivity*. Springer, New Delhi, India, pp. 253–269.

Boyno, G., Ansari, R. A., Durak, E. D., Güneş, H., Çevik, R. & Demir, S. (2024). Arbuscular mycorrhizal technology in sustainable agriculture: Current knowledge and challenges in agroforestry. In: Ansari, R. A., Rizvi, R. & Mahmood, I.(eds) *Mycorrhizal Symbiosis and Agroecosystem Restoration*,, Springer Nature, Singapore, pp. 173–195.

Campos-Herrera, R., Pathak, E., El-Borai, F. E., Schumann, A., Abd-Elgawad, M. M. M. & Duncan, L. W. (2013). New citriculture system suppresses native and augmented entomopathogenic nematodes. *Biological Control*, *66*, 183–194.

Cao, J., Guenther, R. H., Sit, T. L., Lommel, S. A., Opperman, C. H. & Willoughby, J. A. (2015). Development of abamectin loaded plant virus nanoparticles for efficacious plant parasitic nematode control. ACS *Applied Materials & Interfaces*, *7*(18), 9546–9553.

Carrillo-López, L. M., Zavaleta-Mancera, H. A., Vilchis-Nestor, A., Soto-Hernández, R. M., Arenas-Alatorre, J., Trejo-Téllez, L. I. & Gómez-Merino, F. (2014). Biosynthesis of silver nanoparticles using *Chenopodium ambrosioides*. *Journal of Nanomaterials*, 951746.

Chariou, P. L., Steinmetz, N. F. (2017). Delivery of pesticides to plant parasitic nematodes using tobacco mild green mosaic virus as a nanocarrier. *ACS Nano*, *11*(5), 4719–4730.

Chitwood, D. J. (2003). Nematicides. In: Plimmer, J. R. (ed) *Encyclopedia of Agrochemicals*, Vol. 3. John Wiley & Sons, New York, pp. 1104–1115.

Choudhary, G., Kumar, J., Waliam S., Parsad, R. & Parmar, B. S. (2006). Development of controlled release formulations of carbofuran and evaluation of their efficacy against *Meloidogyne incognita*. *Journal of Agricultural* and *Food Chemistry*, *54*, 4727–4733.

Christian, P., Von der Kammer, F., Baalousha, M. & Hofmann, T. (2008). Nanoparticles: Structure, properties, preparation and behaviour in environmental media. *Ecotoxicology*, *17*(5), 326–343.

Coupland, J., Abd-Elgawad, M. M. M. & Askary, T. H. (2017). Beneficial nematodes and the changing scope of crop protection. In: Abd-Elgawad, M. M. M., Askary, T. H. & Coupland, J. (eds) *Biocontrol Agents: Entomopathogenic and Slug Parasitic Nematodes*. CAB International, Wallingford, UK, pp. 26–42.

Cromwell, W. A., Yang, J., Starr, J. L. & Jo, Y. K. (2014). Nematocidal effects of silver nanoparticles on root-knot nematode in bermudagrass. *Journal of Nematology*, *46*, 261–266.

Dang, T. M. D., Le, T. T. T., Fribourg-Blanc, E. & Dang, M. C. (2011). Synthesis and optical properties of copper nanoparticles prepared by a chemical reduction method. *Advances in Natural Sciences: Nanoscience and Nanotechnology*, *2*(6), 015009.

de Freitas Bueno, R. C. O., Ansari, R. A., Lima, G. P. P. & Sakate, R. K. (2019). Phytosanitation: A novel approach toward disease management. In: Ansari, R. A. & Mahmood, I. (eds) *Plant Health Under Biotic Stress: Volume 1: Organic Strategies*, Springer Nature, Singapore Pte Ltd, pp. 73–90.

Devi, T. S., Das, D., Ansari, R. A., Rizvi, R., Sumbul, A. & Mahmood, I. (2020). Role of organic additives in the sustainable management of phytoparasitic nematodes. In: Ansari, R. A., Rizvi, R. & Mahmood, I. (eds) *Management of Phytonematodes: Recent Advances and Future Challenges*, Springer Nature, Singapore Pte Ltd, pp. 279–295.

Dutta, T. K. & Phani, V. (2023). The pervasive impact of global climate change on plant-nematode interaction continuum. *Frontiers in Plant Science*, *14*, 1143889.

Dutta, T. K., Khan, M. R. & Phani, V. (2019) Plant-parasitic nematode management via biofumigation using brassica and non-brassica plants: Current status and future prospects. *Current Plant Biology*, *17*, 17–32.

El-Ansary, M. S. M., Hamouda, R. A. & Elshamy, M. M. (2022). Using biosynthesized zinc oxide nanoparticles as a pesticide to alleviate the toxicity on banana infested with parasitic nematode. *Waste and Biomass Valorization*, *13*(1), 405–415.

El-Sherif, A. G., Gad, S. B., Megahed, A. A. & Sergany, M. I. (2019). Induction of tomato plants resistance to *Meloidogyne incognita* infection by mineral and nano-fertilizer. *Journal of Entomology and Nematology*, *11*(2), 21–26.

Fabiyi, O. A., Alabi, R. O. & Ansari, R. A. (2020). Nanoparticles' synthesis and their application in the management of phytonematodes: An over-view. In: Ansari, R. A., Rizvi, R. & Mahmood, I. (eds) *Management of Phytonematodes: Recent Advances and Future Challenges*. Springer Nature Singapore Pte Ltd., Singapore, pp. 125–140.

Fabiyi, O. A., Alabi, R. O. & Ansari, R. A. (2020). Nanoparticles' synthesis and their application in the management of phytonematodes: An overview. In: Ansari, R. A., Rizvi, R. & Mahmood, I. (eds.) *Management of phytonematodes: Recent Advances and Future Challenges*, Springer Nature, Singapore Pte Ltd, pp. 125–140.

Fabiyi, O. A., Ansari, R. A., Rizvi, R., Sumbul, A. & Mahmood, I. (2024). Influence of arbuscular mycorrhizal fungi on soil health amelioration and plant fitness under hostile environment. In: Ansari, R. A., Rizvi, R. & Mahmood, I. (eds.) *Mycorrhizal Symbiosis and Agroecosystem Restoration*. Springer Nature Singapore, , pp. 227–248.

FAO. (2017). *The Future of Food and Agriculture – Trends and Challenges*. Rome. https://www.fao.org/publications/en. Accessed on 17 Aug 2024.

Forghani, F. & Hajihassani, A. (2020). Recent advances in the development of environmentally benign treatments to control root-knot nematodes. *Frontiers in Plant Science*, *11*, 1125.

Forster, H., Wolfrum, C. & Peukert, W. (2012). Experimental study of metal nanoparticle synthesis by an arc evaporation/condensation process. *Journal of Nanoparticle Research*, *14*, 926–941.

Fu, Z., Chen, K., Li, L., Zhao, F., Wang, Y., Wang, M., Shen, Y., Cui, H., Liu, D. & Guo, X. (2018). Spherical and spindle-like abamectin-loaded nanoparticles by flash nanoprecipitation for Southern root-knot nematode control: Preparation and characterization. *Nanomaterials*, *8*(6), 449.

Fullana, A. M., Expósito, A., Escudero, N., Cunquero, M., Loza-Alvarez, P., Giné, A. & Sorribas, F. J. (2023). Crop rotation with *Meloidogyne*-resistant germplasm is useful to manage and revert the (a)virulent populations of Mi1.2 gene and reduce yield losses. *Frontiers in Plant Science*, *14*, 1133095.

Ghareeb, R. Y., Alfy, H., Fahmy, A. A., Ali, H. M. & Abdelsalam, N. R. (2020). Utilization of *Cladophora glomerata* extract nanoparticles as econematicide and enhancing the defense responses of tomato plants infected by *Meloidogyne javanica*. *Scientific Reports*, *10*, 19968.

Gkanatsiou, C. H., Ntalli, N., Menkissoglu-Spiroudi, U. & Dendrinou-Samara, C. (2019). Essential metal-based nanoparticles (copper/iron NPs) as potent nematocidal agents against *Meloidogyne* spp. *Journal of Nanotechnology Research*, 1(2), 44–58.

Hachicho, N., Hoffmann, P., Ahlert, K. & Heipieper, H. J. (2014). Effect of silver nanoparticles and silver ions on growth and adaptive response mechanisms of *Pseudomonas putida* mt-2. *FEMS Microbiology Letters*, *355*, 71–77.

Hamed, S. M., Hagag, E. S. & El-Raouf, N. A. (2019). Green production of silver nanoparticles, evaluation of their nematocidal activity against *Meloidogyne javanica* and their impact on growth of *faba bean*. *Beni-Suef University Journal: Basic and Applied Science*, *8*(9), 1–12.

Handoo, Z. A., Mowery, J. D. & Chitwood, D. J. (2013). The history of the USDA nematode collection and its database: Valuable resources for taxonomic research and identification. Proceedings of the 10th Int. Nematological Symposium, Russian Society of Nematologists, Golitsyno – Bolshie Vyazemy/ Moscow, 1–5 July 2013, pp. 110–113.

Haq, I. U. & Ijaz, S. (2019). Use of metallic nanoparticles and nanoformulations as nanofungicides for sustainable disease management in plants. In: Prasad, R., Kumar, V., Kumar, M. & Choudhary, D. (eds) *Nanobiotechnology in bioformulations*. Springer, Cham, Switzerland, pp. 289–316.

Hardman, R. (2006). Toxicologic review of quantum dots: Toxicity depends on physicochemical and environmental factors. *Environmental Health Perspective*, *114*, 165–172.

Hassan, M. E. M., Zawam, H. S., El-Nahas, S. E. M. & Desoukey, A. F. (2016). Comparative study between silver nanoparticles and two nematicides against *Meloidogyne incognita* on tomato seedlings. *Plant Pathology Journal*, *15*, 144–151.

Heflish, A. A., Hanfy, A. E., Ansari, M. J., Dessoky. E. S., Attia, A. O., Elshaer, M. M., Gaber, M. K., Kordy, A., Doma, A. S., Abdelkhalek, A. & Behiry, S. I. (2021). Green biosynthesized silver nanoparticles using *Acalypha wilkesiana* extract control root-knot nematode. *Journal of King Saud University–Science*, *33*(6), 101516.

Ibrahim, D. S. S., Elderiny, M. M., Ansari, R. A., Rizvi, R., Sumbul, A. & Mahmood, I. (2020). Role of *Trichoderma* spp. in the management of plant-parasitic nematodes infesting important crops. In: Ansari, R.A., Rizvi, R. & Mahmood, I. (eds.) *Management of Phytonematodes: Recent Advances and Future Challenges*, Springer Nature, Singapore, pp. 259–278.

Ibrahim, D. S. S., Mostafa, M., Abdellatif, A. A. & Ansari, R. A. (2022). Sustainable management of soil-borne plant pathogens. In: Furze, J.N., Eslamian, S., Raafat, S.M. and Swing, K. (eds.) *Earth Systems Protection and Sustainability: Volume 1*. Springer International Publishing, Cham, pp. 35–56.

Ibrahim, H. M. M., Ahmad, E. M., Martínez-Medina, A. & Aly, M. A. M. (2019). Effective approaches to study the plant-root knot nematode interaction. *Plant Physiology and Biochemistry*, *141*, 332–342.

Jeevanandam, J., Barhoum, A., Chan, Y. S., Dufresne, A. & Danquah, M. K. (2018). Review on nanoparticles and nanostructured materials: History, sources, toxicity and regulations. *Beilstein Journal of Nanotechnology*, *9*, 1050–1074.

Jo, Y., Starr, J. L. & Deng Y. (2014) Use of silver nanoparticles for nematode control on the bermudagrass putting green. *Turfgrass and Environmental Research Online*, *13*(2), 16–18.

Jones, J. T., Haegeman, A., Danchin, E. G. J., Gaur, H. S., Helder, J., Jones, M. G. K., Kikuchi, T., Manzanilla-Lopez, R., Palomares-Rius, J. E., Wesemael, W. M. L. & Perry, R. N. (2013). Review: Top 10 plant-parasitic nematodes in molecular plant pathology. *Molecular Plant Pathology*, *14*(9), 946–961.

Kabeil, S. A., Hammam, M. M. A., Mohamed, M. M. M. & Abd-Elgawad, M. M. M. (2016). Biocontrol of potato brown rot and root-knot diseases in Egypt. *Egyptian Journal of Biological Pest Control*, *26*, 209–215.

Kalaiselvi, D., Sundararaj, P., Premasudha, P. & Hafez, S. L. (2017). Nematocidal activity of green synthesized silver nanoparticles using plant extracts against root-knot nematode *Meloidogyne incognita*. *International Journal of Nematology*, *27*, 81–94.

Kalimuthu, K., Babu, R. S., Venkataraman, D., Mohd, B. & Gurunathan, S. (2008). Biosynthesis of silver nanocrystals by *Bacillus licheniformis*. *Colloids and Surfaces B: Biointerfaces*, *65*, 150–153.

Keshari, N., Kranti, K. V. V. S. K., Gunda, N. K. & Ansari, R. A. (2024). Arbuscular mycorrhizal fungi: A potential agent for phytonematodes management in diverse agro-climatic zones. In: Ansari, R.A., Rizvi, R. and Mahmood, I. (eds.) *Mycorrhizal Symbiosis and Agroecosystem Restoration*. Springer Nature Singapore, Singapore, pp. 147–169.

Khan, A., Khan, A. A., Jameel, M., Khan, M. F., Khan, M., Khan, A., Ahmad, F. & Alam, M. (2023). Grass-shaped zinc oxide nanoparticles synthesized by the sol-gel process and their antagonistic properties towards the biotrophic parasite, *Meloidogyne incognita*. *Bioinorganic Chemistry and Applications*, *2023*, 1–14.

Khan, A., Rashid, R., Murtaza, G. & Zahra, A. (2014). Gold nanoparticles: Synthesis and applications in drug delivery. *Tropical Journal of Pharmaceutical Research*, *13*, 1169–1177.

Li, P., Xu, T., Wu, S., Lei, L. & He, D. (2017). Chronic exposure to graphene-based nanomaterials induces behavioural deficits and neural damage in *Caenorhabditis elegans*. *Journal of Applied Toxicology*, *37*(10), 1140–1150.

Lim, D., Roh, J. Y., Eom, H. J., Hyun, J. W. & Choi, J. (2012). Oxidative stress–related PMK-1 P38 MAPK activation as a mechanism for toxicity of silver nanoparticles to reproduction in the nematode *Caenorhabditis elegans*. *Environmental Toxicology and Chemistry*, *31*(3), 585–592.

Loha, K. M., Shakil, N. A., Kumar., J., Singh, M. K., Adak, T. & Jain, S. (2011). Release kinetics of beta-cyfluthrin from its encapsulated formulations in water. *Journal of Environmental Science and Health, Part B*, *46*, 201–206.

Luc, M., Bridge, J. & Sikora, R. A. (2005). Reflections on nematology in sub-tropical and tropical agriculture. In: Luc, M., Sikora, R. A. & Bridge, J. (eds) *Plant Parasitic Nematodes in Sub-tropical and Tropical Agriculture*. CAB International, Wallingford, UK, pp. 1–10.

Mahmood, I., Rizvi, R., Sumbul, A. & Ansari, R. A. (2019). Potential role of plant growth promoting rhizobacteria in alleviation of biotic stress. In: Ansari, R.A. & Mahmood, I. (eds.) *Plant Health Under Biotic Stress: Volume 2: Microbial Interactions*, Springer Nature Singapore, Singapore, pp. 177–188.

Malandrakis, A. A., Kavroulakis, N. & Chrysikopoulos, C. V. (2019). Use of copper, silver and zinc nanoparticles against foliar and soil-borne plant pathogens. *Science of the Total Environment*, *670*, 292–299.

Malik, M. A., Wani, M. Y. & Hashim, M. A. (2012). Microemulsion method: A novel route to synthesize organic and inorganic nanomaterials. *Arabian Journal of Chemistry*, *5*(4), 397–417.

Mie, R., Samsudin, M. W., Din, L. B., Ahmad, A., Ibrahim, N. & Adnan, S. N. (2014). Synthesis of silver nanoparticles with antibacterial activity using the lichen *Parmotrema praesorediosum*. *International Journal Nanomedicine*, *9*, 121–127.

Mittal, D., Kaur, G., Singh, P., Yadav, K. & Ali, S. A. (2020). Nanoparticle-based sustainable agriculture and food science: Recent advances and future outlook. *Frontiers in Nanotechnology*, *2*, 579954.

Moodley, J. S., Krishna, S. B. N., Pillay, K. & Govender, P. (2018). Green synthesis of silver nanoparticles from *Moringa oleifera* leaf extracts and its antimicrobial potential. *Advances in Natural Sciences:* Nanoscience *and Nanotechnology*, *9*(1), 015011.

Moon, J., Kwak, J. I. & An, Y. (2019). The effects of silver nanomaterial shape and size on toxicity to *Caenorhabditis elegans* in soil media. *Chemosphere*, *215*, 50–56.

Moosavi, M. R. & Zare, R. (2015). Factors affecting commercial success of biocontrol agents of phytonematodes. In: Askary, T. H. & Martinelli, P. R. P. (eds) *Biocontrol Agents of Phytonematodes*. CAB International, Wallingford, UK, pp. 423–445.

Nassar, A. M. K. (2016). Effectiveness of silver Nano-particles of extracts of *Urtica urens* (Urticaceae) against root-knot nematode *Meloidogyne incognita*. *Asian Journal Nematology*, *5*, 14–19.

Noling, J. W. (2019). *Nematode Management in Tomatoes, Peppers, and Eggplant*. University of Florida publication Series no. ENY-032, USA, p. 16.

Nour El-Deen, A. H. & El-Deeb, B. A. (2018). Effectiveness of silver nanoparticles against root-knot nematode, *Meloidogyne incognita* infecting tomato under greenhouse conditions. *Journal of Agricultural Science*, *10*(2), 148–156.

Ntalli, N. & Caboni, P. (2017). A review of isothiocyanates biofumigation activity on plant parasitic nematodes. *Phytochemistry Reviews*, *16*, 827–834.

Ntalli, N., Adamski, Z., Doula, M. & Monokrousos, N. (2020). Nematocidal amendments and soil remediation. *Plants*, 9, 429.

Nuruzzaman, M., Rahman, M. M., Liu, Y. J. &, Naidu, R. (2016). Nanoencapsulation, nano-guard for pesticides: A new window for safe application. *Journal of Agricultural and Food Chemistry*, *64*, 1447–1483.

Pankaj, P., Shakil, N. A., Kumar, J., Singh, M. K. &, Singh, K. (2012). Bioefficacy evaluation of controlled release formulations based on amphiphilic nano-polymer of carbofuran against *Meloidogyne incognita* infecting tomato. *Journal of Environmental Science and Health, Part B*, *47*, 520–528.

Patil, S., Ansari, R. A., Gurikar, C., Bhavan, S. & Hariprasad, K. (2024). Propagation of mycorrhizal fungi and their strength in plant health ameliorations. In: Ansari R.A., Rizvi, R. & Mahmood, I. (eds.) *Mycorrhizal Symbiosis and Agroecosystem Restoration*. Springer Nature Singapore, Singapore, pp. 77–98.

Prasad, B. N., Subramanyam, B., Lakshmipathi, R. N., Ansari, R. A., Rizvi, R., Sumbul, A., ... & Rachmi, C. M. (2020). Utilization of beneficial microorganisms in sustainable control of phytonematodes. In: Ansari R.A., Rizvi, R. & Mahmood, I. (eds.) *Management of Phytonematodes: Recent Advances and Future Challenges*, Springer Nature, Singapore, pp. 317–337.

Priyanka, P., Kumar, D., Yadav, K. & Yadav, A. (2019). Nanopesticides: Synthesis, formulation and application in agriculture. *Nanobiotechnology Applications in Plant Protection*, *7*, 129–143.

Rai, M., Bonde, S., Golinska, P., Trzcinska-Wencel. J., Gade, A., Abd-Elsalam, K. A., Shende, S., Gaikwad, S. & Ingle, A. P. (2021). *Fusarium* as a novel fungus for the synthesis of nanoparticles: Mechanism and applications. *Journal of Fungi*, *7*, 139.

Rejeb, I. B., Pastor, V. & Mauch-Mani, B. (2014). Plant responses to simultaneous biotic and abiotic stress: Molecular mechanisms. *Plants*, *3*, 458–475.

Roh, J. Y., Sim, S. J., Yi, J., Park, K., Chung, K. H., Ryu, D. Y. & Choi, J. (2009). Ecotoxicity of silver nanoparticles on the soil nematode *Caenorhabditis elegans* using functional ecotoxicogenomics. *Environmental Science & Technology*, *43*(10), 3933–3940.

Sabry, A. H. (2019). Role of nanotechnology applications in plant-parasitic nematode control. In: Abd-Elsalam, K. A. & Prasad, R. (eds) *Nanobiotechnology Applications in Plant Protection, Nanotechnology in the Life Sciences*. Springer Nature, Switzerland AG, pp. 223–240.

Sabry, H., Nashaat, Y., Omar, N., Negm, A. & Shaheen, N. (2019). Comparative study of the antibacterial effect of nano-silver irrigant, sodium hypochlorite and chlorhexidine against enterococcus faecalis biofilm. *Egyptian Dental Journal*, *65*(2), 1503–1509.

Safeena, M. I. S. & Zakeel, M. C. M. (2020). Nanobiotechnology-driven management of phytonematodes. In: Ansari, R. A., Rizvi, R. & Mahmood, I. (eds) *Management of Phytonematodes: Recent Advances and Future Challenges*. Springer Nature Singapore Pte Ltd., Singapore, pp. 1–33.

Saifuddin, N., Wong, C. W. & Nur yasumira, A. A. (2009). Rapid biosynthesis of silver nanoparticles using culture supernatant of bacteria with microwave irradiation. *European Journal of Chemistry*, *6*, 61–70.

Salama, H. S. & Abd-Elgawad, M. M. M. (2003). Quarantine problems: An analytical approach with special reference to palm weevils and phytonematodes. *Archives of Phytopathology and Plant Protection*, *36*, 41–46.

Shang, Y., Kamrul Hasan, M., Ahammed, G. J., Li, M., Yin, H. & Zhou, J. (2019). Applications of nanotechnology in plant growth and crop protection: A review. *Molecules*, *24*, 2558.

Sharma, H., Dhirta, B. & Shirkot, P. (2017). Evaluation of biogenic iron nano formulations to control *Meloidogyne incognita* in okra. *International Journal of Chemical Studies*, *5*, 1278–1284.

Shehata, I. E., Hammam, M. M. A. & Abd-Elgawad, M. M. M. (2021). Effects of inorganic fertilizers on virulence of the entomopathogenic nematode *Steinernema glaseri* and peanut germination under field conditions. *Agronomy*, 11, 945.

Shehata, I. E., Hammam, M. M. A., El-Borai, F. E., Duncan, L. W. & Abd-Elgawad, M. M. M. (2019). Comparison of virulence, reproductive potential, and persistence among local *Heterorhabditis indica* populations for the control of *Temnorhynchus baal* (Reiche & Saulcy) (Coleoptera: Scarabaeidae) in Egypt. *Egyptian Journal of Biological Pest Control*, 29, 32.

Shekoohi, S. S., Charehgani, H., Abdollahi, M. & Rajabi, H. R. (2021). Combined effect of β-aminobutyric acid and silver nanoparticles on eggplants, *Solanum melongena*, infected with *Meloidogyne javanica*. *Nematology*, *23*(9), 1–16.

Sherly, E. D., Madgula, K. & Kakkar, R. (2011). Green nanotechnology for cleaner environment present and future research needs. *Current World Environment*, *6*, 177–181.

Sikora, R. A., Bridge, J. & Starr, J. L. (2005). Management practices: An overview of integrated nematode management technologies. In: Luc, M., Sikora, R. A. & Bridge, J. (eds) *Plant Parasitic Nematodes in Subtropical and Tropical Agriculture*, 2nd ed. CAB International, Wallingford, UK, pp. 793–825.

Solanki, M. K., Kashyap, P. L., Ansari, R. A. & Kumari, B. (2020). *Microbiomes and Plant Health: Panoply and Their Applications*, Academic Press, Elsevier, The Netherland. ISBN: 9780128197158. www.elsevier.com/books/microbiomes-and-plant-health/solanki/978-0-12-819715-8.

Solanki, M. K., Kashyap, P. L., Kumari, B., Ansari, R. A., Sumbul, A., Rizvi, R. & Mahmood, I. (2021). Mycorrhizal fungi and its importance in plant health amelioration. In: Solanki, M.K., Ansari, R.A., Kashyap, P.L., & Kumari, B. (eds.) *Microbiomes and Plant Health*. Academic Press, Amsterdam, pp. 205–223.

Sumbul, A., Ansari, R. A., Rizvi, R. & Mahmood, I. (2020). Azotobacter: A potential bio-fertilizer for soil and plant health management. *Saudi Journal of Biological Sciences*, *27*(12), 3634–3640.

Taha, E. H. & Abo-Shady, N. M. (2016). Effect of silver nanoparticles on the mortality pathogenicity and reproductivity of entomopathogenic nematodes. *International Journal of Zoological Research*, *12*, 47–50.

Taha, R. A. (2016). Nano carbon applications for plant. *Advances in Plants & Agriculture Research*, *5*(2), 483–484.

Taskinen, H., Lindbohm, M. & Sallmén, M. (2011). Occupational exposure to chemicals and reproductive health. In: Gupta, R. C. (Ed.) *Reproductive and Developmental Toxicology*. Academic Press, London, UK, pp. 949–959.

Thakur, R. K. & Shirkot, P. (2017). Potential of biogold nanoparticles to control plant pathogenic nematodes. *Journal of Bioanalysis & Biomedicine*, *9*, 220–222.

Thiruvenkataswamy, S., Arul, N. & Paulpandi, M. (2021). Biosynthesis, characterization, nematocidal efficacy of silver nanoparticles synthesized using *Solanum nigrum* fruit against root knot nematode *Meloidogyne incognita*. *Journal of Nanoscience & Nanotechnology*, *4*(2), 243–246.

Tien, D-C., Tseng, K-H., Liao, C-Y., Huang, J-C. & Tsung, T. T. (2008). Discovery of ionic silver in silver nanoparticle suspension fabricated by arc discharge method. *Journal of Alloys* and *Compounds*, *463*, 408–411.

Timper, P. (2014). Conserving and enhancing biological control of nematodes. *Journal of Nematology*, *46*, 75–89.

Tran, Q. H., Nguyen, V. Q. & Le, A. (2013). Silver nanoparticles: Synthesis, properties, toxicology, applications and perspectives. *Advances in Natural Sciences: Nanoscience and Nanotechnology*, 4, 033001.

Ur Rahim, H., Qaswar, M., Uddin, M., Giannini, C., Herrera, M. L. & Rea, G. (2021). Nano-enable materials promoting sustainability and resilience in modern agriculture. *Nanomaterials*, 11, 2068.

Usman, M. S., Zowalaty, M. E. E., Shameli, K., Zainuddin, N., Salama, M. & Ibrahim, N. A. (2013). Synthesis, characterization, and antimicrobial properties of copper nanoparticles. *International Journal of Nanomedicine*, *8*, 4467–4479.

Wang, C., Cui, B.,Wang, Y., Wang, M., Zeng, Z., Gao, F., Sun, C., Guo, L., Zhao, X. & Cui, H. (2021). Preparation and size control of efficient and safe nanopesticides by anodic aluminum oxide templates-assisted method. *International Journal of Molecular Sciences*, 22, 8348.

Yin, Y-h., Guo, Q-m., Han, Y., Wang, L-j. & Wan, S-q. (2012). Preparation, characterization and nematocidal activity of lansiumamide B nano-capsules. *Journal of Integrative Agriculture*, *11*, 1151–1158.

Yu, D., Chu, Y., Dong, L. H., Zhuo, Y. J. (2010). Controllable synthesis of $CaCO_3$ micro/nanocrystals with different morphologies in microemulsion. *Chemical Research: Chinese University*, *26*(5), 678–682.

Zamiri, R., Azmi, B. Z., Ahangar, H. A., Zamiri, G., Husin, M. S. & Wahab, Z. A. (2012). Preparation and characterization of silver nanoparticles in natural polymers using laser ablation. *Bulletin of Materials Science*, *35*(5), 727–731.

Zhao, Y-L., Wu, Q-L., Li, Y-P. & Wang, D-Y. (2013). Translocation, transfer, and *in vivo* safety evaluation of engineered nanomaterials in the non-mammalian alternative toxicity assay model of nematode *Caenorhabditis elegans*. *RSC Advances*, *3*, 5741–5757.

Zheng, Q., Putker, V. & Goverse, A. (2021). Molecular and cellular mechanisms involved in host-specific resistance to cyst nematodes in crops. *Frontiers in Plant Science*, 12, 641582.

3 Exploring the Potential of Nanomaterials in a Crusade against Plant Disease Management

Arsh Rehan Siddiqui, Rizwan Ali Ansari, Aisha Sumbul, Rose Rizvi, Noor Fatima, Safiuddin Safiuddin, Atul Upadhyay, Uma Shankar Rajput, and Irshad Mahmood

3.1 INTRODUCTION

As per recent projections, there is a urgent need to augment global food production by 2050 to alleviate hunger. The severity of this serious issue increases since climate abruptions are anticipated to alter the sequence of food production (Elmer & White, 2018). The current agricultural system faces a sizable challenge from crop losses caused by biotic and abiotic factors worldwide. Plant diseases can inflict a significant impact on crop yield, with losses of up to 25% attributable to these pathogenic invaders (Malandrakis et al., 2019). Traditional approaches are recognized as a leading and economically viable strategy for sustainable management of plant diseases (Pandey et al., 2018; Ansari & Mahmood 2017a; Ansari & Mahmood 2019a; Ansari and Mahmood 2019b; Ansari et al., 2020). Crop plants are given agrochemicals as a means of safeguarding themselves against various plant pathogens. The way active ingredients are delivered can increase costs, harm beneficial microbes, and negatively affect soil and water quality. It also harms beneficial organisms and important pollinators, and reduces soil qualities such as pH, soil texture, organic matter, and absorption rate (Ansari et al., 2020; Ansari et al., 2019; Mahmood et al., 2019; Ansari & Khan 2012;). Synthetic compounds can contribute to the persistence of diseases and environmental harm by leaving behind residues in the soil and water (Meena et al. 2020; Fabiyi et al., 2020; Patil et al., 2024; Solanki et al., 2021; Sumbul et al., 2020). This has given rise to significant concerns regarding the safety and sustainability of various agrochemicals.

Nanoparticles (NPs) offer a sustainable approach to agrochemical synthesis, leveraging their unique characteristics for the gradual release of active ingredients, low risk in developing of pathogens resistance, cost-effectiveness, regulated release, unique mode of action, and enhanced efficacy even at lower concentrations. NPs present a promising avenue for improving agricultural productivity by preventing the pathogens' growth (Kumar et al., 2021a). NPs possess the ability to transform into a complex system consisting of several active constituent agents, enabling them to effectively function within a target organism. The engineering process increases the performance of the target organism on multiple levels, making it more valuable in terms of functionality and efficacy (Singh et al. 2020; Saied et al. 2021). The NPs are the materials that can be either synthetic or naturally occurring, however, synthetic NPs have a variety of shapes including rods,

DOI: 10.1201/9781003256762-3

sheets, spherical balls, dendrimers, and carbon based multiple layered nanotubes (Husen and Jawaid 2020). Natural NPs, on the other hand, have a more variable and irregular shape. As per the report of Elmer and White (2018), Zheng et al. (2011) and Alghuthaymi et al. (2021), the potential of a wide range of NPs including those composed of metalloids (such as boron and silicon), non-metals (such as fullerenes, graphene oxide, and quantum dots), and metals (such as titanium, nickel, manganese, iron, copper, cerium, gold, aluminium, and silver) in suppressing plant pathogens have been ensured. NPs possess a significant surface area-to-volume ratio, enhancing their ability to bind and absorb nanomolecules such as proteins, RNA, DNA, high-affinity probes, and nanodrugs, facilitating their transport (Yin et al., 2012; Elmer and White 2018). NPs exhibit distinct properties in comparison to their counterparts with larger volumes. To synthesize NPs, researchers use a variety of techniques, which are regularly updated with new platforms emerging in the twenty-first century. One method for synthesizing NPs that has gained popularity is the use of green sources, which is a less harmful and environmentally friendly approach. NPs exhibit distinct properties in comparison to their counterparts with larger volumes. Single-step green synthesis is another method that allows for rapid synthesis at high pressure and temperature and is easier to scale up (Dreizin, 2009). The paradox of employing bacteria, fungi, and plants to impede disease activity simplifies the process of synthesizing NPs through environmentally friendly, green organic methods. If regulators and legislators decide to use these methods in plant pathology, they could play an important role in disease identification and control.

In the current climatic context, different abiotic and biotic stress entities present practical food security tasks (Ansari et al., 2024; Ansari and Mahmood 2017b; Boyno et al., 2024; Fabiyi et al., 2024;). NPs-based smart agriculture offers many advantages over traditional methods for ensuring necessary farming functions. NPs are crucial to the early pathogens' diagnosis and their treatment (Bhattacharya et al., 2016). The nanomaterial driven formulations have considerably enhanced the specificity, solubility, stability, and permeability with targets, slow-releasing qualities that ease plant nutrient management at pre and post-harvest stages, or prevent the early breakdown of active chemicals. Such target-driven delivery not only increases global food supply but also mitigates adverse environmental effects (Prasad et al., 2017; Ansari et al., 2017; Ansari et al., 2023). This chapter focuses on the synthesis, characterization, and application of NPs in the management plant diseases caused by a diverse array of pathogens. It delves into the challenges inherent in nanoparticle-related research, providing insights into potential solutions. The concluding section offers avenues for redressal, empowering future researchers to address and resolve these issues effectively.

3.2 BIOSYNTHESIS OF NPS

The biosynthesis of NPs has surfaced as a promising area of investigation owing to its environmentally friendly and cost-effective characteristics. This approach employs diverse biological sources, including bacteria, fungi, plants, and algae, to create NPs with desired properties. The use of green synthesis methods not only reduces the harmful impact on the environment but also enhances the biocompatibility and stability of the NPs. Studies have shown that biosynthesis of NPs may be utilized in various fields including the agricultural industry (Singh et al., 2020; Saied et al., 2021; Husen and Jawaid, 2020). NPs have unique optical, chemical, photoelectrochemical, and electronic properties, making their biosynthesis using biological entities more interesting (Peto et al. 2002). Biosynthesis of nanomaterials has emerged as a new and unexplored field of study, as natural processes can create inorganic materials at both nano- and micron scales (Sastry et al., 2004). To create environmentally friendly and non-toxic processes for the synthesis and assembly of NPs, researchers have investigated various natural sources such as microbes, fungi, and plants (Sastry et al., 2004).

The biomass of *Verticillium* sp. was observed to synthesize intracellular silver NPs in the presence of an aqueous $AgNO_3$ solution, whereas the biomass of *Fusarium oxysporum* was found to generate extracellular silver NPs (Dias et al 2002; Ahmad et al 2003; Bansal et al. 2004; Birla et al. 2013).

Moreover, the application of microorganisms, especially bacteria, yeasts, fungi, and actinomycetes, has been judiciously exploited for the enhanced production of NPs (Mandal et al., 2006; Gericke and Pinches 2006a). Various variables like pH, temperature, concentration of substrate and time of exposure to the substrate can influence the formation of particles considerably (Gericke and Pinches, 2006a). Furthermore, endeavours have been undertaken to adjust crucial growth parameters to regulate the pattern of gold NPs which are produced extracellularly by microorganisms (Gericke and Pinches, 2006b). The research is currently focused on producing NPs with precise control over their size, monodispersity, and chemical composition. By understanding and replicating this natural process, researchers anticipate creating NPs with a high level of control over their physical and chemical properties (Zhang and Xu 2013).

3.3 NP-MEDIATED PLANT DISEASE MANAGEMENT

3.3.1 Fungal Disease Management through NPs

Plants are a vital source of food, fuel, and raw materials for many industries. However, plant health is damaged by various pathogens resulting in a considerable yield reduction worldwide (de Freitas Bueno et al., 2019; Ibrahim et al., 2020; Devi et al., 2020; Ibrahim et al., 2022). Traditional chemical-based methods of controlling plant diseases have several drawbacks, including environmental pollution, health hazards, and the development of resistance. To tackle these challenges, scientists are investigating the capabilities of nanomaterials in crop health improvement by reducing the pathogen population. NPs have emerged as an alternative to traditional methods for effectively managing plant diseases. NPs are tiny particles with unique physical and chemical properties that make them effective against various plant pathogens. Studies have shown that NPs can inhibit the growth and proliferation of fungi, bacteria, and viruses that cause plant diseases. For instance, NPs of silver have shown antimicrobial properties against *Fusarium oxysporum* and *Alternaria solani* (Singh et al. 2020). These NPs function by disrupting the pathogen's cell membrane, ultimately causing its death. Similarly, CuONPs have also shown the antifungal properties of *Phytophthora infestans*, the pathogen that causes potato late blight (Ghormade et al. 2011). Furthermore, NPs can also enhance the plant's natural defence mechanisms against pathogens. For example, ZnONPs may enhance disease resistance ability against two different pathogens (Shah et al. 2023). Efforts are being made to optimize the application of NPs in plant disease management by exploring different synthesis methods, particle sizes, and surface modifications. With further research and development, NPs have the potential to revolutionize plant disease management and contribute to sustainable agriculture.

3.3.1.1 AgNPs: Biosynthesis and Their Use in the Management of Plant Disease

A pivotal facet of current nanotechnology research involves establishing reliable methods for synthesizing AgNPs. AgNPs enjoy antimicrobial properties and can serve as good agents for sustainable and eco-friendly management of plant diseases (Aboelfetoh et al., 2017). Gardea-Torresdey et al. (2003) reported that different plant parts are utilized for producing NPs by utilizing the different plant species. The production of AgNPs has been achieved effectively through the utilization of natural sources such as *Ananas cosmosus* (Ahmad et al., 2012), *Ziziphora tenuior* (Sadeghi and Gholamhoseinpoor, 2015), *Coriandrum sativum* (Nazeruddin et al., 2014), and *Pelargonium graveolens* (Shankar et al., 2003). These AgNPs have demonstrated significant antifungal properties when synthesized through environmentally friendly methods. Krishnaraj et al. (2012) reported the antimicrobial activity of *Acalypha indica* leaf extract driven AgNPs using different concentrations. The study demonstrated considerable inhibitory activity against many microorganisms causing significant yield losses to the plants. In a study, *Curvularia lunata* causing leaf spot disease in maize, was found to be inhibited by 95% after incubation for 24 hours in sunlight and using an autoclave. This approach demonstrates the potential for using AgNPs as an effective tool in combating fungal plant diseases. (Swamy and Nargund 2017). An additional investigation was undertaken to explore the antimicrobial capabilities of AgNPs synthesized through green methods. Combining bark extracts from *Boswellia ovalifoliolata* and *Shorea tumbuggaia* with the leaf extract of *Svensonia hyderobadensis* facilitated the synthesis of these NPs. The study aimed to evaluate the effectiveness of these NPs against many micro fungi causing diseases on plants. The synthesis of AgNPs involved the utilization of *Svensonia hyderabadensis* leaf extract with $AgNO_3$ (1 mM) solution, showcasing notable antifungal efficacy against *Rhizopus arrhizus, Fusarium oxysporum, Curvularia lunata,* and *Aspergillus niger*. The NPs were synthesized through a green approach using olive seed extract, exhibiting effectiveness in combating plant pathogenic fungi across various groups including *Cylindrocarpon, Corynespora, Botrytis, Alternaria, Fusarium, Pythium* and *Stemphylium* (Kim et al., 2012). The utilization of AgNPs for the management of plant diseases or the reduction of pathogenic fungi inoculum load has been an area of active investigation. The application of AgNPs as agents for controlling plant pathogenic fungi shows considerable promise, and additional work is required to be done to elucidate their mechanism of action and refine their synthesis for potential commercial applications.

3.3.1.2 CuNPs: Biosynthesis and Their Use in Plant Disease Management

The investigation relating to application of CuNPs has recently gained momentum, however there are still few studies that reveal how these NPs are biologically synthesized. CuNPs have demonstrated significant potential in effectively managing various plant diseases owing to their potent antifungal and antibacterial properties. The use of CuNPs as plant protection agents has been shown to reduce the use of conventional fungicides, which are toxic to the environment. The CuNPs work by disrupting the plasmalemma of the pathogen, ultimately resulting in its death (Kim et al., 2008). The production of these NPs can be achieved through a biological process utilizing the extract obtained from magnolia leaves (Valodkar et al. 2011). The CuNPs can be produced using various environmentally friendly methods. CuNPs can be produced using basil extract from *Ocimum sanctum* and pomegranate peel extract from *Punica granatum* in a $CuSO_4$ solution. Dulta et al. (2021) illustrated the synthesis of CuNPs employing extract of *Carica papaya*, while Gunalan et al. (2012) used *Aloe Vera* (*Aloe barbadensis* Miller) leaf extract for the same purpose. In another study, blackberry fruit and leaf extract (*Rubus glaucus*) were used for the biosynthesis of CuNPs (Kumar et al. (2017). CuNPs synthesized using citron juice (*Citrus medica*) demonstrated antimicrobial activity against multiple species of fusaria (Kumar et al., 2017). CuNPs exhibit remarkable antifungal potential against a diverse array of phytopathogens including *Curvularia lunata, Alternaria alternata, Fusarium oxysporum, Phoma destructiva, Phytophthora cinnamon* and *Pseudomonas syringae*. This innovative approach harnesses the unique properties of Cu to combat plant diseases effectively, offering promising solutions for sustainable agriculture. Among these fungi, *Fusarium culmorum* demonstrated higher susceptibility to CuNPs in comparison to *Fusarium oxysporum* and *Fusarium graminearum* (Shende et al., 2015). CuNPs synthesized through clove bud (*S. aromaticum*) extract were found to effectively inhibit *Aspergillus flavus, Aspergillus niger*, and *Penicillium* spp., showcasing their potential for application in plant pathology to manage fungal infections in plants (Kanhed et al., 2014; Banik and Perez-de-luque, 2017). Furthermore, CuNPs produced with citron juice (*Citrus medica*) showed antimicrobial effects against many species of *Fusarium* (Kamis et al., 2017). CuNPs have shown better suppression of numerous fungal infections compared to other readily available commercial fungicides such as bavistin (Shende et al. 2015).

3.3.1.3 AuNPs: Biosynthesis and Their Use in Plant Disease Management

Gold (Au) NPs are promising candidates for plant disease management due to their different unique properties, and are synthesized through different plant extracts such as *Aloe vera*, garlic and green tea (Kulabhusan et al. 2023). The production of AuNPs through biosynthesis has exhibited notable antimicrobial efficacy against a diverse array of fungal pathogens. Utilizing AuNPs is regarded as a safe and environmentally friendly alternative to chemical fungicides in managing crop diseases. Therefore, the use of AuNPs holds great potential in plant health management and could contribute to the development of sustainable agriculture. The bioproduction of AuNPs can be achieved using different fresh leaf extracts of *Mentha piperita*. To enhance their compatibility, non-toxic reducing agents should be used, such as sodium citrate and sodium borohydride which are commonly used (Ghosh et al. 2008). Plant-derived reducing and stabilizing agents are also commonly used in the production of AuNPs (Mittal et al. 2013). These AuNPs have demonstrated efficacy in managing various fungal infections. For example, *Abelmoschus esculentus* seed extract synthesizes AuNPs with antifungal properties against *Aspergillus niger, Aspergillus flavus, Puccinia graminis tritici* and *Candida albicans* exhibiting the most significant inhibitory zones for *P. graminis* (17 mm) and *C. albicans* (18 mm) (Jayaseelan et al., 2013). Similarly, *Mentha piperita* derived NPs impede *Sclerotium rolfsii* (Thanighaiarassu et al., 2014). AuNPs crafted from an edible mushroom (*Agaricus bisporus*) extract and $HAuCl_4$ solution also exhibit antimicrobial activity against *Aspergillus terreus* and *Aspergillus flavus* (Eskandari-Nojedehi et al., 2018). Furthermore, *Salix alba* leaf extract can generate AuNPs when exposed to aqueous gold ions, which inhibit many fungal species (*Alternaria solani, Aspergillus flavus,* and *Aspergillus niger*). Among these fungi, *A. niger* and *A. solani* show the most robust inhibition zones, while *A. flavus* displays the lowest (Islam et al. 2019).

3.3.1.4 ZnONPs: Biosynthesis and Their Use in Plant Disease Management

Various plant extracts, including those obtained from *Azadirachta indica, Aloe vera*, and *Camellia sinensis*, facilitate the biological synthesis of ZnO nanoparticles (Basnet et al., 2018). Utilizing plant extracts as both reducing and stabilizing agents is recognized as a straightforward, effective, and environmentally friendly method for ZnONP synthesis. The ZnONPs produced from *Parthenium hysterophorus* plant extracts were assessed for their efficacy against various fungal infections. The parthenium-derived ZnONPs exhibited a significant inhibitory effect on *Aspergillus flavus* and *A. niger* (Rajiv et al. 2013). The antifungal activity of ZnONPs, synthesized through a green method using extracts from *Syzygium aromaticum* buds was assessed against *Fusarium graminearum* resulting in a significant suppression of mycelial development and mycotoxin formation, such as zearalenone and deoxy-nivalenol. The treatment also led to an increase in ROS generation, lipid peroxidation, and a reduction in ergosterol content in the fungal membrane causing serious harm to plant pathogens (Senthilkumar and Sivakumar 2014). The morphology of macroconidia underwent changes, displaying a rough, wrinkled, and shrinking texture, indicative of the antimicrobial activity. Similarly, ZnONPs made from *Eucalyptus globulus* phyto-extract were tested to detect the fungus responsible for apple orchid disease. *Alternaria mali, Botryosphaeria dothidea*, and *Diplodia seriata* exhibited maximum inhibition at 100 ppm concentrations, demonstrating the potential use of these NPs in the management of fungal infections in fruit crops. These environmentally-friendly and cost-effective ZnONPs made from *Parthenium hysterophorus* plant extracts were also effective against various fungal infections such as *Aspergillus flavus* and *A. niger*.

3.3.2 Bacte rial Disease Management through Various NP Applications

The utilization of NPs has demonstrated significant potential in controlling various bacterial diseases, however, with varied rate of effectiveness. Considerable research is necessary to optimize the application of NPs for sustainable bacterial disease management. Jiang et al. (2021) applied metal oxide NPs in tomato bacterial wilt management induced by *Ralstonia solanacearum*. The study showed that metal NPs, including CuO, ZnO, and FeO, decreased the occurrence of bacterial wilt and modified the composition and structure of the rhizosphere bacterial community. Moreover, applying metal oxide NPs enhanced the morphological and physiological characteristics of tomato plants.

The application of CuONPs markedly elevated the Chao1 and Shannon indices. In the initial stages, they substantially decreased the relative abundance of pathogens. Concurrently, there was a significant increase in the relative abundance of beneficial *Streptomyces* bacteria, illustrating an inverse relationship with the prevalence of pathogenic bacteria. This study offers valuable evidence and strategies for mitigating tomato bacterial wilt through the utilization of metal oxide NPs. Parveen and Siddiqui (2021) investigated the efficacy of ZnONPs against various bacterial strains and fungal pathogens that impact tomato plants.

Their findings revealed that the most substantial improvements in plant growth, encompassing enhancements in photosynthetic pigments and proline content, as well as the most significant reduction in disease index were achieved through foliar application and as a seed priming. Ogunyemi et al. (2019) reported the biofabrication of ZnONPs using phytoextracts from olive leaves, chamomile flowers, and fruits of red tomato. Upon phytochemical analysis, the extracts were found to contain a wide array of organic components. These compounds reduce and stabilize the NPs during the biosynthesis process. Characterization of the synthesized nanoparticles involved UV-Vis spectroscopy and FTIR, enabled confirmation of their formation and enabled identification of characteristic absorption bands. The study demonstrated that the synthesized NPs exhibited antibacterial activity against rice bacterial leaf blight disease.

3.3.3 Management of Plant-Parasitic Nematodes through NPs

Many nanomaterials, such as AgNPs and SNPs have been found to be reducing the nematode population thus can help us control them effectively (Hardman 2006). NPs have shown potential as an effective means of controlling nematodes with various studies demonstrating that NPs can suppress the nematode population considerably, and improve plant growth and overall biomass of the crop plants. The use of NPs for nematode management is yet to be fully explored to optimize their efficacy, safety, and environmental impact. However, the potential benefits of NP based approaches include improved crop yields, reduced chemical inputs, and enhanced soil health. The effect of NPs on nematode management is an exciting area of research which is required to be carried out for sustainable agriculture. Developing effective and environmentally friendly strategies for nematode control is possible (Keshari et al., 2024; Prasad et al., 2020; Solanki et al., 2020). Heflish et al. (2021) approved the biosynthesis of AgNPs through *Acalypha wilkesiana* leaf extract. The NPs are considerably effective against plant nematode (root-knot nematode), and have been found to be instrumental in reducing nematode activity, mortality, egg hatchability, and movement of larvae. The biosynthesis method is cost-effective, eco-friendly, and biocompatible with nature. Ghareeb et al. (2022) examined the nematocidal efficacy of extracts derived from two marine algae (*Colpomenia sinuosa* and *Corallina mediterranea*) as well as AgNPs synthesized from these extracts. The target nematode species under investigation was *Meloidogyne incognita*, a common nematode affecting tomato plants. The findings indicated that NPs sourced from *Colpomenia sinuosa* showcased the most potent nematocidal activity against *M. incognita* compared to other treatments. This particular treatment notably reduced nematode-related factors, resulting in improved plant growth and yield components. However, methylene chloride driven from *C. sinuosa* exhibited more effectiveness as compared to that obtained from *Corallina mediterranea*. Moreover, a highly effective solvent combination for *C. sinuosa* extract was found to be a mixture of hexane, methylene chloride, and ethyl acetate in a 1:0.5:0.5 ratio. Within the eluent fractions, the third fraction showed exceptional effectiveness, leading to 87.5% mortality within 12 hours of exposure to *M. incognita*, followed by complete mortality after 24 and 72 hours. The silver NPs produced from *Colpomenia sinuosa* exhibit potential as an alternative strategy for controlling root-knot nematodes, providing a viable substitute for chemical nematicides. Mohammad et al. (2022) assessed the nematicidal activity of aqueous extracts from both plants, including their nano-forms, against *Meloidogyne incognita* in tomatoes grown in infected soil. Phenolic compounds were identified in the extracts, with chlorogenic acid being predominant in sweet annie and p-hydroxybenzoic acid in garden cress.

Nano-formulations exhibited higher efficacy, leading to increased mortality of *M. incognita*. Additionally, treated tomato plants showed improved growth and also significantly enhanced the activity of defensive enzymes, indicating systemic resistance. Ghareeb et al. (2020) investigated the effects of *C. glomerata* extract and biosynthesized Green Silver NanoParticles (GSNPs) on root-knot nematodes. They used Q-PCR to analyze enzyme expression in tomato plants treated with GSNPs, revealing significant eco-nematicidal activity against *M. javanica*. GSNPs influenced egg hatchability and nematode mortality, offering a promising eco-friendly nematode management solution. Additionally, GSNPs outperformed the chemical nematicide Rugby 60%, and reduced nematode reproduction. Enzyme activity increased in treated plants compared to the control, peaking at 14 days post-inoculation. Kalaiselvi et al. (2019) proposed the utilization of *Euphorbia tirucalli* L. latex for efficient biofabrication of AgNPs. The AgNPs exhibited notable nematicidal activity against *Meloidogyne incognita*, emphasizing the potential of plant biomass in NP fabrication for nematode management. The study encompassed the processing of latex, AgNPs synthesis, and their characterization. The optimal concentration for green AgNPs synthesis was determined to be 3% latex extract concentration. This concentration demonstrated effective nematicidal activity against *Meloidogyne incognita*. Cromwell et al. (2014) explored the nematicidal effects of AgNPs on root-knot nematodes in Bermudagrass. They found that applying AgNPs led to a notable decrease in nematode populations and stimulated plant growth. The AgNPs effectively eradicated nematodes and hindered their ability to reproduce.

3.4 NPS PENETRATION, TRANSPORT AND MECHANISM OF ACTION

AgNPs exhibit the capability to disrupt various cellular mechanisms, including membrane permeability, ATP synthesis, and responses to oxidative stress in both eukaryotic (Roh et al., 2009; Lim et al., 2012) and prokaryotic cells (Sondi and Salopek-Sondi, 2004; Choi and Hu, 2008). This broad-spectrum antimicrobial property makes AgNPs effective against plant-pathogenic bacteria and fungi (Park et al., 2006; Jo et al., 2009). In the context of warm-season turfgrass, AgNP treatment may also exhibit antifungal effects against common root-associated fungal pathogens. By protecting turfgrass from additional weaknesses and stresses caused by these pathogens, AgNP enhances tolerance to root-knot nematode damage and contributes to overall turf quality improvement (Cromwell, 2014). AgNPs give promising results as compared to nematicides, which are risky if used non-judiciously, and biocontrol agents, which have shown inconsistent results in different agroclimatic zones. To effectively manage root-knot nematode issues in turfgrass using AgNPs, one may need to apply them frequently (biweekly) at high doses ($\geq 90.4\,mg/m^2$). Improving the applicability of AgNPs can be achieved by incorporating them into an irrigation system, such as fertigation, or by combining them with compatible chemicals that enhance their nematicidal effect (Cromwell, 2014). Remarkably, the AgNPs bind to bacterial cell walls, penetrate them, and induce structural modifications in the cell membrane, such as increased permeability and cell death. This interaction results in the formation of surface "pits" on the bacterial cells and the accumulation of nanoparticles. Additionally, AgNPs generate free radicals upon contact with bacteria thus contributing to cell death. The interaction between AgNPs and bacteria confirmed by studies using electron spin resonance spectroscopy, forms free radicals, rendering the cell membrane porous and leading to cell death. (Danilcauk et al., 2006; Kim et al., 2007). There is a proposed mechanism where AgNPs may release silver ions, which can interact with crucial enzyme thiol (-SH) groups, resulting in their inactivation (Matsumura et al., 2003). Upon contacting silver ions, they inhibit various cellular functions, causing damage and generating Reactive Oxygen Species (ROS). This ROS production can be attributed to the inhibition of respiratory enzymes by silver ions, initiating an attack on bacterial cells. The impact of AgNPs on bacterial cells triggers these reactions, ultimately causing cell death. Importantly, bacterial DNA, a vital component, contains sulfur and phosphorus. AgNPs interact with these components leading to DNA destruction and subsequent cell death (Hatchett and Henry 1996).

A wide range of NPs possess antifungal and antibacterial characters as well that can help reduce nematode population. NPs such as silver, gold, silica carbide, and copper have been found to have a nematicidal effect against the root-knot nematodes (Sabry 2019). The nematocidal properties of AgNPs, in particular, have been well-documented (Tuncsoy 2021). Silicon NPs also possess nematocidal properties against an array of plant parasitic nematodes including root-knot (El-Ashry 2021). The exact mechanism is yet to be fully explored, but it is thought to involve the disruption of the nematodes' cellular membranes, leading to their death (Sabry 2019). Some types of NPs, such as zinc oxide nanoparticles, have been found to disrupt the behaviour of plant nematodes, leading to a reduction in their populations (Bhau et al., 2016). Other NPs such as titanium dioxide have been found to induce plant defence mechanisms against nematodes. The exact mechanism by which titanium dioxide NPs induce plant defence mechanisms has not yet been explored, but it is thought to involve the activation of plant signalling pathways. Similarly, the exact mechanism by which zinc oxide nanoparticles disrupt nematode behaviour is yet to be explored, but it involves the disruption of the nematode's sensory organs (Bhau et al., 2016).

Silicon (Si) NPs shows nematicidal activity in mitigating the population of *Meloidogyne incognita* infecting eggplants. The collaborative application of silicon NPs alongside nematicides resulted in a reduction in various nematode parameters (El-Ashry et al., 2022).

Chitosan-derived AgNPs also contain nematicidal activity against the root-knot nematode (*Meloidogyne incognita*). Exposure to AgNPs can cause significant mortality in nematodes leading to reduced nematode populations (Bernard et al. 2019). Application of silicon NPs can affect the reproduction, development, and behaviour of nematodes. Silicon NPs can cause oxidative stress and DNA damage in nematodes that reduce the potential population of plant parasitic nematodes (Li et al. 2021).

NPs have exhibited effectiveness against a range of harmful fungi including *Rhizoctonia, Fusarium, Verticillium, Sclerotinia, Phythium, Phytophtora*, and *Alternaria* (Mansoor et al., 2021; Jian et al., 2022). AgNPs impede the growth of *Rhizoctonia solani* by causing damage to fungal hyphae and inhibiting conidial germination. ZnONPs induce oxidative stress and disrupt the fungal cell membrane, inhibiting the growth of *R. solani* (Mansoor et al., 2021). Additionally, CuONPs exhibit inhibitory effects on *R. solani* by inducing oxidative stress and mycelial disruption. AuNPs disrupt the fungal cell membrane and induce oxidative stress, effectively inhibiting the growth of *R. solani* (Al-Khattaf, 2021). Likewise, TiO_2NPs induce oxidative stress and mycelial disruption leads to the inhibition of *R. solani* growth (Dutta et al., 2023). Plant extracts mediated AgNPs inhibited the growth of *Fusarium graminearum* by suppressing conidium germination and germ tube growth (Jijan et al., 2022). Additionally, ZnONPs inhibit the growth of *Fusarium oxysporum* through the induction of oxidative stress and disruption of the fungal cell membrane (Dutta et al., 2023). Similarly, CuONPs exhibit inhibitory effects on *Fusarium oxysporum* by inducing oxidative stress and disrupting the fungal cell membrane. Similarly, AuNPs effectively impede the growth of *Fusarium oxysporum* by disrupting the fungal cell membrane and inducing oxidative stress (Al-khattaf, 2021). Likewise, TiO_2NPs induce oxidative stress and disrupt the fungal cell membrane, leading to the inhibition of *Fusarium oxysporum* growth. NPs exhibit inhibitory effects on plant pathogenic bacteria through diverse mechanisms, with their size playing a crucial role in their antibacterial activity. Research has demonstrated that smaller NPs possess enhanced penetration capabilities within bacteria, thereby heightening their efficacy. One of the mechanisms attributed to the inhibition of pathogens involves the direct physical interaction between the sharp edges of NPs and the cell walls of these microorganisms. This interaction has been recognized as a contributing factor to the observed inhibition (More et al., 2023). Utilizing NP sprays on plants not only promotes plant growth but also acts as a preventive measure against pathogenic infections (Li et al., 2023).

AgNPs synthesized through biological means have demonstrated substantial inhibitory effects on the pathogen that causes wildfire disease (*Pseudomonas syringae* pv. *tabaci*) on tobacco. This inhibition occurs when the bacteria are directly targeted, inducing plant resistance in *Nicotiana benthamiana* (Jiang et al., 2022). Several metallic NPs have demonstrated potential as antimicrobial agents against plant pathogens. Harnessing the unique properties of NPs offers promising opportunities for combating plant pathogenic bacteria, with their diverse mechanisms of action and potential for promoting plant growth. By further exploring the capabilities of NPs and their tailored applications, we can unlock innovative strategies to enhance plant health and disease management.

3.5 UNDERSTANDING POTENTIAL RISKS IN NANOPARTICLE-BASED PLANT DISEASE MANAGEMENT

NPs used in fertilizers can harm soil biota and fertility by negatively impacting soil microbes, microfauna, and the digestive systems of earthworms. These detrimental effects can disrupt the delicate balance of the soil ecosystem (Singh et al., 2021). NPs have potential toxicological implications for both humans and animals. It is imperative to assess the possible hazards linked to NPs exposure to safeguard the health of both humankind as well as the surroundings (Iavicoli et al., 2017). The absence of comprehensive regulations and guidelines for assessing the risks of nanotechnology poses challenges and uncertainties. The lack of a unified framework makes it difficult to establish clear safety standards for NPs, creating confusion and ambiguity surrounding their usage and potential impacts (Mwaanga, 2017). The NPs and their applications in agriculture present occupational hazards for workers handling these materials. Occupational exposure to NPs can pose health risks, emphasizing the need for adequate safety measures and protective protocols to safeguard the well-being of those involved in nanoparticle-related activities (Iavicoli et al., 2014). Whether nano-based or not, agrochemicals must undergo rigorous evaluation to ensure their safety before being deployed in agricultural practices. Thorough assessments are essential to mitigate any potential risks to both environment and humankind (Iavicoli et al., 2017).

3.6 CONCLUSIONS AND FUTURE OUTLOOK

NPs exhibit significant potential in the control of plant diseases caused by various pathogens. Their distinct physical and chemical properties render them highly effective to a wide array of plant pathogenic propagules. Heterogenous groups of biosynthesized NPs have demonstrated antimicrobial properties and have also enhanced plant growth and yield attributes considerably. NPs may help us achieve sustainable and eco-friendly solutions for controlling soil-borne plant pathogens and improving overall plant health and productivity. However, while utilizing NPs for plant pathogen management, it is crucial to consider their potential impact on environmental variables. NPs have the potential to interact with soil components, water systems, and beneficial organisms in the ecosystem. These interactions can influence the physicochemical properties of soil, water quality, and the overall balance of the ecological system. It is essential to evaluate the potential risks associated with the application of NPs, considering factors such as their persistence in the environment, potential toxicity to untargeted microorganism, and the risk of accumulation in the food chain.

REFERENCES

Aboelfetoh, E. F., El-Shenody, R. A., & Ghobara, M. M. (2017). Eco-friendly synthesis of silver nanoparticles using green algae (Caulerpa serrulata): Reaction optimization, catalytic and antibacterial activities. *Environmental Monitoring and Assessment*, *189*, 1–15.

Ahamed, M., Posgai, R., Gorey, T. J., Nielsen, M., Hussain, S. M., & Rowe, J. J. (2010). Silver nanoparticles induced heat shock protein 70, oxidative stress and apoptosis in Drosophila melanogaster. *Toxicology and Applied Pharmacology*, *242*(3), 263–269.

Ahmad, A., Mukherjee, P., Senapati, S., Mandal, D., Khan, M. I., Kumar, R., & Sastry, M. (2003). Extracellular biosynthesis of silver nanoparticles using the fungus Fusarium oxysporum. *Colloids and Surfaces B: Biointerfaces*, *28*(4), 313–318.

Ahmad, K., Bhatti, I. A., Muneer, M., Iqbal, M., & Iqbal, Z. (2012). Removal of heavy metals (Zn, Cr, Pb, Cd, Cu and Fe) in aqueous media by calcium carbonate as an adsorbent. *International Journal of Chemical and Biochemical Sciences*, *2*, 48–53.

Alghuthaymi, M. A., Kalia, A., Bhardwaj, K., Bhardwaj, P., Abd-Elsalam, K. A., Valis, M., & Kuca, K. (2021). Nanohybrid antifungals for control of plant diseases: Current status and future perspectives. *Journal of Fungi*, *7*(1), 48.

Ali, M., Tahir, M. N., Siwy, Z., Neumann, R., Tremel, W., & Ensinger, W. (2011). Hydrogen peroxide sensing with horseradish peroxidase-modified polymer single conical nanochannels. *Analytical Chemistry*, *83*(5), 1673–1680.

Al-Khattaf, F. S. (2021). Gold and silver nanoparticles: Green synthesis, microbes, mechanism, factors, plant disease management and environmental risks. *Saudi Journal of Biological Sciences*, *28*(6), 3624–3631.

Ansari, R. A., & Khan, T. A. (2012). Parasitic association of root-knot nematode, *Meloidogyne incognita* on guava. *e-Journal of Science Technology*, *5*(12), 65–67.

Ansari, R. A., & Mahmood, I. (2017a). Determination of disease incidence caused by *Meloidogyne* spp. and or *Fusarium udum* on pigeonpea in Aligarh district: A survey. *Trends in Biosciences*, *10*(24), 5239–5243.

Ansari, R. A., & Mahmood, I. (2017b). Optimization of organic and bio-organic fertilizers on soil properties and growth of pigeon pea. *Scientia Horticulturae*, *226*, 1–9.

Ansari, R. A., & Mahmood, I. (2019a). *Plant Health Under Biotic Stress – Volume II, Microbial Interactions*. Springer Nature, Singapore. ISBN: 978-981-13-6040-4.

Ansari, R. A., & Mahmood, I. (2019b). *Plant Health Under Biotic Stress – Volume I, Organic Strategies*, Springer Nature, Singapore. ISBN: 978-981-13-6043-5.

Ansari, R. A., Mahmood, I., Rizvi, R., Sumbul, A., & Safiuddin. (2017). Siderophores: Augmentation of soil health and crop productivity. In: *Probiotics in Agroecosystem*, (Editors: Kumar, V., Kumar, M. Sharma, S. and Prasad R.), Springer Nature, Singapore, pp. 291–312.

Ansari, R. A., Rizvi, R., & Mahmood, I. (2020). *Management of Phytonematodes: Management of Phytonematodes: Recent Advances and Future Challenges*, Springer Nature, Singapore. ISBN: 978-981-15-4087-5.

Ansari, R. A., Rizvi, R., & Mahmood, I. (2023). *Mycorrhizal Symbiosis and Agroecosystem Restoration*, Springer Nature Singapore, Imprint: Springer.

Ansari, R. A., Rizvi, R., Sumbul, A., & Mahmood, I. (2020). Plant-growth-promoting rhizobacteria (PGPR)-based sustainable management of phytoparasitic nematodes: Current understandings and future challenges. In: *Management of Phytonematodes: Recent Advances and Future Challenges*, (Editors: Ansari, R.A., Rizvi, R. and Mahmood, I.), Springer Nature, Singapore, pp. 51–71.

Ansari, R. A., Sumbul, A., Rizvi, R., & Mahmood, I. (2019). Organic soil amendments: Potential tool for soil and plant health management. In: *Plant Health Under Biotic Stress: Volume 1: Organic Strategies*, (Editors: Ansari, R.A. and Mahmood, I.), Springer Nature, Singapore, pp. 1–35.

Ansari, R. A., Sumbul, A., Rizvi, R., & Mahmood, I. (2024). Impact of environmental gases on mycorrhizal symbiosis and its influence on ecosystem functioning under the current climate change scenario. In: *Mycorrhizal Symbiosis and Agroecosystem Restoration*. ((Editors: Ansari, R.A., Rizvi, R., and Mahmood, I.), Springer Nature Singapore, Singapore, pp. 51–76.

Banik, S., & Pérez-de-Luque, A. (2017). In vitro effects of copper nanoparticles on plant pathogens, beneficial microbes and crop plants. *Spanish Journal of Agricultural Research*, *15*(2), e1005.

Bansal, V., Rautaray, D., Ahmad, A., & Sastry, M. (2004). Biosynthesis of zirconia nanoparticles using the fungus *Fusarium oxysporum*. *Journal of Materials Chemistry*, *14*(22), 3303–3305.

Basnet, P., Chanu, T. I., Samanta, D., & Chatterjee, S. (2018). A review on bio-synthesized zinc oxide nanoparticles using plant extracts as reductants and stabilizing agents. *Journal of Photochemistry and Photobiology B: Biology*, *183*, 201–221.

Bernard, G. C., Fitch, J., Min, B., Shahi, N., Egnin, M., Ritte, I., ... & Bonsi, C. (2019). Potential nematicidal activity of silver nanoparticles against the root-knot nematode (Meloidogyne incognita). *Online Journal of Complementary & Alternative Medicine*, *2*, 000531.

Bhattacharya, D., & Rajinder, G. (2005). Nanotechnology and potential of microorganisms. *Critical Reviews Biotechnology*, *25*, 199–204.

Bhattacharyya, A., Prasad, R., Buhroo, A. A., Duraisamy, P., Yousuf, I., Umadevi, M., ... & Khanday, A. L. (2016). One-pot fabrication and characterization of silver nanoparticles using Solanum lycopersicum: An eco-friendly and potent control tool against rose aphid, *Macrosiphum rosae*. *Journal of Nanoscience*, Article ID 4679410, 7 pages.

Bhau, B. S., Phukon, P., Ahmed, R., Gogoi, B., Borah, B., Baruah, J., ... & Wann, S. B. (2016). A novel tool of nanotechnology: Nanoparticle mediated control of nematode infection in plants. In: *Microbial Inoculants in Sustainable Agricultural Productivity: Vol. 2: Functional Applications*, (Editors: Singh, D.P., Singh, H.B. and Prabha, R.), Springer Nature, Singapore, pp. 253–269.

Birla, S. S., Gaikwad, S. C., Gade, A. K., & Rai, M. K. (2013). Rapid synthesis of silver nanoparticles from *Fusarium oxysporum* by optimizing physicocultural conditions. *The Scientific World Journal*, Article ID 796018, 12 pages.

Boyno, G., Ansari, R. A., Durak, E. D., Güneş, H., Çevik, R., & Demir, S. (2024). Arbuscular mycorrhizal technology in sustainable agriculture: Current knowledge and challenges in agroforestry. In: *Mycorrhizal Symbiosis and Agroecosystem Restoration*, (Editors: Ansari, R.A., Rizvi, R. and Mahmood, I.), Springer Nature, Singapore, 173–195.

Chandrasekharan, N., & Kamat, P. V. (2000) Improving the photoelectrochemical performance of nanostructured TiO_2 films by adsorption of gold nanoparticles. *Journal of Physical Chemistry B*, *104*, 10851–10857.

Choi, O., & Hu, Z. (2008). Size dependent and reactive oxygen species related nanosilver toxicity to nitrifying bacteria. *Environmental Science & Technology*, *42*(12), 4583–4588.

Cromwell, W. A., Yang, J., Starr, J. L., & Jo, Y. K. (2014). Nematicidal effects of silver nanoparticles on root-knot nematode in bermudagrass. *Journal of Nematology*, *46*(3), 261.

Danilczuk, M., Lund, A., Sadlo, J., Yamada, H., & Michalik, J. (2006). Conduction electron spin resonance of small silver particles. *Spectrochimica Acta Part A: Molecular and Biomolecular Spectroscopy*, *63*(1), 189–191.

de Freitas Bueno, R. C. O., Ansari, R. A., Lima, G. P. P., & Sakate, R. K. (2019). Phytosanitation: A novel approach toward disease management. In: *Plant Health Under Biotic Stress: Volume 1: Organic Strategies*, (Editors: Ansari, R.A. and Mahmood, I.), Springer Nature, Singapore, pp. 73–90.

Devi, T. S., Das, D., Ansari, R. A., Rizvi, R., Sumbul, A., & Mahmood, I. (2020). Role of organic additives in the sustainable management of phytoparasitic nematodes. In: *Management of Phytonematodes: Recent Advances and Future Challenges*, (Editors: Ansari, R.A., Rizvi, R. and Mahmood, I.), Springer Nature, Singapore, pp. 279–295.

Dias, M. A., Lacerda, I. C. A., Pimentel, P. F., De Castro, H. F., & Rosa, C. A. (2002). Removal of heavy metals by an Aspergillus terreus strain immobilized in a polyurethane matrix. *Letters in Applied Microbiology*, *34*(1), 46–50.

Dreizin, E. L. (2009). Metal-based reactive nanomaterials. *Progress in Energy and Combustion Science*, *35*(2), 141–167.

Dulta, K., Ağçeli, G. K., Chauhan, P., & Chauhan, P. K. (2021). Biogenic production and characterization of CuO nanoparticles by Carica papaya leaves and its biocompatibility applications. *Journal of Inorganic and Organometallic Polymers and Materials*, *31*, 1846–1857.

El-Ashry, R. M., El-Saadony, M. T., El-Sobki, A. E., El-Tahan, A. M., Al-Otaibi, S., El-Shehawi, A. M., ... & Elshaer, N. (2022). Biological silicon nanoparticles maximize the efficiency of nematicides against biotic stress induced by Meloidogyne incognita in eggplant. *Saudi Journal of Biological Sciences*, *29*(2), 920–932.

Elmer, W., & White, J. C. (2018). The future of nanotechnology in plant pathology. *Annual Review of Phytopathology*, *56*, 111–133.

Eskandari-Nojedehi, M., Jafarizadeh-Malmiri, H., & Rahbar-Shahrouzi, J. (2018). Hydrothermal green synthesis of gold nanoparticles using mushroom (*Agaricus bisporus*) extract: Physico-chemical characteristics and antifungal activity studies. *Green Processing and Synthesis*, *7*(1), 38–47.

Fabiyi, O. A., Alabi, R. O., & Ansari, R. A. (2020). Nanoparticles' synthesis and their application in the management of phytonematodes: An overview. In: *Management of phytonematodes: Recent Advances and Future Challenges*, (Editors: Ansari, R.A., Rizvi, R. and Mahmood, I.), Springer Nature, Singapore, pp. 125–140.

Fabiyi, O. A., Ansari, R. A., Rizvi, R., Sumbul, A., & Mahmood, I. (2024). Influence of arbuscular mycorrhizal fungi on soil health amelioration and plant fitness under hostile environment. In: *Mycorrhizal Symbiosis and Agroecosystem Restoration*. (Editors: Ansari, R.A., Rizvi, R. and Mahmood, I.), Springer Nature Singapore, Singapore, pp. 227–248.

Fox, J. P., Capen, J. D., Zhang, W., Ma, X., & Rossi, L. (2020). Effects of cerium oxide nanoparticles and cadmium on corn (*Zea mays* L.) seedlings physiology and root anatomy. *NanoImpact*, *20*, 100264.

Gardea-Torresdey, J. L., Gomez, E., Peralta-Videa, J. R., Parsons, J. G., Troiani, H., & Jose-Yacaman, M. (2003). Alfalfa sprouts: A natural source for the synthesis of silver nanoparticles. *Langmuir*, *19*(4), 1357–1361.

Gericke, M., & Pinches, A. (2006a). Biological synthesis of metal nanoparticles. *Hydrometallurgy*, *83*, 132–140.

Gericke, M., & Pinches, A. (2006b). Microbial production of gold nanoparticles. *Gold Bulletin*, *39*, 22–28.

Ghareeb, R. Y., Alfy, H., Fahmy, A. A., Ali, H. M., & Abdelsalam, N. R. (2020). Utilization of Cladophora glomerata extract nanoparticles as eco-nematicide and enhancing the defense responses of tomato plants infected by Meloidogyne javanica. *Scientific Reports*, *10*(1), 19968.

Ghareeb, R. Y., Shams El-Din, N. G. E. D., Maghraby, D. M. E., Ibrahim, D. S., Abdel-Megeed, A., & Abdelsalam, N. R. (2022). Nematicidal activity of seaweed-synthesized silver nanoparticles and extracts against *Meloidogyne incognita* on tomato plants. *Scientific Reports*, *12*(1), 3841.

Ghormade, V., Deshpande, M. V., & Paknikar, K. M. (2011). Perspectives for nano-biotechnology enabled protection and nutrition of plants. *Biotechnology Advances*, *29*(6), 792–803.

Ghosh, B., Das, M., Banerjee, P., & Das, S. (2008). Fabrication of SnS thin films by the successive ionic layer adsorption and reaction (SILAR) method. *Semiconductor Science and Technology*, *23*(12), 125013.

Hardman, R. (2006). A toxicologic review of quantum dots: Toxicity depends on physicochemical and environmental factors. *Environmental Health Perspectives*, *114*(2), 165–172.

Hatchett, D. W., & White, H. S. (1996). Electrochemistry of sulfur adlayers on the low-index faces of silver. *The Journal of Physical Chemistry*, *100*(23), 9854–9859.

Heflish, A. A., Hanfy, A. E., Ansari, M. J., Dessoky, E. S., Attia, A. O., Elshaer, M. M., ... & Behiry, S. I. (2021). Green biosynthesized silver nanoparticles using Acalypha wilkesiana extract control root-knot nematode. *Journal of King Saud University-Science*, *33*(6), 101516.

Husen, A., & Iqbal, M. (2019). *Nanomaterials and Plant Potential: An Overview*. Springer Nature Switzerland AG , pp. 3–29.

Husen, A., & Jawaid, M. (Eds.). (2020). *Nanomaterials for Agriculture and Forestry Applications*. Elsevier, Netherland.

Iavicoli, I., Leso, V., Beezhold, D. H., & Shvedova, A. A. (2017). Nanotechnology in agriculture: Opportunities, toxicological implications, and occupational risks. *Toxicology and Applied Pharmacology*, *329*, 96–111.

Iavicoli, I., Leso, V., Ricciardi, W., Hodson, L. L., & Hoover, M. D. (2014). Opportunities and challenges of nanotechnology in the green economy. *Environmental Health*, *13*(1), 1–11.

Ibrahim, D. S. S., Elderiny, M. M., Ansari, R. A., Rizvi, R., Sumbul, A., & Mahmood, I. (2020). Role of *Trichoderma* spp. in the management of plant-parasitic nematodes infesting important crops. In: *Management of Phytonematodes: Recent Advances and Future Challenges*, (Editors: Ansari, R.A., Rizvi, R. and Mahmood, I.), Springer Nature, Singapore, pp. 259–278.

Ibrahim, D. S. S., Mostafa, M., Abdellatif, A. A., & Ansari, R. A. (2022). Sustainable management of soil-borne plant pathogens. In: *Earth Systems Protection and Sustainability: Volume 1*. (Editors: Furze, J.N., Eslamian, S., Raafat, S.M. and Swing, K.), Springer International Publishing, Cham, pp. 35–56.

Islam, N. U., Jalil, K., Shahid, M., Rauf, A., Muhammad, N., Khan, A., ... & Khan, M. A. (2019). Green synthesis and biological activities of gold nanoparticles functionalized with Salix alba. *Arabian Journal of Chemistry*, *12*(8), 2914–2925.

Jayaseelan, C., Ramkumar, R., Rahuman, A. A., & Perumal, P. (2013). Green synthesis of gold nanoparticles using seed aqueous extract of *Abelmoschus esculentus* and its antifungal activity. *Industrial Crops and Products*, *45*, 423–429.

Jian, Y., Chen, X., Ahmed, T., Shang, Q., Zhang, S., Ma, Z., & Yin, Y. (2022). Toxicity and action mechanisms of silver nanoparticles against the mycotoxin-producing fungus Fusarium graminearum. *Journal of Advanced Research*, *38*, 1–12.

Jiang, H., Lv, L., Ahmed, T., Jin, S., Shahid, M., Noman, M., ... & Li, B. (2021). Effect of the nanoparticle exposures on the tomato bacterial wilt disease control by modulating the rhizosphere bacterial community. *International Journal of Molecular Sciences*, *23*(01), 414.

Jiang, L., Xiang, S., Lv, X., Wang, X., Li, F., Liu, W., ... & Sun, X. (2022). Biosynthesized silver nanoparticles inhibit Pseudomonas syringae pv. tabaci by directly destroying bacteria and inducing plant resistance in Nicotiana benthamiana. *Phytopathology Research*, *4*(1), 43.

Jo, Y. K., Kim, B. H., & Jung, G. (2009). Antifungal activity of silver ions and nanoparticles on phytopathogenic fungi. *Plant Disease*, *93*(10), 1037–1043.

Kalaiselvi, D., Mohankumar, A., Shanmugam, G., Nivitha, S., & Sundararaj, P. (2019). Green synthesis of silver nanoparticles using latex extract of *Euphorbia tirucalli*: A novel approach for the management of root knot nematode, *Meloidogyne incognita*. *Crop Protection*, *117*, 108–114.

Kamis, Y., Hashim, A., Margarita, R., Algothaymi, M. A., & Abd-Elsalam, K. A. 2017. Fungicidal efficacy of chemically produced copper nanoparticles against *Penicillium digitatum* and *Fusarium solani* on citrus fruit. *Philippine Agriculture Science*, 100(1), 69–78.

Kanhed, P., Birla, S., Gaikwad, S., Gade, A., Seabra, A. B., Rubilar, O., ... & Rai, M. (2014). In vitro antifungal efficacy of copper nanoparticles against selected crop pathogenic fungi. *Materials Letters*, *115*, 13–17.

Keshari, N., Kranti, K. V. V. S. K., Gunda, N. K., & Ansari, R. A. (2024). Arbuscular mycorrhizal fungi: A potential agent for phytonematodes management in diverse agro-climatic zones. In: *Mycorrhizal Symbiosis and Agroecosystem Restoration*. Springer Nature Singapore, Singapore, pp. 147–169.

Kim, D. H., Kim, H. S., Choi, C. W., Kim, E. K., Kim, B. I., & Choi, J. H. (2012). Risk factors for pulmonary artery hypertension in preterm infants with moderate or severe bronchopulmonary dysplasia. *Neonatology*, *101*(1), 40–46.

Kim, J. S., Kuk, E., Yu, K. N., Kim, J. H., Park, S. J., Lee, H. J., ... & Cho, M. H. (2007). Antimicrobial effects of silver nanoparticles. *Nanomedicine: Nanotechnology, Biology and Medicine*, *3*(1), 95–101.

Kim, S. S., Na, S. I., Jo, J., Kim, D. Y., & Nah, Y. C. (2008). Plasmon enhanced performance of organic solar cells using electrodeposited Ag nanoparticles. *Applied Physics Letters*, *93*, 073307.

Kouzegaran, V. J., & Farhadi, K. (2017). Green synthesis of sulphur nanoparticles assisted by a herbal surfactant in aqueous solutions. *Micro & Nano Letters*, *12*(5), 329–334.

Krishnaraj, C., Jagan, E. G., Ramachandran, R., Abirami, S. M., Mohan, N., & Kalaichelvan, P. T. (2012). Effect of biologically synthesized silver nanoparticles on Bacopa monnieri (Linn.) Wettst. plant growth metabolism. *Process Biochemistry*, *47*(4), 651–658.

Krolikowska, A., Kudelski, A., Michota, A., & Bukowska, J. (2003). SERS studies on the structure of thioglycolic acid monolayers on silver and gold. *Surface Science,* 532, 227–232.

Kulabhusan, P. K., Tripathi, A., & Kant, K. (2022). Gold nanoparticles and plant pathogens: An overview and prospective for biosensing in forestry. *Sensors*, *22*(3), 1259.

Kumar, A., Mandal, S., Selvakannan, P. R., Parischa, R., Mandale, A. B., & Sastry, M. (2003). Investigation into the interaction between surface-bound alkylamines and gold nanoparticles. *Langmuir*, *19*, 6277–6282.

Kumar, S., Soukup, M., & Elbaum, R. (2017). Silicification in grasses: Variation between different cell types. *Frontiers in Plant Science*, *8*, 438.

Kumar, Y., Tiwari, K. N., Singh, T., & Raliya, R. (2021). Nanofertilizers and their role in sustainable agriculture. *Annals of Plant and Soil Research*, *23*(3), 238–255.

Laware, S. L., & Raskar, S. (2014). Influence of zinc oxide nanoparticles on growth, flowering and seed productivity in onion. *International Journal of Current Microbiology Science*, *3*(7), 874–881.

Li, Y., Zhang, P., Li, M., Shakoor, N., Adeel, M., Zhou, P., ... & Rui, Y. (2023). Application and mechanisms of metal-based nanoparticles in the control of bacterial and fungal crop diseases. *Pest Management Science*, *79*(1), 21–36.

Li, Y., Zhong, L., Zhang, L., Shen, X., Kong, L., & Wu, T. (2021). Research advances on the adverse effects of nanomaterials in a model organism, *Caenorhabditis elegans*. *Environmental Toxicology and Chemistry*, *40*(9), 2406–2424.

Lim, D., Roh, J. Y., Eom, H. J., Choi, J. Y., Hyun, J., & Choi, J. (2012). Oxidative stress-related PMK-1 P38 MAPK activation as a mechanism for toxicity of silver nanoparticles to reproduction in the nematode Caenorhabditis elegans. *Environmental Toxicology and Chemistry*, *31*(3), 585–592.

Lok, C. N., Ho, C. M., Chen, R., He, Q. Y., Yu, W. Y., Sun, H., ... & Che, C. M. (2006). Proteomic analysis of the mode of antibacterial action of silver nanoparticles. *Journal of Proteome Research*, *5*(4), 916–924.

Mahmood, I., Rizvi, R., Sumbul, A., & Ansari, R. A. (2019). Potential role of plant growth promoting rhizobacteria in alleviation of biotic stress. In: *Plant Health Under Biotic Stress: Volume 2: Microbial Interactions*, (Editors: Ansari, R.A. and Mahmood, I.), Springer Nature, Singapore, pp. 177–188.

Mahmoodzadeh, H., Nabavi, M., & Kashefi, H. (2013). Effect of nanoscale titanium dioxide particles on the germination and growth of canola (*Brassica napus*).

Malandrakis, A. A., Kavroulakis, N., & Chrysikopoulos, C. V. (2019). Use of copper, silver and zinc nanoparticles against foliar and soil-borne plant pathogens. *Science of the Total Environment*, *670*, 292–299.

Mandal, D., Bolander, M. E., Mukhopadhyay, D., Sarkar, G., & Mukherjee, P. (2006). The use of microorganisms for the formation of metal nanoparticles and their application. *Applied Microbiology & Biotechnology*, *69*, 485–492.

Mann, S. (ed.) (1996). *Biomimetic Materials Chemistry*. VCH Publishers, New York.

Mansoor, S., Zahoor, I., Baba, T. R., Padder, S. A., Bhat, Z. A., Koul, A. M., & Jiang, L. (2021). Fabrication of silver nanoparticles against fungal pathogens. *Frontiers in Nanotechnology*, *3*, 679358.

Matsumura, Y., Yoshikata, K., Kunisaki, S. I., & Tsuchido, T. (2003). Mode of bactericidal action of silver zeolite and its comparison with that of silver nitrate. *Applied and Environmental Microbiology*, *69*(7), 4278–4281.

Meena, R. S., Kumar, S., Datta, R., Lal, R., Vijayakumar, V., Brtnicky, M., ... & Marfo, T. D. (2020). Impact of agrochemicals on soil microbiota and management: A review. *Land*, *9*(2), 34.

Mittal, A. K., Chisti, Y., & Banerjee, U. C. (2013). Synthesis of metallic nanoparticles using plant extracts. *Biotechnology Advances*, *31*(2), 346–356.

More, P. R., Pandit, S., Filippis, A. D., Franci, G., Mijakovic, I., & Galdiero, M. (2023). Silver nanoparticles: Bactericidal and mechanistic approach against drug resistant pathogens. *Microorganisms*, *11*(2), 369.

Morones, J. R., Elechiguerra, J. L., Camacho, A., Holt, K., Kouri, J. B., Ramírez, J. T., & Yacaman, M. J. (2005). The bactericidal effect of silver nanoparticles. *Nanotechnology*, *16*(10), 2346.

Mwaanga, P. (2018). Risks, uncertainties, and ethics of nanotechnology in agriculture. In: *New Visions in Plant Science* (Editor: Çelik, O.) InTech.

Nazeruddin, G. M., Prasad, N. R., Waghmare, S. R., Garadkar, K. M., & Mulla, I. S. (2014). Extracellular biosynthesis of silver nanoparticle using Azadirachta indica leaf extract and its anti-microbial activity. *Journal of Alloys and Compounds*, *583*, 272–277.

Ogunyemi, S. O., Abdallah, Y., Zhang, M., Fouad, H., Hong, X., Ibrahim, E., ... & Li, B. (2019). Green synthesis of zinc oxide nanoparticles using different plant extracts and their antibacterial activity against Xanthomonas oryzae pv. oryzae. *Artificial cells, Nanomedicine, and Biotechnology*, *47*(1), 341–352.

Pandey, A. K., & Gelin, R. (2018). A mass-produced sociable humanoid robot: Pepper: The first machine of its kind. *IEEE Robotics & Automation Magazine*, *25*(3), 40–48.

Park, H. J., Kim, S. H., Kim, H. J., & Choi, S. H. (2006). A new composition of nanosized silica-silver for control of various plant diseases. *The Plant Pathology Journal*, *22*(3), 295–302.

Parveen, A., & Siddiqui, Z. A. (2021). Zinc oxide nanoparticles affect growth, photosynthetic pigments, proline content and bacterial and fungal diseases of tomato. *Archives of Phytopathology and Plant Protection*, *54*(17–18), 1519–1538.

Patil, S., Ansari, R. A., Gurikar, C., Bhavan, S., & Hariprasad, K. (2024). Propagation of mycorrhizal fungi and their strength in plant health ameliorations. In: *Mycorrhizal Symbiosis and Agroecosystem Restoration*. (Editors: Ansari, R.A., Rizvi, R., and Mahmood, I.) Springer Nature Singapore, Singapore, pp. 77–98.

Peto, G., Molnar, G. L., Paszti, Z., Geszti, O., Beck, A., & Guczi, L. (2002). Electronic structure of gold nanoparticles deposited on SiOx/Si. *Materials Science and Engineering C*, *19*, 95–99.

Prasad, B. N., Subramanyam, B., Lakshmipathi, R. N., Ansari, R. A., Rizvi, R., Sumbul, A., ... & Rachmi, C. M. (2020). Utilization of beneficial microorganisms in sustainable control of phytonematodes. In: *Management of Phytonematodes: Recent Advances and Future Challenges*, (Editors: Ansari, R.A., Rizvi, R., and Mahmood, I.) Springer Nature Singapore, pp. 317–337.

Prasad, R., Bhattacharyya, A., & Nguyen, Q. D. (2017). Nanotechnology in sustainable agriculture: Recent developments, challenges, and perspectives. *Frontiers in Microbiology*, *8*, 1014.

Rajiv, P., Rajeshwari, S., & Venckatesh, R. (2013). Bio-fabrication of zinc oxide nanoparticles using leaf extract of Parthenium hysterophorus L. and its size-dependent antifungal activity against plant fungal pathogens. *Spectrochimica Acta Part A: Molecular and Biomolecular Spectroscopy*, *112*, 384–387.

Rezaei, V., Rabiee, A., & Khademi, F. (2020). Glioblastoma multiforme: A glance at advanced therapies based on nanotechnology. *Journal of Chemotherapy*, *32*(3), 107–117.

Roh, J. Y., Sim, S. J., Yi, J., Park, K., Chung, K. H., Ryu, D. Y., & Choi, J. (2009). Ecotoxicity of silver nanoparticles on the soil nematode Caenorhabditis elegans using functional ecotoxicogenomics. *Environmental Science & Technology*, *43*(10), 3933–3940.

Sabry, A. K. H. (2019). Role of nanotechnology applications in plant-parasitic nematode control. In: *Nanobiotechnology Applications in Plant Protection: Volume 2*, (Editors: Abd-Elsalam K. A. and Prasad, R.), Springer Nature Switzerland AG, pp. 223–240.

Sadeghi, B., & Gholamhoseinpoor, F. (2015). A study on the stability and green synthesis of silver nanoparticles using Ziziphora tenuior (Zt) extract at room temperature. *Spectrochimica Acta Part A: Molecular and Biomolecular Spectroscopy*, *134*, 310–315.

Saied, E. M., El-Naggar, N. E.-A., & Abdelwahed, N. A. M. (2021). Biosynthesis of nanoparticles: A review. *Arabian Journal of Chemistry, 14*(3), 103155.

Saied, E., Eid, A. M., Hassan, S. E. D., Salem, S. S., Radwan, A. A., Halawa, M., ... & Fouda, A. (2021). The catalytic activity of biosynthesized magnesium oxide nanoparticles (Mgo-nps) for inhibiting the growth of pathogenic microbes, tanning effluent treatment, and chromium ion removal. *Catalysts*, *11*(7), 821.

Salem, N. M., Albanna, L. S., & Awwad, A. M. (2016). Green synthesis of sulfur nanoparticles using *Punica granatum* peels and the effects on the growth of tomato by foliar spray applications. *Environmental Nanotechnology, Monitoring & Management*, *6*, 83–87.

Sastry, M., Ahmad, A., Khan, M. I., & Kumar, R. (2004). Microbial nanoparticle production. In: Niemeyer, C. M., & Mirkin, C. A. (eds) *Nanobiotechnology*. Wiley-VCH, Weinheim, Germany, pp. 126–135.

Senthilkumar, S. R., & Sivakumar, T. (2014). Green tea (Camellia sinensis) mediated synthesis of zinc oxide (ZnO) nanoparticles and studies on their antimicrobial activities. *International Journal of Pharmacy and Pharmaceutical Sciences*, *6*(6), 461–465.

Shah, Z. M., Naz, R., Naz, S., Zahoor, S., Nosheen, A., Shahid, M., ... & Keyani, R. (2023). Incorporation of zinc sulfide nanoparticles, Acinetobacter pittii and Bacillus velezensis to improve tomato plant growth, biochemical attributes and resistance against Rhizoctonia solani. *Plant Physiology and Biochemistry*, *202*, 107909.

Shankar, V., Smith, A. K., & Rangaswamy, A. (2003). Customer satisfaction and loyalty in online and offline environments. *International Journal of Research in Marketing*, *20*(2), 153–175.

Shende, S., Ingle, A. P., Gade, A., & Rai, M. (2015). Green synthesis of copper nanoparticles by Citrus medica Linn.(Idilimbu) juice and its antimicrobial activity. *World Journal of Microbiology and Biotechnology*, *31*, 865–873.

Siddiqui, Z. A., Khan, M. R., Abd_Allah, E. F., & Parveen, A. (2019). Titanium dioxide and zinc oxide nanoparticles affect some bacterial diseases, and growth and physiological changes of beetroot. *International Journal of Vegetable Science*, *25*(5), 409–430.

Singh, P., Kim, Y. J., Wang, C., Mathiyalagan, R., El-Agamy Farh, M., Yang, D. C., & Agastian, P. (2020). Biosynthesis, characterization, and antimicrobial applications of silver nanoparticles. *International Journal of Nanomedicine*, *15*, 61–83.

Singh, S. K., Patra, A., Verma, Y., Chattopadhyay, A., Rakshit, A., & Kumar, S. (2021). Potential and risk of nanotechnology application in agriculture vis-à-vis nanomicronutrient fertilizers. In: *Soil Science: Fundamentals to Recent Advances*, (Editors: Rakshit, A., Singh, S., Abhilash, P., Biswas, A.), Springer Singapore, Singapore, pp. 513–552.

Solanki, M. K., Kashyap, P. L., Ansari, R. A., & Kumari, B. (2020). *Microbiomes and Plant Health: Panoply and Their Applications*, Academic Press, Elsevier, The Netherland. ISBN: 9780128197158. www.elsevier.com/books/microbiomes-and-plant-health/solanki/978-0-12-819715-8.

Solanki, M. K., Kashyap, P. L., Kumari, B., Ansari, R. A., Sumbul, A., Rizvi, R., & Mahmood, I. (2021). Mycorrhizal fungi and its importance in plant health amelioration. In: *Microbiomes and Plant Health: Panoply and Their Applicationsth*. (Editors: Solanki, M.K., Ansari, R.A., Kashyap, P.L., and Kumari, B.), Academic Press, Amsterdam, pp. 205–223.

Soleimani, M., Aflatouni, F., & Khani, A. (2013). A new and simple method for sulfur nanoparticles synthesis. *Colloid Journal*, *75*(1), 112–116.

Sondi, I., & Salopek-Sondi, B. (2004). Silver nanoparticles as antimicrobial agent: A case study on E. coli as a model for Gram-negative bacteria. *Journal of Colloid and Interface Science*, *275*(1), 177–182.

Sumbul, A., Ansari, R. A., Rizvi, R., & Mahmood, I. (2020). Azotobacter: A potential bio-fertilizer for soil and plant health management. *Saudi Journal of Biological Sciences*, *27*(12), 3634–3640.

Swamy, C., & Nargund, V. B. (2017). Sunlight induced mediated silver nanoparticles from seeds of Thevetia peruviana L., characterization and their antifungal efficacy against Curvularia lunata (Wakker) Boedijn. *International Journal of Current Microbiology and Applied Sciences*, *6*(11), 1008–1013.

Team, C. W. (2010, January). Good practice guidance paper on assessing and combining multi model climate projections. In: *IPCC Expert Meeting on Assessing and Combining Multi Model Climate Projections,* p. 1.

Thanighaiarassu, R. R., Nambikkairaj, B., Devika, R., Raghunathan, D., & Sivamani, P. (2014). Green synthesis of gold nanoparticles mediated by plant essential oil and its antifungal activity against human pathogenic fungi. *World Journal of Pharmacy and Pharmaceutical Sciences (WJPPS)*, *3*(12), 1752–1768.

Tudi, M., Daniel Ruan, H., Wang, L., Lyu, J., Sadler, R., Connell, D., ... & Phung, D. T. (2021). Agriculture development, pesticide application and its impact on the environment. *International Journal of Environmental Research and Public Health, 18*(3), 1112.

Tuncsoy, B. (2021). Nematicidal activity of silver nanomaterials against plant-parasitic nematodes. In: *Silver Nanomaterials for Agri-Food Applications,*(Editor: Abd-Elsalam, K.A.) Elsevier, pp. 527–548.

Valodkar, M., Modi, S., Pal, A., & Thakore, S. (2011). Synthesis and anti-bacterial activity of Cu, Ag and Cu–Ag alloy nanoparticles: A green approach. *Materials Research Bulletin*, *46*(3), 384–389.

Yin, Q., Yang, W., Sun, C., & Di, M. (2012). Preparation and properties of lignin-epoxy resin composite. *BioResources*, *7*(4), 5737–5748.

Zhang, G., & Xu, J. (2013). From colloidal nanoparticles to a single crystal: New insights into the formation of nacre's aragonite tablets. *Journal of Structural Biology*, *182*(1), 36–43.

Zheng, G., Yang, Y., Cha, J. J., Hong, S. S., & Cui, Y. (2011). Hollow carbon nanofiber-encapsulated sulfur cathodes for high specific capacity rechargeable lithium batteries. *Nano Letters*, *11*(10), 4462–4467.

4 Nanosensors

An Emerging Tool for Early Detection of Plant Pathogens in Sustainable Agriculture

Sozan E. El-Abeid, Rizwan Ali Ansari, Mohamed A. Mosa, Ahmed G. Soliman, Ahmed M. Saleh, Jamila Boudaden, and Dina S. S. Ibrahim

4.1 INTRODUCTION

Nanosensors are sophisticated devices designed to detect and quantify a wide array of physical, chemical or biological signals with exceptional precision and sensitivity, operating at the nanoscale level (Rolfe, 2012). By the way, materials in their nano form usually show different activity as compared to their ionic state, which enhances their activity as pesticide agents or carriers. They have a significant ability to influence the agriculture sector enormously. This aims to increase the growth and subsequently, yield components, by economic as well as societal empowerment (Parisi et al., 2015). NanoParticles (NPs) have been exploited in agriculture for sensors and crop protection; they have unique features permeability, solubility, thermal stability and degradability. Additionally, the high surface area of nanoparticles makes them more attractive to the target organism (Li et al., 2011). NPs exhibit sizes within the range of 1 nm to less than 100 nm and exist in diverse forms, compositions, nano-encapsulates, nano-emulsions and being used for plant disease management (Das et al., 2017). NP applications play a crucial role in reducing the usage of pesticides and fertilizers (Mittal et al., 2020). Peng and Miller, (2011) reported that nanoparticles exhibit varied dimensions, which contribute to their suitability for constructing electrochemical analytical sensors and other sensors types.

Several nano-sensor technologies have been specifically engineered to detect toxins such as heavy metals like mercury in water sources (Selid et al., 2009). Different types of NPs have been utilized in creating nano-sensors for detecting crop viruses, pests, pathogens, stressors in the environment, and nutrient levels in the soil (Kashyap and Kumar, 2021). As an illustration, nanosensors have been developed to measure oxygen levels and detect auxin distribution (Koren et al., 2015). NPs have also been used in the detection of devastating several plant pests, nematodes and pathogens in agriculture (Fabiyi et al., 2020). Nanosensors serve as diagnostic tools to identify specific molecules (effectors) in plants that are only present in diseased plants or that attract plant-invading nematodes (Fabiyi et al., 2020).

Biosensors are seen as analytical tools that utilize living organisms or biological materials to produce an electrochemical signal that can be measured with a processor (Prasad et al., 2014). The primary functional components consist of a three-part initial interactive bio-recognition element, a second transducer, and a third accessibility signal. A variety of biosensors, including immune sensors, utilize the interaction between a bio-molecular recognition element like an antibody (Ab) and a specific analyte (Ag) (Anusha et al., 2022). Advanced biosensors could provide

DOI: 10.1201/9781003256762-4

benefits for food production research and development, food quality assurance organizations, food manufacturers, and regulators in charge of ensuring food safety and security (Parsad et al., 2017). The conventional methods which are being utilized to detection are indeed effective, but they come with inherent challenges. These methods tend to be costly and time-consuming involving several steps for sample preparation based on certain protocols. Furthermore, they depend on the skills of specialized personnel and sophisticated instruments that are not readily available (Koedrith et al., 2014).

The presence of nanomaterials in biosensors allows for the integration of numerous new signal transmission methods (Otles and Yalcin 2012). Nanomaterial-enhanced biosensors can be used to assess antibiotic resistance, soil conditions, biotic and abiotic stress of plants, and nutritional content, as well as food quality (Prasad et al., 2017). Biological sensitive probes are components that mimic biology and allow biological materials to interact with analytes, giving an analogue signal indicating the concentration of solute. This includes specificity, sensitivity, and the surface of nanoparticles acting as a carrier system in molecular interactions (Algieri et al., 2014). The transducer serves as a critical component facilitating the conversion of one form of energy or signal into another, essential for numerous sensing principles. These principles encompass electrochemical techniques like conductometry, amperometry, and potentiometry, as well as optical phenomena including surface plasmon resonance and luminescence (Philip and Kumar, 2022).The surface plasmon resonance permits the detection of Events that are sensitive to mass, such as the detection of signals indicating the presence of chemicals or biological substances. Luminescence is the result of light energy stimulating chemicals in substances to emit visible light waves. The data recording equipment consists of an amplifier, a signal processor, and a display to present the final output (Ahmad, et al., 2023). Some nanomaterials used in creating nanomaterial-based biosensors include gold nanoparticles, carbon nanotubes, and nano-magnetic materials, offering promising applications in enhancing soil health through the facilitation of mineralization processes (Malekzad et al., 2017; Mitchell et al., 2005).

Moreover, conventional techniques are accurate and trustworthy, but they require extensive processes and bulky machinery, along with the need to transport samples to labs (Wang, 2005; Rai et al., 2012). Dyk and Pletschke (2011) conducted a study on the detection of organochlorine, carbamate, and organophosphate residues through the use of biosensors. The sensing technologies and NPs may be integrated together to enhance the reliability of diagnostics. These possess wide surfaces with high conductivity characteristics, and are responsive to colour changes. Additionally, nanoparticles, nanorods and nanowires are effective instruments in enhancing detection sensitivity. Carbon nanotubes, which have a significant level of electrical and thermal conductivity along with a large surface area, are also of interest in the production of nano-biosensors, alongside quantum dots nanomaterial.

To effectively tackle the challenges related to detecting organochlorine, carbamate, and organophosphate residues, Dyk and Pletschke (2011) have explored the use of biosensors that rely on enzyme detection techniques. However, to overcome these limitations and improve detection capabilities, a combination of sensing technologies and nanotechnology has also been presented. Nanomaterials, characterized by their varying dimensions and broad surface areas with strong conductivity properties, offer promising solutions. NPs and nano-wires possess enhanced resilience and sensitivity, while nano-rods provide powerful detection capabilities (Kumar and Kumbhat, 2018). The carbon nanotubes are well versed with high electrical and thermal conductivity, offer large surface area for improved sensing. Quantum dots, with their colour tenability, have also gained interest in the development of nanobiosensors (Pandey, et al., 2021).

A sensor represents an indispensable tool capable of identifying and discerning the presence of an analyte within a given matrix (Jurs et al. 2000). The function of a nanosensor involves the detection or perception of gases, chemicals, light, electric fields, and heat through the utilization of at least one nanostructure (Liao et al., 2019). The success of the biosensor relies on how the bioreceptor layer is chosen and connected to the transducer surface (Grieshaber et al., 2008). The primary goal of the sensor is to establish a firm and dependable link between the biological component and the sensor surface. A perfect situation includes the creation of self-assembled monolayers incorporated onto the sensor surface, guaranteeing excellent performance and sensitivity (Milvich et al., 2018). The inclusion of nanoparticles results in the creation of self-assembled monolayers, which greatly enhances the system's sensitivity (Murugan et al., 2015).

Additionally, the functional layer in biosensors, which helps with the binding of a particular analyte for precise detection, is often a biological component such as an antibody, enzyme, nucleic acid (DNA/RNA), or sometimes an entire bacterial cell. Nanoparticles are commonly utilized in these systems to connect the converter and the bio-receptor at the nanoscale level (Kaushal and Wani 2017; Huang et al., 2021). Frequently used electrochemical techniques in sensors include potentiometric, chronometric, voltammetry, electrical resistance measurement, and field-effect transistor (Huang et al., 2021).

This chapter discusses the nanosensors, their types, characteristic features and usage in detection of plant pathogens infesting different crop commodities. The outcome of the chapter will shed light on current status of nanosensors being utilized for the detection of a myriad of microorganisms which causes significant crop yield losses.

4.2 TYPES OF NANOBIOSENSORS FOR PATHOGEN DETECTION

A sensor or biosensor is a tool that detects the response of a specific substance, known as the 'analyte', in a medium such as gas or liquid. An integrated device in a medium could allow for continuous monitoring of a specific analysis with full reversibility without causing any harm to the medium being studied (Naresh and Lee, 2021). Various types of nanosensors have been employed for this purpose, an overview of some of these are presented in the following section.

Biosensors work on biological recognition and transduction of the signal mechanisms as well. These sensors have diverse categories like piezoelectric, electrochemical, optical and thermal sensors each offering unique advantages in detecting and quantifying analytes with precision and sensitivity (Kwak et al., 2017). Their diverse range of detection capabilities makes biosensors a powerful tool for precise and reliable measurements in various fields. By harnessing the principles of bio-recognition and leveraging innovative signal transduction mechanisms, biosensors offer immense potential for advancing research, enhancing diagnostics, and enabling efficient monitoring in numerous applications.

4.2.1 Fluorescence Resonance Energy Transfer Nanosensor

The Fluorescence Resonance Energy Transfer (FRET) nanosensor represents a highly advanced optical sensing technology that exploits the principle of fluorescence resonance energy transfer for precise measurements (Algar and Krull 2008). This cutting-edge technology utilizes two distinct fluorescent dyes, referred to as the donor and the acceptor, which interact through intermolecular dipole-dipole interactions. The transfer of energy takes place as soon as the spectrum of the donor lines up with the excitation of the spectrum of the acceptor dye. This initiates the process as the distance between dyes ranges from 1 to 10 nm (Stanisavljevic et al., 2015). The domains, which are substrate-specific and located between appropriate donor and acceptor dyes at opposite ends, serve as a vital element in enabling efficient FRET pairing (Ellinger and Voigt 2014). The sensor experiences a discernible shift in energy transfer due to an interaction with the substrate binding, primarily caused by changes in the spatial arrangement or conformation of the two fluorophores (Kwak et al., 2017).

4.2.2 Surface Enhanced Raman Scattering Nanosensor

This is a remarkable spectroscopy detection tool, has the extraordinary ability to identify analytes down to individual molecules. Fabricated predominantly with metal nanoparticle surfaces it enhances the Raman scattering signatures of adsorbed molecules (Kahraman et al., 2017). Importantly, Raman signals can be enhanced up to 10^{14} times, matching or even exceeding those of fluorescent organic dyes (Stiles et al., 2008). The primary thing responsible for inducing SERS effects involve two main factors: electromagnetic enhancement and chemical enhancement (Sun et al., 2008). The electromagnetic phenomenon boosts the electromagnetic field near the surface of nanoparticles within a substantial range of 10–200 nm. The chemical phenomenon involves short-range charge transfers between guest molecules and NPs, occurring specifically when an analyte adheres to a surface coated with NPs (Wei et al., 2015). Gold and silver NPs, with their powerful and adjustable (plasmon resonance in the near-infrared region), are commonly employed for this purpose. Thanks to the inherent possibilities of the near-infrared region, various sensors have been designed for detecting pathogens within agriculture (Chen and Park 2016). Chen et al. (2021) introduced a cutting-edge aptasensor utilizing surface-enhanced Raman scattering technology, enabling the concomitant identification of zearalenone and ochratoxin A in two crops (wheat and corn). Their assessment of the sensor's performance revealed impressive levels of accuracy, precision, and specificity. This pioneering approach allows for the identification of both mycotoxins, and also enables their differentiation with precision.

Furthermore, the versatility of this platform extends to the simultaneous identification of various mycotoxin types through the modification of aptamer and signal probes.

4.2.3 Electrochemical Nanosensors

The ElectroChemical Nanosensors (ECN) present a groundbreaking and powerful approach for detecting crop diseases in their early stages. They are very simple, affordable, highly sensitive, easy miniaturization, and analyse data directly. The versatility of ECNs allows for seamless examination of any biological sample, as they can effortlessly convert a biological occurrence into an electronic signal (Zhu et al., 2015). These state-of-the-art nanosensors comprise three distinct electrode types: counter, reference, and working electrodes, each playing a crucial role in the sensor's functionality.

(Barry and O'Riordan 2015). The sophisticated instruments (amperometry, potentiometry, voltammetry, conductometry) records the biological/chemical indications into electrical outputs (Qureshi et al., 2009). Electrochemical electrodes designed with NPs possess a remarkable active surface area, which enhances sensitivity and broadens the detection range for redox and ion species relevant to plants (Haroon and Stine 2023) (Clausmeyer and Schuhmann, 2016). ECNs possess significant potential to revolutionize pathogen detection methods for field and horticultural crops, offering improved accuracy and efficiency in disease management (Srivastava et al., 2018).

4.2.4 Piezoelectric Nanosensors

Piezoelectric nanosensors are advanced devices that utilize piezoelectric materials at the nanometre scale to detect minuscule mechanical vibrations and convert them into electrical signals (Tavker & Sharma, 2022; Swain et al., 2023; Zhu et al., 2020). These nanomaterials exhibit high sensitivity to weak, low-frequency vibrations, making them ideal for harvesting energy from the environment and converting it into electrical energy for applications like nanogenerators and self-powered devices (Ji et al., 2023). Additionally, nanosensors, including piezoelectric ones are utilized in agriculture to detect soil nutrients, pests, pathogens, and environmental changes to enhance crop yield and protect plants from external factors (Kamari et al., 2023).

4.3 DESIGN AND FABRICATION OF NANOBIOSENSORS FOR PATHOGEN DETECTION

The rapid advancements in nanotechnology have revolutionized the field of sensing by enabling the design and development of nanosensors with remarkable capabilities (Lichtenstein et al., 2014; Walsh et al., 2015). These pioneering nanosensors possess the capability to investigate the spatio-temporal dynamics of specific analytes, facilitating precise environmental sensing (Reuel et al., 2012; Wise and Brasuel, 2011), and monitor cellular functions (Howes et al., 2014; Sun et al., 2017). In order to fulfil the requirements of various applications, these innovative nanosensors are ideally characterized by essential features including rapid dynamics, exceptional sensitivity, precision, stability and reproducibility (Sagadevan and Periasamy, 2014). The recent strides in nanotechnology have led to the emergence of incredibly adaptable and sensitive nanoscale sensors, offering unmatched reliability and temporal resolution. These sensors, when fused with disciplines like electronics, biology, and computer science, demonstrate exceptional performance. In agriculture, nanosensors offer vast potential across a spectrum of applications. They serve as indispensable tools for detecting pathogens, precisely managing soil nutrients, and mitigating environmental pollution by optimizing fertilizer doses (Rai et al., 2012). The nanosensors have demonstrated efficacy in detecting contaminants in vegetables such as organophosphorus (Amine et al., 2006). Additionally, intracellular sensors are very instrumental in elucidating the intricate molecular mechanisms within plant systems by monitoring metabolic intermediates and signalling molecules. The rational design and fabrication of these sensors contribute to a better understanding of plant biology (Medintz, 2006).

4.4 APPLICATION OF NANOBIOSENSORS IN PATHOGEN DETECTION

Incorporating NPs as sensing elements significantly boosts the detection sensitivity across various bacterial, fungal, and viral diseases, expanding the diagnostic capabilities in disease management (Castillo-Henríquez et al., 2020; Mocan et al., 2017). The history of using nanosensors to detect plant diseases is relatively recent, starting around 2009 (Kashyap et al., 2019). One of the first examples of nanosensors for detecting plant pathogens was reported by Wang in 2010 (Wang et al. 2010). Furthermore, they devised a lateral flow immunoassay utilizing AuNPs that enable rapid and highly sensitive detection of tobacco mosaic virus. The assay can detect TMV (0.1 ng/mL) within 25 minutes (Drygin et al., 2009). Other nanosensors have been also developed for detecting bacteria, fungi and viruses, using different sensing principles, such as optical, electrochemical, and piezo-electric (Dyussembayev et al., 2021). NPs either individually or in conjunction with other particles improve the techniques of direct pathogen detection or the identification of specific compounds which is linked to disease progression. For instance, Au nanorods functionalized with antibodies have demonstrated efficacy in rapidly detecting viral infections (Jian et al., 2018; Lin et al., 2022).

4.4.1 Electrochemical Nanobiosensor

These sensors use electrical signals to measure changes in the electrical properties of nanomaterials or biomolecules when they interact with plant pathogens. For example, carbon nanotubes can be used to detect bacterial cells (Kashyap et al., 2019). Franco et al. (2019) unpicked the progress of DNA driven nanobiosensors targeting *Phytophthora palmivora* infections in cacao. These nanobiosensors utilize sandwich hybrids formed by pairing two oligonucleotide probes with the genomic DNA of *P. palmivora*. These hybrids are carefully designed based on the ITS sequence of field isolates. This design strategy enhances the specificity and sensitivity of the nanobiosensors for detecting *P. palmivora*. Through electrochemical detection, the nanobiosensor demonstrates both sensitivity and selectivity towards *P. palmivora*, distinguishing it from other fungal pathogens associated with cacao plants.

4.4.2 Piezoelectric Nanobiosensors

These devices use mechanical vibrations to measure changes in the mass or stress of nanomaterials or biomolecules when they interact with plant pathogens. For example, zinc oxide nanowires can be used to detect viral particles by measuring the resonant frequency change (Chandna et al., 2022; Iyer et al., 2014).

AgNPs are minute particles endowed with distinctive optical, electrical, and catalytic attributes. Leveraging these properties, they serve as highly effective sensors for detecting plant pathogens, offering unparalleled precision and sensitivity in disease surveillance. Pathogen binding changes the colour, fluorescence, or electrical properties of AgNPs, which electrochemical methods or optical methods can measure. The utilization of AgNPs as signal amplifiers in molecular assays, such as polymerase chain reaction or loop-mediated isothermal amplification, significantly enhances the sensitivity and accuracy of detection methods for diverse pathogens and biomolecules. By augmenting the amount of DNA or RNA produced in the reaction, AgNPs contribute to heightened sensitivity and specificity in detection. AgNPs conjugated with aptamers were used to detect Ralstonia solanacearum, a bacterial wilt pathogen, in tomato plants (Opiyo, 2022). Similarly, AgNPs conjugated with DNA probes were used to detect rice tungro bacilliform virus, a viral disease agent, in rice plants (Uda et al., 2017; Uda et al., 2019). The AgNPs conjugated with antibodies were employed to detect *Fusarium graminearum*, a fungal pathogen responsible for Fusarium head blight in wheat plants. This innovative approach showcases the potential of AgNPs in enhancing the specificity and sensitivity of pathogen detection methods, thereby contributing to improved disease management (Carbone et al., 2019). AgNP driven sensors offer numerous advantages, including high sensitivity, specificity, rapidity, portability, and cost-effectiveness. They can help in the early diagnosis and management of plant diseases, as well as monitoring plant health and growth. Paniel et al. (2010) utilized magnetic NPs for detecting the aflatoxin-M1, a mycotoxin synthesized by the fungus *Aspergillus flavus*, in food products. This pioneering method underscores the potential of magnetic NPs for enhancing toxin detection, thereby contributing to food safety measures and public health protection. Additionally, an immunosensor was also found to help perceive the toxin at a very low (0.01 ppb) concentration (Kashyap et al., 2017).

4.5 CONCLUSIONS

Nanosensors have emerged as groundbreaking tools in agricultural practices, offering unparalleled capabilities for inspecting the nature of myriad pathogens. Their integration into crop pathology provides enhanced disease management strategies and improved crop health. The ability of nanosensors to detect specific plant pathogens enables rapid and accurate identification of diseases even at their early stages. This early detection empowers farmers with the knowledge to implement targeted treatment strategies such as judicious pesticide, biological control agent applications or the deployment of resistant cultivars. By intervening early, we can minimize the spread of diseases, prevent significant yield losses and ensure sustainable agriculture practices. Furthermore, careful assessment of the environmental impact and long-term stability of nanosensors is crucial to guarantee their safe and sustainable utilization. Nanosensors offer unprecedented sensitivity, speed and portability in the early detection of plant pathogens. Continued research and development are crucial to overcome challenges and unlock the full potential of nanosensors in revolutionizing plant disease diagnosis followed by its eco-friendly management.

REFERENCES

Ahmad, M., Hasan, M., Tarannum, N., & Ahmed, S. (2023). Recent advances in optical and photoelectrochemical nanobiosensor technology for cancer biomarker detection. *Biosensors and Bioelectronics: X*, 1100375.

Algar, W. R., & Krull, U. J. (2008). Quantum dots as donors in fluorescence resonance energy transfer for the bioanalysis of nucleic acids, proteins, and other biological molecules. *Analytical and Bioanalytical Chemistry, Anal Bioanal Chem 391*, 1609–1618.

Algieri, C., Drioli, E., Guzzo, L., & Donato, L. (2014). Bio-mimetic sensors based on molecularly imprinted membranes. *Sensors, 14*(8), 13863–13912.

Amine, A., Mohammadi, H., Bourais, I., & Palleschi, G. (2006). Enzyme inhibition-based biosensors for food safety and environmental monitoring. *Biosensors and Bioelectronics, 21*(8), 1405–1423.

Anusha, T., Bhavani, K. S., Kumar, J. S., Brahman, P. K., & Hassan, R. Y. (2022). Fabrication of electrochemical immunosensor based on GCN-β-CD/Au nanocomposite for the monitoring of vitamin Vitamin D deficiency. *Bioelectrochemistry, 143*, 107935.

Barry, S., & O'Riordan, A. (2015). Electrochemical nanosensors: advances advances and applications. *Report Electrochemistry, 2016*(6), 1–14.

Bellingham, B.K, (2011). *Proximal soil sensing, Vadose Zone J., 10*(4), 1340–1341.

Byrne, B., Stack, E., Gilmartin, N., & O'Kennedy, R. (2009). Antibody-based sensors: principles, problems and potential for detection of pathogens and associated toxins. *Sensors, 9*, 4407–4445.

Carbone, K., Paliotta, M., Micheli, L., Mazzuca, C., Cacciotti, I., Nocente, F., ... & Dell'Abate, M. T. (2019). A completely green approach to the synthesis of dendritic silver nanostructures starting from white grape pomace as a potential nanofactory. *Arabian Journal of Chemistry, 12*(5), 597–609.

Castillo-Henríquez, L., Brenes-Acuña, M., Castro-Rojas, A., Cordero-Salmerón, R., Lopretti-Correa, M., & Vega-Baudrit, J. R. (2020). Biosensors for the detection of bacterial and viral clinical pathogens. *Sensors, 20*(23), 6926.

Chandna, S., Gogde, K., Kaur, R., Sagar, V., & Bhaumik, J. (2022). Nano-biosensors for Plant plant Biomassbiomass: Concept concept and Applicationsapplications. In *The Role of Nanoparticles in Plant Nutrition under Soil Pollution: Nanoscience in Nutrient Use Efficiency* (pp. 199–221). Cham: Springer International Publishing.

Chen, J., & Park, B. (2016). Recent advancements in nanobioassays and nanobiosensors for foodborne pathogenic bacteria detection. *Journal of Food Protection, 79*, 1055–1069.

Chen, R., Li, S., Sun, Y., Huo, B., Xia, Y., Qin, Y., ... & Gao, Z. (2021). Surface-enhanced Raman spectroscopy aptasensor for simultaneous determination of ochratoxin A and zearalenone using Au@ Ag core-shell nanoparticles and gold nanorods. *Microchimica Acta, 188*, 1–9.

Clausmeyer, J., & Schuhmann, W. (2016). Nanoelectrodes: applications in electrocatalysis, single-cell analysis and high-resolution electrochemical imaging. *TrAC Trends in Analytical Chemistry, 79*, 46–59. Return to ref 2016 in article.

Das, R. K., Pachapur, V. L., Lonappan, L., Naghdi, M., Pulicharla, R., Maiti, S., ... & Brar, S. K. (2017). Biological synthesis of metallic nanoparticles: plants, animals and microbial aspects. *Nanotechnology for Environmental Engineering, 2*, 1–21.

Drygin, Y. F., Blintsov, A. N., Osipov, A. P., Grigorenko, V. G., Andreeva, I. P., Uskov, A. I., ... & Atabekov, J. G. (2009). High-sensitivity express immunochromatographic method for detection of plant infection by tobacco mosaic virus. *Biochemistry (Moscow), 74*, 986–993.

Dyussembayev, K., Sambasivam, P., Bar, I., Brownlie, J. C., Shiddiky, M. J., & Ford, R. (2021). Biosensor technologies for early detection and quantification of plant pathogens. *Frontiers in Chemistry, 9*, 636245.

Ellinger, D., & Voigt, C. A. (2014). The use of nanoscale fluorescence microscopic to decipher cell wall modifications during fungal penetration. *Frontiers of Plant Science, 5*, 270.

Fabiyi, O. A., Alabi, R. O., & Ansari, R. A. (2020). Nanoparticles' synthesis and their application in the management of phytonematodes: an overview. In *Management of Phytonematodes: Recent Advances and Future Challenges* (pp. 125–140).

Franco, A. J. D., Merca, F. E., Rodriguez, M. S., Balidion, J. F., Migo, V. P., Amalin, D. M., ... & Fernando, L. M. (2019). DNA-based electrochemical nanobiosensor for the detection of *Phytophthora palmivora* (Butler) Butler, causing black pod rot in cacao (*Theobroma cacao* L.) pods. *Physiological and Molecular Plant Pathology, 107*, 14–20.

Grecco, H. E., & Verveer, P. J. (2011). FRET in cell biology: still shining in the age of super-resolution? *ChemPhysChem, 12*, 484–490.

Grieshaber, D., MacKenzie, R., Vörös, J., & Reimhult, E. (2008). Electrochemical biosensors-sensor principles and architectures. *Sensors, 8*(3), 1400–1458.

Haroon, N., & Stine, K. J. (2023). Electrochemical detection of hormones using nanostructured electrodes. *Coatings, 13*(12), 2040.

Howes, P. D., Chandrawati, R., & Stevens, M. M. (2014). Colloidal nanoparticles as advanced biological sensors. *Science, 346*(6205), 1247390.

Huang, X., Zhu, Y., & Kianfar, E. (2021). Nano Biosensors: properties, applications and electrochemical techniques. *Journal of Materials Research and Technology, 12*, 1649–1672.

Iyer, M. A., Oza, G., Velumani, S., Maldonado, A., Romero, J., Munoz, M. D. L., ... & Yi, J. (2014). Scanning fluorescence-based ultrasensitive detection of dengue viral DNA on ZnO thin films. *Sensors and Actuators B: Chemical, 202*, 1338–1348.

Ji, J., Yang, C., Shan, Y., Sun, M., Cui, X., Xu, L., ... & Li, Z. (2023). Research trends of piezoelectric nanomaterials in biomedical engineering. *Advanced NanoBiomed Research, 3*(1), 2200088.

Jian, Y. S., Lee, C. H., Jan, F. J., & Wang, G. J. (2018). Detection of Odontoglossum ringspot virus infected Phalaenopsis using a nano-structured biosensor. *Journal of the Electrochemical Society, 165*(9), H449.

Jurs, P. C., Bakken, G. A., & McClelland, H. E. (2000). Computational methods for the analysis of chemical sensor array data from volatile analytes. *Chemical Reviews, 100*(7), 2649–2678.

Kahraman, M., Mullen, E. R., Korkmaz, A., & Wachsmann-Hogiu, S. (2017). Fundamentals and applications of SERS-based bioanalytical sensing. *Nanophotonics*.

Kamari, M., Kumar, N., Motaung, D. E., Issaoui, N., Kumar, S., & Rani, G. (2023). Nanosensors for crop protection. In *The Impact of Nanoparticles on Agriculture and Soil* (pp. 323–349). Academic Press.

Kashyap, B., & Kumar, R. (2021). Sensing methodologies in agriculture for monitoring biotic stress in plants due to pathogens and pests. *Inventions, 6*(2), 29.

Kashyap, P. L., Kumar, S., & Srivastava, A. K. (2017). Nanodiagnostics for plant pathogens. *Environmental Chemistry Letters, 15*, 7–13.

Kashyap, P. L., Kumar, S., Jasrotia, P., Singh, D. P., & Singh, G. P. (2019). Nanosensors for plant disease diagnosis: Current current understanding and future perspectives. In: Pudake, R., Chauhan, N., Kole, C. (eds), *Nanoscience for Sustainable Agriculture* (pp. 189–205). Springer, Cham.

Kaushal, M., & Wani, S. P. (2017). Nanosensors: frontiers in precision agriculture. *In Nanotechnology* (pp. 279–291). Springer, Singapore.

Koedrith, P., Thasiphu, T., Tuitemwong, K., Boonprasert, R., & Tuitemwong, P. (2014). Recent advances in potential nanoparticles and nanotechnology for sensing food-borne pathogens and their toxins in foods and crops: current technologies and limitations. *Sensors Mat, 26*(10), 711–736.

Koren, K., Brodersen, K. E., Jakobsen, S. L., & Kühl, M. (2015). Optical sensor nanoparticles in artificial sediments–a new tool to visualize O2 dynamics around the rhizome and roots of seagrasses. *Environmental Science & Technology, 49*(4), 2286–2292.

Kosaka, P. M., Pini, V., Ruz, J. J., Da Silva, R. A., González, M. U., Ramos, D., ... & Tamayo, J. (2014). Detection of cancer biomarkers in serum using a hybrid mechanical and optoplasmonic nanosensor. *Nature Nanotechnology, 9*(12), 1047–1053.

Kumar, N., & Kumbhat, S. (2018). *Concise Concepts of Nanoscience and Nanomaterials*. Scientific Publishers.

Kwak, S.-Y., Wong, M. H., Lew, T. T. S., Bisker, G., Lee, M. A., Kaplan, A., Dong, J., Liu, A. T., Koman, V. B., Sinclair, R., Hamann, C., & Strano, M. S. (2017). Nanosensor technology applied to living plant systems. *Annual Review of Analytical Chemistry, 10*, 113–140.

Li, F., Zhang, Z. P., Peng, J., Cui, Z. Q., Pang, D. W., Li, K., ... & Zhang, X. E. (2009). Imaging viral behaviour in mammalian cells with self-assembled capsid–quantum-dot hybrid particles. *Small, 5*(6), 718–726.

Li, X., Xu, H., Chen, Z. S., & Chen, G. (2011). Biosynthesis of nanoparticles by microorganisms and their applications. *Journal of Nanomaterials, 2011*, 1–16.

Li, Y., Schluesener, H. J., & Xu, S. (2010). Gold nanoparticle-based biosensors. *Gold Bull, 43*(1), 29–41.

Liao, Y., Zhang, R., & Qian, J. (2019). Printed electronics based on inorganic conductive nanomaterials and their applications in intelligent food packaging. *RSC Advances, 9*(50), 29154–29172.

Lichtenstein, A., Havivi, E., Shacham, R., Hahamy, E., Leibovich, R., Pevzner, A., ... & Patolsky, F. (2014). Supersensitive fingerprinting of explosives by chemically modified nanosensors arrays. *Nature Communications, 5*(1), 4195.

Lin, W. P., Wang, W. J., Lee, C. H., Jan, F. J., & Wang, G. J. (2022). A two-in-one immunoassay biosensor for the simultaneous detection of Odontoglossum ringspot virus and Cymbidium mosaic virus. *Sensors and Actuators B: Chemical, 350*, 130875.

Malekzad, H., Sahandi Zangabad, P., Mirshekari, H., Karimi, M., & Hamblin, M. R. (2017). Noble metal nanoparticles in biosensors: recent studies and applications. *Nanotechnology Reviews*, *6*(3), 301–329.

Medintz, I. L. (2006). Recent progress in developing FRET-based intracellular sensors for the detection of small molecule nutrients and ligands. *Trends in Biotechnology*, *24*(12), 539–542.

Milvich, J., Kohler, D., Freude, W., & Koos, C. (2018). Surface sensing with integrated optical waveguides: a design guideline. *Optics Express*, *26*(16), 19885–19906.

Mitchell, J. S., Wu, Y., Cook, C. J., & Main, L. (2005). Sensitivity enhancement of surface plasmon resonance biosensing of small molecules. *Analytical Biochemistry*, *343*(1), 125–135.

Mittal, D., Kaur, G., Singh, P., Yadav, K., & Ali, S. A. (2020). Nanoparticle-based sustainable agriculture and food science: Recent recent advances and future outlook. *Frontiers in Nanotechnology*, *2*, 579954.

Mocan, T., Matea, C. T., Pop, T., Mosteanu, O., Buzoianu, A. D., Puia, C., ... & Mocan, L. (2017). Development of nanoparticle-based optical sensors for pathogenic bacterial detection. *Journal of Nanobiotechnology*, *15*, 1–14.

Murugan, P., Krishnamurthy, M., Jaisankar, S. N., Samanta, D., & Mandal, A. B. (2015). Controlled decoration of the surface with macromolecules: polymerization on a self-assembled monolayer (SAM). *Chemical Society Reviews*, *44*(10), 3212–3243.

Naresh, V., & Lee, N. (2021). A Review review on Biosensors biosensors and Recent recent Development development of Nanostructured Materials-Enabled Biosensors. *Sensors*, *21*(4), 1109.

Opiyo, I. L. A. (2022). *Development and Validation of a Colorimetric Probe for Detection of Bacterial Wilt of Tomato in Kenya* (Doctoral dissertation, JKUAT-AGRICULTURE).

Otles, S., & Yalcin, B. (2012). Review on the application of nanobiosensors in food analysis. *Acta Scientiarum Polonorum Technologia Alimentaria, Acta Sci Pol Technol Aliment*, *11*(1), 7–18.

Pandey, Raja R. RamR., and & Charles C. Chusuei, C. C. (2021). Carbon Nanotubes, Graphene, and Carbon Dots as Electrochemical electrochemical Biosensing biosensing Composites. *Molecules, 26*(21), 6674.

Paniel, N., Radoi, A., & Marty, J. L. (2010). Development of an electrochemical biosensor for the detection of aflatoxin M1 in milk. *Sensors*, *10*(10), 9439–9448.

Parisi, C., Vigani, M., & Rodríguez-Cerezo, E. (2015). Agricultural nanotechnologies: what are the current possibilities? *Nano Today*, *10*(2), 124–127.

Peng, H. I., & Miller, B. L. (2011). Recent advancements in optical DNA biosensors: exploiting the plasmonic effects of metal nanoparticles. *Analyst*, *136*(3), 436–447.

Philip, A., & Kumar, A. R. (2022). The performance enhancement of surface plasmon resonance optical sensors using nanomaterials: A a review. *Coordination Chemistry Reviews*, *458*, 214424.

Prasad, R., Kumar, M., & Kumar, V. (Eds.). (2017). *Nanotechnology: an An Agricultural Paradigm*. Springer.

Prasad, R., Kumar, V., & Prasad, K. S. (2014). Nanotechnology in sustainable agriculture: present concerns and future aspects. *African Journal of Biotechnology*, *13*(6), 705–713.

Prasad, R., Pandey, R., & Barman, I. (2016). Engineering tailored nanoparticles with microbes: quo Vadis. *WIREs Nanomedicine Nanobiotechnology, 8*, 316–330.

Qureshi, A., Kang, W. P., Davidson, J. L., & Gurbuz, Y. (2009). Review on carbon-derived, solid-state, micro and nano sensors for electrochemical sensing applications. *Diamond and Related Material, Diam Relat Mater*, *18*, 1401–1420.

Rai, V., Acharya, S., & Dey, N. (2012). Implications of nanobiosensors in agriculture. *Journal of Biomaterials and Nanobiotechnology,* 3, 315–324.

Rashid, A., Zubair, U., Ashraf, M., Javid, A., Abid, H. A., & Akram, S. (2023). Flexible piezoelectric coatings on textiles for energy harvesting and autonomous sensing applications: a review. *Journal of Coatings Technology and Research*, *20*(1), 141–172.

Ray, M. K., Mishra, A. K., Mohanta, Y. K., Mahanta, S., Chakrabartty, I., Kungwani, N. A., ... & Pudake, R. N. (2023). Nanotechnology as a promising tool against phytopathogens: a futuristic approach to agriculture. *Agriculture*, *13*(9), 1856.

Ray, M., Ray, A., Dash, S., Mishra, A., Gopinath Achary, K., Nayak, S., & Singh, S. (2017). Fungal disease detection in plants: traditional assays, novel diagnostic techniques and biosensors. *Biosensors Bioelectronics, 87*, 708–723.

Reuel, N. F., Mu, B., Zhang, J., Hinckley, A., & Strano, M. S. (2012). Nanoengineered glycan sensors enabling native glycoprofiling for medicinal applications: towards profiling glycoproteins without labeling or liberation steps. *Chemical Society Reviews*, *41*(17), 5744–5779.

Rolfe, P. (2012). Micro- and Nanosensors nanosensors for Medical medical and Biological Measurement. *Sensors and Materials*, *24*(6), 275–302.

Sagadevan, S., & Periasamy, M. (2014). Recent trends in nanobiosensors and their applications-a review. *Reviews on Advanced Materials Science, Rev Adv Mater Sci*, *36*(2014), 62–69.

Selid, P. D., Xu, H., Collins, E. M., Striped Face-Collins, M., & Zhao, J. X. (2009). Sensing mercury for biomedical and environmental monitoring. *Sensors*, *9*(7), 5446–5459.

Srivastava, A.K.,, Dev A, & Karmakar, S. (2018) Nanosensors and nanobiosensors in food and agriculture. *Environ Chem Lett*. *16*(1), 161–182.

Stanisavljevic, M., Krizkova, S., Vaculovicova, M., Kizek, R., & Adam, V. (2015). Quantum dots-fluorescence resonance energy transfer-based nanosensors and their application. *Biosensors Bioelectronics, 74*(2015), 562–574.

Stiles, P. L., Dieringer, J. A., Shah, N. C., & Van Duyne, R. P. (2008). Surface-enhanced Raman spectroscopy. *Annual Review of Analytical Chemistry*, *1*, 601–626.

Sun, J., Lei, L., Tsai, C.M. *et al.* (2017). Engineered proteins with sensing and activating modules for automated reprogramming of cellular functions. *Nat Commun*, *8*, 477.

Sun, L. L., Song, Y. H., Wang, L., Guo, C. L., Sun, Y. J., Liu, Z., & Li, Z. (2008). Ethanol-induced formation of silver nanoparticle aggregates for highly active SERS substrates and application in DNA detection. *Journal of Physical Chemistry C, 112*, 1415–1422.

Swain, J., Swain, S., Singh, D., Jena, A., Samantaray, R., & Sahu, R. (2023). Role of Piezoelectric Biosensors. *Point-of-Care Biosensors for Infectious Diseases* (pp. 129–145), Wiley-VCH GmbH.

Tavker, N., & Sharma, M. (2022). Nanosensors for intelligent food packaging. In *Nanosensors for Smart Agriculture* (pp. 737–756). Elsevier.

Tiwari, M., Krishnamurthy, S., Shukla, D., Kiiskila, J., Jain, A., Datta, R., ... & Sahi, S. V. (2016). Comparative transcriptome and proteome analysis to reveal the biosynthesis of gold nanoparticles in Arabidopsis. *Scientific Reports*, *6*(1), 21733.

Uda, M. N. A., Hasfalina, C. M., Samsuzana, A. A., Hashim, U., Ariffin, S. A., & Gopinath, S. C. (2017, March). Immunosensor development for rice tungro bacilliform virus (RTBV) detection using antibody nano-gold conjugate. In *AIP Conference Proceedings* (Vol. 1808, No. 1). AIP Publishing.

Uda, M. N. A., Hasfalina, C. M., Samsuzanaa, A. A., Faridah, S., Gopinath, S. C., Parmin, N. A., ... & Mat, M. (2019). A disposable biosensor based on antibody-antigen interaction for tungro disease detection. In *Nanobiosensors for Biomolecular Targeting* (pp. 147–164). Elsevier.

Van Dyk, J. S. & Pletschke, B. (2011). Review on the use of enzymes for the detection of organochlorine, organophosphate and carbamate pesticides in the environment, *Chemosphere*, *82*(3), 291–307,

Walsh, R., Morales, J. M., Skipwith, C. G., Ruckh, T. T., & Clark, H. A. (2015). Enzyme-linked DNA dendrimer nanosensors for acetylcholine. *Scientific Reports*, *5*(1), 14832.

Wang, J., (2005). Carbon-nanotube based electrochemical biosensors: a review. *Electroanalysis, 17*(1), 7–14.

Wang, M., & Li, Z. (2008). Nano-composite ZrO2/Au film electrode for voltammetric detection of parathion. *Sensors and Actuators B: Chemical*, *133*(2), 607–612.

Wang, Z., Wei, F., Liu, S. Y., Xu, Q., Huang, J. Y., Dong, X. Y., Yua, J. H., Yang, Q., Zhao, Y. D., & Chen, H. (2010). Electrocatalytic oxidation of phytohormone salicylic acid at copper nanoparticles-modified gold electrode and its detection in oilseed rape infected with fungal pathogen Sclerotinia sclerotiorum. *Talanta, 80*, 1277–1281.

Wei, H., Abtahi, S. M. H., & Vikesland, P. J. (2015). Plasmonic colorcolourimetric and SERS sensors for environmental analysis. *Environ mental Science Nano, 2*, 120–135.

Wise, K., & Brasuel, M. (2011). The current state of engineered nanomaterials in consumer goods and waste streams: the need to develop nanoproperty-quantifiable sensors for monitoring engineered nanomaterials. *Nanotechnology, Science and Applications*, *1*(4), 73–86.

Zhu, C., Yang, G., Li, H., Du, D., & Lin, Y. (2015). Electrochemical sensors and biosensors based on nanomaterials and nanostructures. *Analytical Chemistry, 87*(1), 230–249.

Zhu, R., & Wang, Z. (2020). Piezoelectric one-to two-dimensional nanomaterials for vibration energy harvesting devices. In *Emerging 2D materials and devices for the internet of things* (pp. 221–241). Elsevier.

5 Nanoparticles from Microbes
The Next Generation Tool for Combatting Plant Diseases

Gökhan Boyno, Necmettin Teniz, Emre Demirer Durak, Younes Rezaee Danesh, and Semra Demir

5.1 INTRODUCTION

Plant diseases are a major concern for the agriculture industry, affecting crop yield, quality, and profitability (Teniz 2020; Boyno et al., 2022; Çevik et al., 2022; Keshari et al., 2024). Traditional methods of plant disease control such as the use of chemical pesticides have several limitations including high toxicity, environmental pollution and the development of resistance in pathogens. Therefore, there is a need for alternative strategies that are safe, effective and sustainable. The use of nanoparticles to manage plant disease has become a promising area in recent years (Singh et al., 2016). Nanoparticles are small particles ranging from 1 to 100 nm in diameter (Ealia and Saravanakumar 2017). They possess unique physical and chemical properties compared to their bulk counterparts such as high surface area and reactivity (Saleh 2020). Metals, metal oxides, carbon, polymers, and other materials can all be used for producing nanoparticles (Foresti et al., 2017; Wang et al., 2021). The unique characteristics and potential uses of nanoparticles formed from microorganisms have made them stand out among the several sources of nanoparticles (Tsekhmistrenko et al., 2020; Bhardwaj et al., 2020).

Microbe-synthesized NanoParticles (MNPs) such as bacterial, fungal, and viral nanoparticles have undoubtedly emerged as a promising tool for controlling plant diseases (Bhattacharya et al., 2022; Khanna et al., 2022). These nanoparticles have several advantages over traditional chemical pesticides, including greater efficacy, lower toxicity, and environmental safety (Abdollahdokht et al., 2022; Kannan et al., 2023). Furthermore, MNPs can be produced using eco-friendly and cost-effective methods making them a viable alternative to chemical pesticides (Maity et al., 2022; Chinni et al., 2023). The mechanisms of action of MNPs in controlling plant diseases are diverse and involve various modes of action, including antifungal, antibacterial, and antiviral activities (Malandrakis et al., 2022; Kumar et al., 2022). The nanoparticles can penetrate the cell wall and interfere with the normal physiological processes of the pathogen leading to its death or reduced virulence (Li et al., 2023; Yin et al., 2023). Moreover, some nanoparticles can activate the plant's defence mechanisms by inducing the production of various biomolecules that enhance resistance in crop plants against plant pathogens (Singh et al., 2020). These biomolecules include phytohormones such as Salicylic Acid (SA), Jasmonic Acid (JA), and EThylene (ET), which play crucial roles in regulating plant defence responses (Kouzai et al., 2016; Di et al., 2017). When triggered by nanoparticles, these biomolecules act as signalling molecules, initiating a cascade of defence-related events (Imada et al., 2016; Tripathi et al., 2022).

Despite the potential advantages of MNPs, several challenges remain. The lack of knowledge regarding these nanoparticles' long-term consequences on the environment and human health is one of the main problems (Pietroiusti et al., 2018; Tortella et al., 2020). Additionally, the production and

DOI: 10.1201/9781003256762-5

commercialization of these nanoparticles face regulatory and economic barriers, limiting their widespread adoption. Recent developments in nanotechnology and biotechnology, however, have created new opportunities for the creation of MNPs that are both safer and more effective in controlling plant diseases (Singh et al., 2020; Li et al., 2023; Yin et al., 2023).

This chapter discusses the use of nanoparticles derived from microbes for combatting plant diseases in the agriculture industry. It aims to review the current state of research on this topic and discusses the various modes of action by which these nanoparticles can manage plant diseases. Additionally, the chapter discusses the challenges and limitations of using MNPs and future directions for research and development in this field. The chapter provides important insights into the potential of MNPs for sustainable agriculture and plant disease management.

5.2 MICROBES AS NANOPARTICLE FACTORIES

Nanoparticles are small particles ranging from 1 to 100 nm in diameter that possess unique physical and chemical properties compared to their bulk counterparts (Ealia & Saravanakumar, 2017; Saleh 2020). Due to their potential advantages, nanoparticle utilization for several applications has recently attracted a lot of attention (Boyno 2022). Due to their capacity to manufacture a variety of chemicals, including nanoparticles, microbes such as bacteria and fungus have emerged as possible sources for generating nanoparticles (Purohit et al., 2019; John et al., 2021). This ability has led to the development of a new field perhaps known as microbial nanotechnology.

MNPs can occur through various mechanisms, including extracellular and intracellular synthesis (Ahmad et al., 2019; Khalil et al., 2022). Extracellular synthesis involves the secretion of nanoparticles by a microbe into the surrounding environment, while intracellular synthesis involves the formation of nanoparticles within the microbial cell (Patil & Chandrasekaran 2020). Microbial proteins, including external proteins, cell membrane proteins, and intracellular proteins, as well as enzymes like reductases, oxidases, and nitrate reductases, mediate the creation of nanoparticles in both situations (Lahiri et al., 2021; Banik et al., 2022). The use of microbes as nanoparticle factories has several advantages over traditional methods of nanoparticle synthesis. For instance, microbial synthesis is typically more eco-friendly and cost-effective than traditional chemical methods, as it can occur under ambient conditions using renewable resources (Maity et al., 2022; Chinni et al., 2023). Additionally, microbial synthesis can result in nanoparticles with unique physicochemical properties that are difficult to achieve using traditional methods (Tsekhmistrenko et al., 2020; Bhardwaj et al., 2020).

Microbes can produce nanoparticles that exhibit unique physical and chemical properties such as small size, high surface area and reactivity making them attractive for various applications. For instance, nanoparticles derived from microbes have shown great potential in plant disease control due to their ability to penetrate the cell wall of the pathogen and interfere with its normal physiological processes leading to its death or reduced virulence (Li et al., 2023; Yin et al., 2023). Additionally, some nanoparticles can activate the plant's defence mechanisms, leading to enhanced resistance to pathogen infection (Singh et al., 2020). Moreover, MNPs can be produced using eco-friendly and cost-effective methods, making them a viable alternative to chemical pesticides (Tsekhmistrenko et al., 2020; Bhardwaj et al., 2020). The environmental impact of pesticide use has been a major concern for many years, and the use of MNPs can help reduce this impact (Shafey, 2020). Additionally, the specificity of nanoparticles can be engineered to target specific pathogens, while leaving beneficial microbes and non-target organisms unaffected (Kannan et al., 2023). This can reduce the environmental impact of disease management practices and minimize the risk of developing resistance to the treatment (Singh et al., 2020).

The use of microbes as nanoparticle factories has significant potential for various applications, including plant disease control, drug delivery, and environmental remediation (Singh et al., 2018). The ability to produce nanoparticles using renewable resources and under ambient conditions makes microbial synthesis a sustainable and eco-friendly alternative to traditional methods of nanoparticle synthesis (Saleh, 2020; Bhardwaj et al., 2020). MNPs might be used for a variety of purposes in agriculture and other fields with more study and development, offering a sustainable, efficient, and eco-friendly alternative to conventional methods (Singh et al., 2020).

Overall, using microbes as nanoparticle factories has many potential benefits, including cost-effectiveness, sustainability, and precision in nanoparticle synthesis. This approach is still in its early stages of development, but it has already shown promising results in various applications, such as drug delivery, imaging, and catalysis (Balasubramanian et al., 2020; Dikshit et al., 2021). The ability to produce nanoparticles using renewable resources and under ambient conditions makes microbial synthesis an attractive alternative to traditional methods of nanoparticle synthesis, and the potential applications of MNPs are vast and can revolutionize various fields, providing a sustainable, effective, and environmentally friendly alternative to traditional approaches (Balasubramanian et al., 2020; Singh et al., 2020;).

5.2.1 Types of Microorganisms Involved in Nanoparticle Synthesis

Microorganisms are a diverse group of living organisms that are ubiquitous in nature and can be found in a wide range of environments, including soil, water, air, and living organisms. They include bacteria, fungi, viruses, and archaea. In recent years, microorganisms have gained significant attention in the field of nanotechnology due to their ability to synthesize a wide range of compounds, including nanoparticles (Bhattacharya et al., 2022; Khanna et al., 2022).

Microorganisms such as bacteria, fungi, viruses, and archaea have unique properties that can influence the properties of the synthesized nanoparticles (Khan & Lee 2020). For example, bacterial nanoparticles tend to be smaller and more stable than fungal nanoparticles, while fungal nanoparticles tend to be more diverse in terms of size and shape (Mukherjee et al., 2001; Prasad et al., 2016). Additionally, different types of microbes can synthesize nanoparticles using different mechanisms, such as extracellular or intracellular synthesis (Ahmad et al., 2019; Khalil et al., 2022) (Figure 5.1). Bacteria are single-celled microorganisms that can be found in various environments, including soil, water, and living organisms. Due to their potential uses in a variety of industries, including medical, environmental cleanup, and agriculture, bacterial nanoparticles have been thoroughly explored (Panáček et al., 2018; Iravani & Varma, 2020. Bacterial nanoparticles have unique properties, such as small size, high surface area, and stability, which make them attractive for various applications. *Bacillus subtilis* (Tariq et al., 2020), *Pseudomonas aeruginosa* (Abdo et al., 2021) and *Escherichia coli* (Gholami-Shabani et al., 2015) are some of the bacteria that have been used for nanoparticle synthesis. Fungi are multicellular microorganisms that can be found in various environments, including soil and living organisms. Fungal nanoparticles have unique properties, such as diverse size and shape, that make them attractive for various applications (Boroumand Moghaddam et al., 2015; Guilger-Casagrande and Lima 2019). Fungal nanoparticles can be synthesized by various fungi, including *Aspergillus fumigatus* (Sarsar et al., 2016), *Fusarium oxysporum* (Yin et al., 2016), and *Trichoderma reesei* (Gemishev et al., 2022).

Viruses are small infectious agents that can infect living organisms, including bacteria. Viral nanoparticles have been extensively studied due to their potential applications in medicine, including drug delivery and imaging (Chung et al., 2020). Viral nanoparticles have unique properties, such as uniform size and shape, which make them attractive for various applications. Tobacco mosaic virus (Niehl et al., 2016), bacteriophage MS2 (Park et al., 2014), and cowpea mosaic viruses (Beatty & Lewis, 2019) are some of the viruses that have been used for nanoparticle synthesis. Archaea are single-celled microorganisms that can be found in extreme environments, including hot springs, salt lakes, and deep-sea hydrothermal vents (Delgado-García et al., 2018). Archaeal nanoparticles have unique properties, such as stability in extreme conditions, that make them attractive for various applications (Abdollahnia et al., 2020). *Halobacterium salinarum* and *Sulfolobus acidocaldarius*

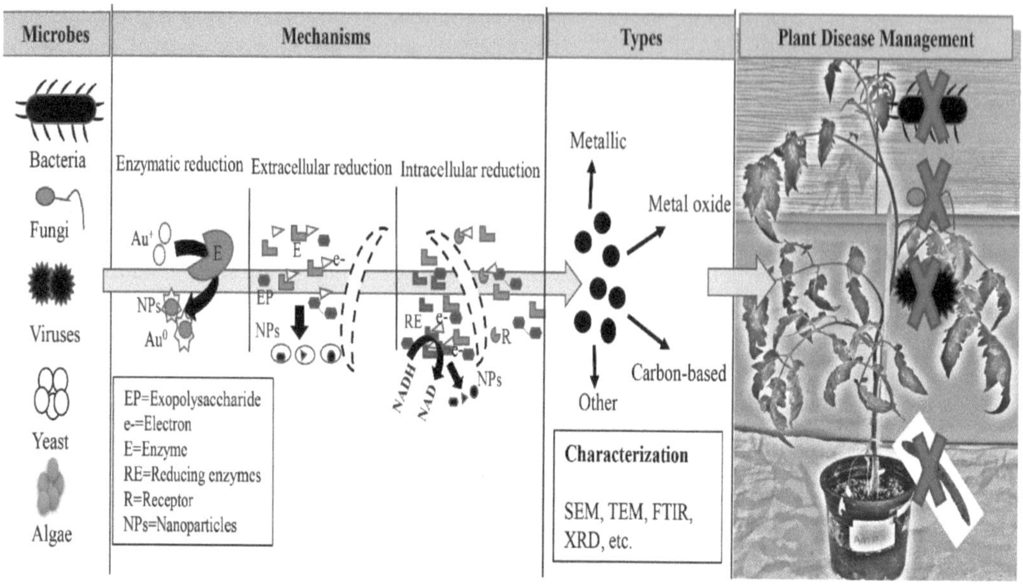

FIGURE 5.1 General scheme of nanoparticle synthesis by microbes and their types and application in plant disease management: Some microorganisms have the natural ability to convert metallic ions such as gold (Au) or other compounds into nanoparticles through a biological process. This process occurs within the cells of the microorganism or using extracellular enzymes or molecules produced by the microorganism. During biosynthesis, microorganisms act as biological catalysts and reduce metal ions or other precursor molecules to their nanoscale forms. They usually use enzymes or special proteins to facilitate this reduction process. The resulting nanoparticles are stabilised through interactions with specific biomolecules such as proteins or polysaccharides produced by microorganisms. The synthesised nanoparticles are subjected to characterisation processes to determine their chemical and physical properties. These nanoparticles find applications in various fields such as plant disease management.

are some of the archaea that have been used for nanoparticle synthesis (Kalabegishvili et al., 2015; Moghimipour et al., 2021).

5.2.2 Mechanisms of Nanoparticle Synthesis by Microbes

There are several mechanisms by which microorganisms can synthesize nanoparticles. Some of the common mechanisms are:

5.2.2.1 Enzymatic Reduction

This reduction is a process used by certain microorganisms to synthesize nanoparticles using enzymes as reducing agents (Barapatre et al., 2016). This method involves the use of microorganism-produced enzymes to convert metal ions into nanoparticles by transferring electrons, which produces stable nanoparticles (Barapatre et al., 2016). The process of enzymatic reduction involves several steps. Firstly, the microorganisms produce enzymes, such as reductases, which are capable of reducing metal ions (Kuroda & Ueda, 2010; Xu et al., 2016). These enzymes have specific binding sites that can recognize and bind to metal ions (Kuroda & Ueda, 2010; Xu et al., 2016). Once the enzyme has bound to the metal ion, it initiates the transfer of electrons from the surrounding molecules to the metal ion. This electron transfer process causes the metal ion to gain electrons and subsequently form a stable nanoparticle (Kuroda & Ueda 2010; Xu et al., 2016). The enzymatic reduction process has several advantages over other methods of nanoparticle synthesis. Firstly, it is a green and eco-friendly process as it uses natural enzymes and does not require harsh chemicals (Gholami-Shabani et al., 2015). Secondly, it is a relatively simple and cost-effective process as it requires minimal equipment and can be carried out under mild reaction conditions (Ma et al., 2017). Finally, it produces nanoparticles with a narrow size distribution and high stability, such as gold nanoparticles (Kitching et al., 2015), quantum dots (Borovaya et al., 2015), magnetic iron oxide nanoparticles (Laurent et al., 2008) and silica nanoparticles (Qin et al., 2016), making it appropriate for a variety of applications (Hulkoti & Taranath 2014).

5.2.2.2 Extracellular Reduction

This reduction is a mechanism in which microorganisms synthesize nanoparticles outside their cell membrane (Prakash et al., 2010). This process involves the secretion of enzymes and other biomolecules by microorganisms that react with metal ions and convert them into nanoparticles (Prakash et al., 2010; Durán et al., 2011). A number of variables, including temperature, pH, the concentration of metal ions, and the presence of electron donors, can start the extracellular reduction process (Elangovan et al., 2010; Li et al., 2011). For example, the presence of electron donors such as glucose or fructose can stimulate the secretion of enzymes that can reduce metal ions into nanoparticles (Elangovan et al., 2010). The mechanism of extracellular reduction is dependent on the type of microorganism involved in the process (Prakash et al., 2010). Some microorganisms secrete enzymes that can chelate metal ions, leading to the formation of nanoparticle-protein complexes (Ravindran et al., 2013; Ponnuvel et al., 2015). These complexes can further interact with other biomolecules and form stable nanoparticles. Another mechanism involved in extracellular reduction is the secretion of enzymes that can reduce metal ions directly into nanoparticles (Hulkoti & Taranath, 2014). The enzymes can transfer electrons from the electron donors to the metal ions, leading to the formation of stable nanoparticles.

5.2.2.3 Biomineralization

Biomineralization is a process where living organisms use organic molecules to create minerals (Yao et al., 2017; Fabiyi et al., 2024; Fabiyi et al., 2020; Patil et al., 2024; Solanki et al., 2021). In the context of nanoparticle synthesis, some microorganisms secrete organic molecules that act as templates for the formation of nanoparticles. These organic molecules can be proteins, peptides, polysaccharides, or other biomolecules (Saallah & Lenggoro, 2018). The process of biomineralization typically involves three stages: nucleation (Zhang et al., 2020), growth (Gal et al., 2015), and stabilization (Benzerara et al., 2011). In the nucleation stage, the metal ions bind to the organic molecules and form small clusters (Zhang et al., 2020). These clusters then grow and aggregate to form larger nanoparticles in the growth stage (Gal et al., 2015). Finally, in the stabilization stage, the nanoparticles are stabilized by further binding with the organic molecules (Benzerara et al., 2011). Biomineralization is a highly regulated and complex process that requires specific interactions between the organic molecules and metal ions (Yao et al., 2017). The resulting nanoparticles are often highly uniform in size and shape, making them ideal for a variety of applications. One example of biomineralization in nanoparticle synthesis is the production of magnetite nanoparticles by the bacterium *Magnetospirillum magneticum* (Jones et al., 2015). These bacteria use a protein called magnetosome to synthesize magnetite nanoparticles for use in magnetotaxis, a process where the bacteria orient themselves in a magnetic field (Jones et al., 2015). Overall, biomineralization is a fascinating mechanism for nanoparticle synthesis by microorganisms that has potential for various applications in medicine, materials science, and other fields (Yao et al., 2017).

5.2.2.4 Intracellular Reduction

This reduction is a mechanism of nanoparticle synthesis in which microorganisms take up metal ions in their cells and reduce them to form nanoparticles (Sanghi and Verma, 2009). In this process, the metal ions are transported across the cell membrane by various transporters and are then reduced by intracellular enzymes or other biomolecules (Yan et al., 2013). The production of gold nanoparticles by the bacteria *Shewanella oneidensis* is an illustration of intracellular reduction (Suresh et al., 2011). In this process, the bacterium takes up gold ions from the surrounding environment and reduces them to form nanoparticles inside its cytoplasm (Suresh et al., 2011). The reduction is believed to be mediated by the activity of certain intracellular enzymes, such as reductases, which transfer electrons to the gold ions to reduce them to nanoparticles (Yan et al., 2013). Intracellular reduction is a relatively less studied mechanism of nanoparticle synthesis as it requires complex cellular machinery and metabolic pathways (Thakkar et al., 2010). However, it has the advantage of producing nanoparticles with a high degree of purity and uniformity, as the synthesis occurs within the controlled environment of the cell (Ovais et al., 2018).

5.2.3 Advantages of Microbe-Synthesized Nanoparticles over Chemical Synthesis

MNPs have gained considerable attention in recent years as a viable alternative to chemical synthesis (Parveen et al., 2016). This is primarily due to the numerous advantages offered over traditional chemical synthesis methods (Figure 5.2). One of the significant advantages of microbial synthesis is its cost-effectiveness (Iravani & Varma, 2020; Chmykhalo et al., 2021). Microbial synthesis is a relatively low-cost process as it requires minimal infrastructure and the microorganisms used can be grown in large quantities using inexpensive growth media. Moreover, the microorganisms used in microbial synthesis are usually easy to cultivate and do not require any special equipment or infrastructure, making an attractive option for industries looking to produce large quantities of nanoparticles. Another significant advantage of microbial synthesis is its environmental friendliness (Saravanan et al., 2021). Unlike chemical synthesis, MNPs is a green process that does not involve the use of toxic chemicals or hazardous solvents. This makes it an environmentally friendly option that is safe for both humans and the environment (Devi & Joshi, 2015). This aspect is particularly important as industries are looking to reduce their environmental impact and comply with stringent environmental regulations. The production of nanoparticles with precise shapes and sizes is achieved by the great degree of specificity and selectivity offered by microbial synthesis. This is not possible with chemical synthesis, which often produces nanoparticles with a broad size distribution (Duan et al., 2015). The ability to produce nanoparticles with specific properties makes microbial

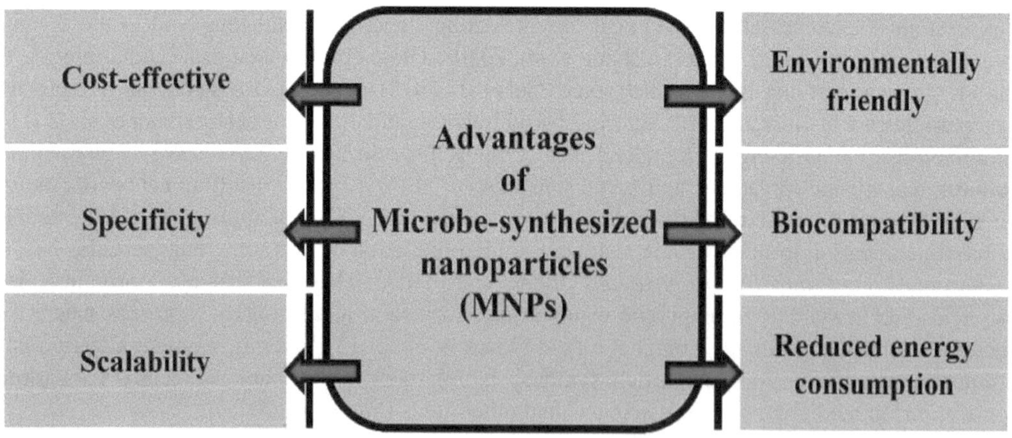

FIGURE 5.2 Advantages of microbe-synthesized nanoparticles: Microbial synthesis of nanoparticles offers advantages such as cost-effectiveness, environmental friendliness, specificity, biocompatibility, scalability, and reduced energy consumption compared to chemical synthesis methods. These advantages make microbial synthesis a promising and sustainable approach for producing nanoparticles for various applications.

synthesis particularly attractive for use in biomedical applications such as drug delivery, imaging, and therapy (Stark et al., 2015). The nanoparticles produced by microbial synthesis are also biocompatible and non-toxic (Thiyagarajan et al., 2018), making them suitable for use in various biomedical applications (Rojas et al., 2015). They are also less likely to cause adverse reactions in the human body, making them a safer option than chemically synthesized nanoparticles (Thiyagarajan et al., 2018). Microbial synthesis is easily scalable, which makes it a practical approach for the production of nanoparticles (Lateef et al., 2021). The use of microorganisms that grow rapidly and produce large quantities of nanoparticles makes the process easily scalable to meet the increasing demand for nanoparticles. Finally, MNPs require less energy input compared to chemical synthesis (Park et al., 2016). This results in lower energy consumption and reduces the carbon footprint of the process, making it an attractive option for industries looking to reduce their energy consumption and carbon footprint.

5.3 TYPES OF MICROBIAL NANOPARTICLES

Microbial nanoparticles refer to nanoparticles that are synthesized by microorganisms, such as bacteria, fungi, and viruses (Bhattacharya et al., 2022; Khanna et al., 2022). There are different types of microbial nanoparticles that can be synthesized by microorganisms, and some of them include:

5.3.1 Metallic Nanoparticles

A intriguing and important class of nanoparticles, metallic nanoparticles have special physicochemical characteristics that make them suitable for a variety of commercial and biological applications (Ovais et al., 2018). The size and shape of the nanoparticles can be controlled by manipulating the reaction conditions and the microbial strain used for synthesis. This allows for the production of nanoparticles with specific properties tailored to their intended application. A promising and flexible method for producing nanoparticles with particular characteristics for diverse commercial and medicinal uses is the microbial synthesis of metallic nanoparticles (Ovais et al., 2018). They are synthesized by microorganisms and can be tailored with specific properties for various applications.

Silver nanoparticles have remarkable antimicrobial properties and are used in wound dressings, coatings, and water treatment applications (Durán et al., 2016). Their small size allows them to penetrate bacterial cells and disrupt their metabolic processes, leading to cell death (Durán et al., 2016). They can also interact with the cell membranes of microorganisms and cause increased permeability, leading to the leakage of intracellular components and ultimately causing cell death (Tang & Zheng, 2018). These unique properties make silver nanoparticles an effective and versatile antimicrobial agent (Tang & Zheng, 2018). Gold nanoparticles have unique optical properties and are used in biological imaging and drug delivery applications (Cabuzu et al., 2015). These properties are due to the surface plasmon resonance effect that causes the collective oscillation of free electrons on the surface of the nanoparticle in response to incident light. This effect gives gold nanoparticles a distinctive colour that varies depending on the shape and size of the particle (Cabuzu et al., 2015). They can be functionalized with biomolecules such as antibodies or peptides to target specific cells or tissues, making them useful in targeted drug delivery applications (Goddard et al., 2020; Zong et al., 2017). These unique optical properties make gold nanoparticles an attractive option for a wide range of biological and biomedical applications (Carabineiro 2017). Copper nanoparticles exhibit unique physicochemical properties such as high surface area-to-volume ratio and high reactivity, which make them effective in a wide range of industrial and biomedical applications (Din et al., 2017). One of the most notable applications of copper nanoparticles is their use as antimicrobial agents (Kaur et al., 2016). They have potent antimicrobial activity against a variety of microorganisms, including bacteria, viruses, and fungi (Rai et al., 2018). The antimicrobial activity of copper nanoparticles is due to their ability to release copper ions, which can disrupt the cell membranes and cellular processes of microorganisms (Kaur et al., 2016). They have also been studied for their potential in cancer therapy due to their ability to induce apoptosis in cancer cells by disrupting mitochondrial function and inducing oxidative stress (Wang et al., 2019). Copper nanoparticles have a high surface area to volume ratio and are highly reactive, making them suitable catalysts (Din et al., 2017).

Platinum nanoparticles are highly effective catalysts due to their unique electronic and geometric structures, which make them suitable for a variety of catalytic reactions (Jameel et al., 2020). They are commonly used in catalytic converters to convert toxic gases emitted from vehicles into less harmful substances. Platinum nanoparticles also have potential applications in the field of drug delivery and cancer therapy (Pedone et al., 2017). They can be used as carriers for drugs due to their small size and large surface area, which allows them to easily penetrate cells and deliver drugs directly to the target site (Pedone et al., 2017; Fahmy et al., 2020).

5.3.2 Metal Oxide Nanoparticles

Metal oxide nanoparticles are another type of nanoparticle synthesized by microorganisms (Falcaro et al., 2016; Jeevanandam et al., 2016). Zinc oxide, titanium dioxide, iron oxide, and cerium oxide are a few of the metal oxide nanoparticles that are frequently examined (Agarwal et al., 2018; Subhapriya & Gomathipriya, 2018). Metal oxide nanoparticles have unique physicochemical properties, such as high surface area-to-volume ratio, thermal stability, and photocatalytic activity, that make them suitable for a wide range of industrial and biomedical applications (Falcaro et al., 2016; Jeevanandam et al., 2016).

Zinc oxide nanoparticles are among the most commonly studied metal oxide nanoparticles synthesized by microorganisms (Agarwal et al., 2018). They have unique physicochemical properties that make them suitable for a wide range of applications, including in sunscreens, coatings, and antibacterial agents (Lallo da Silva et al., 2019). They have also demonstrated positive antibacterial capabilities and helped in the creation of fresh antimicrobial substances (Agarwal et al., 2018; Lallo da Silva et al., 2019). Their small size and high surface area-to-volume ratio allow them to penetrate bacterial cell walls and interfere with cellular processes, leading to bacterial death (Agarwal et al., 2018; Lallo da Silva et al., 2019).

Titanium dioxide (TiO_2) nanoparticles are widely used in various applications such as pigments, sunscreens, coatings and photocatalysts due to their unique properties (Subhapriya & Gomathipriya 2018). Due to these unique properties and potential applications, these have recently attracted attention in the field of agriculture and plant disease management (Owolade & Ogunleti 2008). Its use in the control of ailments including bacterial, fungal, and nematode diseases that significantly reduce plant yields is gaining ground every day (Owolade & Ogunleti 2008; Prasad et al., 2014; Kotliarevski et al., 2022).

TiO_2 nanoparticles exhibit antibacterial properties and can be effective against various plant pathogenic bacteria. When exposed to ultraviolet (UV) light, TiO_2 nanoparticles produce Reactive Oxygen Species (ROS) such as hydroxyl radicals with strong oxidative potential (Ma et al., 2012). These ROS can damage bacterial cell walls, proteins and DNA, leading to inactivation or death of the bacteria (Li et al., 2011). TiO2 nanoparticles can be incorporated into plant coatings or applications to help control bacterial diseases such as Xanthomonas leaf spot, *Ralstonia solanacearum* and bacterial blight (Khan et al., 2022; Younis et al., 2023). In addition, Parveen & Siddiqui (2022) found that TiO_2 nanoparticles were more effective against *Xanthomonas campestris* pv. *vesicatoria*, *Pseudomonas syringae* pv. tomato, *Pectobacterium carotovorum* subsp. *carotovorum* and *Ralstonia solanacearum* bacterial diseases in tomato by foliar spray application.

TiO_2 nanoparticles show promise in the management of fungal diseases affecting plants (Alabdallah et al., 2023). Similar to their antibacterial mechanism, TiO_2 nanoparticles produce ROS when exposed to UV light, leading to oxidative stress and disruption of fungal cell membranes and structures (Liu et al., 2016; Nurdin et al., 2023). This inhibits fungal growth and can help control diseases caused by fungal pathogens such as *Puccinia striiformis*, *Fusarium solani*, *Alternaria solani*, *Phomopsis vexans* and *Venturia inaequalis* (Khan et al., 2022; Parveen & Siddiqui 2022). In addition, TiO_2 nanoparticles can act as a carrier for antifungal agents and enhance their efficacy against plant fungal pathogens (Garg et al., 2020).

Plant parasitic nematodes are a significant threat to crop productivity. TiO_2 nanoparticles have shown potential in the management of nematode infections in plants (Kotliarevski et al., 2022). For example, Khan et al. (2022) revealed that titanium dioxide nanoparticles were effective against *Meloidogyne incognita*, the root-knot nematode of aubergine. Moreover, TiO_2 nanoparticles can inhibit the hatching of nematode eggs, disrupt the nematode life cycle and reduce nematode populations (Dong et al., 2018; Hu et al., 2018). This can lead to reduced damage to plant roots and improved plant health.

The use of TiO_2 nanoparticles in plant disease management offers several advantages. They are environmentally friendly as they are easily degradable and have low toxicity compared to conventional chemical pesticides (Roh et al., 2010). In addition, their nanoscale size increases their efficacy against pathogens by providing more surface area and better penetration into plant tissues (Worrall et al., 2018).

5.3.3 Carbon-Based Nanoparticles

Carbon-based nanoparticles are another type of nanoparticle synthesized by microorganisms (Dizaj et al., 2015). Carbon-based nanoparticles include fullerenes, carbon nanotubes, graphene, and carbon quantum dots (Dizaj et al., 2015). These nanoparticles have unique properties that make them suitable for various industrial and biomedical applications (Gupta et al., 2019). Fullerenes are spherical molecules made entirely of carbon atoms arranged in a pattern resembling a soccer ball (Dizaj et al., 2015; Rahman et al., 2022). They have unique physicochemical properties, such as high stability, low reactivity, and good electron-accepting ability, which make them useful in drug delivery, imaging, and electronics (Rahman et al., 2022). Carbon nanotubes are cylindrical structures made of carbon atoms arranged in a tube-like structure (Dizaj et al., 2015). They have unique mechanical and electronic properties, such as high strength, high electrical conductivity, and good thermal conductivity, which make them useful in electronics, sensors, and drug delivery (Pardo et al., 2018). Graphene is a two-dimensional sheet of carbon atoms arranged in a hexagonal pattern. It has excellent electrical conductivity, high mechanical strength, and high surface area, making it suitable for applications in electronics, energy storage, and catalysis (Dizaj et al., 2015). Carbon quantum dots are small carbon nanoparticles with a size range of a few nanometres to tens of nanometres (Dizaj et al., 2015). They have excellent optical properties, including strong fluorescence and good photostability, making them useful in biological imaging, sensing, and drug delivery (Pardo et al., 2018).

Carbon-based nanoparticles have gained attention in recent years for their potential applications in the management of plant diseases (Mohamed et al., 2018). Carbon-based nanoparticles, such as carbon nanotubes and graphene-based nanoparticles, have shown antimicrobial properties against plant pathogens (Alimardani et al., 2019; Patel et al., 2019). They can disrupt the cell walls and membranes of bacteria and fungi, leading to their inactivation or death (Shoala 2020). Carbon-based nanoparticles can be incorporated into coatings (Erkoc & Ulucan-Karnak 2021), or nanocomposites (Díez-Pascual 2021) to provide antimicrobial protection against plant pathogens, reducing disease incidence and severity (Shoala 2020). For example, Corredor et al. (2009) reported that carbon-coated iron nanoparticles applied to diseased pumpkin (*Cucurbita pepo*) plants showed a potential effect in the treatment of the plant. Furthermore, González-García et al. (2022) reported that carbon-based nanomaterials significantly reduced the incidence and severity of *F. oxysporum*.

Carbon-based nanoparticles have the potential to induce plant defence responses, thereby enhancing plant resistance against pathogens. For example, Adeel et al. (2021) suggested that carbon-based nanomaterials suppress Tobacco Mosaic Virus (TMV) infection by increasing the resistance of tobacco (*Nicotiana benthamiana*) plants. Carbon-based nanoparticles can stimulate the plant's innate immune system, triggering the production of defence-related compounds, such as phytohormones, enzymes, and antimicrobial peptides (Shoala 2018). This can strengthen the plant's defence mechanisms and improve its ability to resist pathogen attacks. In addition, carbon-based nanoparticles, particularly carbon nanotubes, have been explored for their effects on plant growth and development (Mohamed et al., 2018). They can modulate plant hormone levels, regulate gene expression, and enhance photosynthesis efficiency (Jordan et al., 2018; Husen 2020). By promoting plant growth and vigour, carbon-based nanoparticles can indirectly contribute to disease management by increasing the plant's resilience and ability to withstand pathogenic attacks (Mohamed et al., 2018; Adeel et al., 2021).

Carbon-based nanoparticles can be employed in the development of nanosensors for the detection and monitoring of plant diseases (Zhu et al., 2022). Functionalized carbon-based nanoparticles can interact with specific pathogen biomarkers or plant metabolites associated with disease presence (Shoala 2020). This enables the rapid and sensitive detection of pathogens, facilitating early disease diagnosis and timely intervention (Kashyap et al., 2019).

It's worth noting that the use of carbon-based nanoparticles in plant disease management is still an active area of research, and their practical applications in agriculture may require further development and evaluation (Abd-Elsalam 2019). Considerations regarding carbon-based nanoparticle synthesis, characterization, dose optimization, and potential environmental impacts are crucial for their safe and effective implementation.

5.3.4 Other Types of Microbial Nanoparticles

In addition to metallic, metal oxide and carbon-based nanoparticles, microorganisms can also synthesize other types of nanoparticles including:

5.3.4.1 Quantum Dots

These are semiconductor nanoparticles that have unique optical and electronic properties due to their size and composition (Pardo et al., 2018). Microorganisms have been shown to synthesize quantum dots of different sizes and materials, such as cadmium sulfide, zinc sulfide, and indium phosphide, which have potential applications in imaging, sensing, and solar cells (Shivaji et al., 2018).

The use of quantum dots in plant disease management is still in its early stages. For example, quantum dots have been explored as a potential tool for the detection and management of Citrus Tristeza Virus (CTV), a bacterial disease that affects citrus crops (Safarnejad et al., 2017). Furthermore, although the active ingredients of pesticides have been replaced with quantum dots to treat various bacterial and fungal diseases, they have been more effective against nematodes and insects (Khan & Lee 2020; Khan et al., 2022). It's important to note that while these examples demonstrate the potential applications of quantum dots in plant disease management, further research is needed to optimize their formulations, evaluate their effectiveness, and ensure their safe and sustainable use in agricultural settings.

5.3.4.2 Magnetic Nanoparticles

These are magnetic nanoparticles that are employed in a number of biological applications, including medication administration and magnetic resonance imaging (MRI) (Cristea et al., 2017). Microorganisms such as magnetotactic bacteria are capable of synthesizing magnetic nanoparticles with controlled size, shape, and magnetic properties (Jacob and Suthindhiran 2016).

The potential use of magnetic nanoparticles in plant disease management is among the current researches. Mohamed et al. (2019) suggested that there is also the possibility that magnetic nanoparticles can be adapted on a large scale by integrating them into existing agricultural applications, thus preventing crop loss due to pests and diseases. Ali et al. (2021) revealed that MNPs can be used as soil fertilisers to increase production as well as plant disease management.

5.3.4.3 Chitosan Nanoparticles

Chitosan is a biopolymer derived from chitin, which is found in the exoskeleton of crustaceans (Chandrasekaran et al., 2020). Chitosan nanoparticles have shown potential in drug delivery, wound healing, and tissue engineering. Microorganisms have been used to synthesize chitosan nanoparticles with different sizes and properties (Chandrasekaran et al., 2020).

Chitosan nanoparticles have demonstrated promise in the treatment of bacterial, fungal, and nematode plant diseases. Antibacterial efficacy against several plant pathogenic bacteria has been demonstrated using chitosan nanoparticles. In some studies, it was revealed that chitosan nanoparticles showed antibacterial activity against plant bacterial pathogens *Xanthomonas campestris* pv. *campestris* (Sreelatha et al., 2022a, b) and *Xanthomonas oryzae* pv. *oryzae* (Abdallah et al., 2020). They can disrupt bacterial cell walls and membranes, leading to bacterial inactivation or death (Abdallah et al., 2020; Sreelatha et al., 2022a, b). Similarly, chitosan nanoparticles have antifungal properties that make them effective against plant fungal pathogens (Oh et al., 2019). For example, it has been determined that it suppresses diseases by showing antifungal effect against *F. solani* in tomato (Abd-Ellatif et al., 2022) and *Botrytis cinerea* in strawberry (El-Naggar et al., 2022). They can inhibit fungal growth, disrupt fungal cell structures and interfere with basic metabolic processes (Abd-Ellatif et al., 2022; El-Naggar et al., 2022). Moreover, chitosan nanoparticles have shown potential in the management of plant parasitic nematode infections. For example, Alfy et al. (2020) revealed that chitosan nanoparticles are effective against *Spodoptera littoralis*, *Locusta migratoria* and *Meloidogyne incognita*. Mouniga et al. (2022) revealed that chitosan nanoparticles protect plants against *Meloidogyne incognita* and induce systemic resistance against nematodes by increasing phenol content, peroxidase and polyphenol oxidase activity in tomato. Moreover, it was determined that chitosan nanoparticles increased the resistance of the plant against Tobacco Mosaic Tobamovirus (TMV) as well as *Meloidogyne incognita* in tomato (Khalil et al., 2022).

The efficacy of chitosan nanoparticles in disease management may vary depending on factors such as nanoparticle size, concentration and formulation (Mehta et al., 2022). Furthermore, specific plant-pathogen interactions and target pathogen species may affect their efficacy (Mehta et al., 2022).

5.3.4.4 Lipid Nanoparticles

These are nanoparticles composed of lipids and are used in drug delivery and gene therapy (Shah et al., 2015). Microorganisms have been used to synthesize lipid nanoparticles with different compositions and sizes (Shah et al., 2015).

Lipid nanoparticles have gained attention for their potential applications in managing plant pathogens, including bacterial, fungal, and nematode diseases (McDaniel et al., 2019). For example, lipid nanoparticles can protect the antimicrobial agents, improve their stability, and facilitate their penetration into bacterial cells (Pinilla et al., 2021). This approach can help control bacterial infections and reduce disease severity. Furthermore, lipid nanoparticles protect the antifungal agents from degradation, improve their solubility, and enhance their uptake by fungal cells (McDaniel et al., 2019). This enables more effective targeting of fungal pathogens, inhibiting their growth and reducing disease development. Similarly, lipid nanoparticles can be utilized to deliver nematicidal compounds or biocontrol agents to target nematodes (Ojha et al., 2018). Lipid nanoparticles protect the active ingredients, improve their stability, and facilitate their delivery to nematode-infested soil or plant tissues (Djiwanti & Kaushik 2019). This approach enhances the efficacy of nematode management strategies, reducing nematode populations and minimizing plant damage. In addition, Madalosso et al. (2017) reported that lipid nanoparticles were effective against *Meloidogyne* sp.

The specific formulation and composition of lipid nanoparticles may vary depending on the intended application and target pathogen. Despite their promise, it is important to note that their practical application in plant disease management is still evolving and factors such as dosage, application methods and potential environmental effects need to be considered for their safe and effective use (Lai et al., 2006).

5.3.4.5 Protein Nanoparticles

These are nanoparticles composed of proteins and are used in drug delivery, imaging, and vaccine development (López-Sagaseta et al., 2016). Microorganisms such as viruses and bacteriophages can be genetically engineered to produce protein nanoparticles with specific properties (López-Sagaseta et al., 2016).

Protein nanoparticles have shown promise in managing plant pathogens (Resham et al., 2015). Protein nanoparticles can be engineered to carry antibacterial peptides or proteins, enhancing their stability and targeted delivery to bacterial pathogens (Carratalá et al., 2020). These nanoparticles can disrupt the bacterial cell membrane, inhibit bacterial growth, or interfere with essential bacterial processes (Carratalá et al., 2020). Furthermore, protein nanoparticles can be functionalized to deliver antifungal proteins, enzymes, or bioactive peptides to fungal pathogens. These nanoparticles can interact with the fungal cell wall or membrane, leading to disruption of fungal structures or inhibition of fungal growth. In this context, Narendrakumar et al. (2018) found that protein nanoparticles effectively reduced the biomass of the plant pathogenic fungus *Alternaria alternata* and highly damaged fungal hyphae fragments.

The development and application of protein nanoparticles for plant pathogen management are still areas of active research. Factors such as nanoparticle design, formulation, and delivery strategies need to be optimized for specific pathogens and target plants. Additionally, safety and environmental considerations should be addressed to ensure the responsible use of protein nanoparticles in agriculture.

5.4 CHARACTERIZATION OF MICROBE-SYNTHESIZED NANOPARTICLES

Characterization of MNPs is crucial for determining their physical and chemical properties, which can affect their behaviour and performance in various applications (Ealia and Saravanakumar 2017). The characterization process involves a series of analytical techniques to evaluate the nanoparticles' size, shape, composition, and surface properties.

5.4.1 Size and Shape Characterization

The size and shape of nanoparticles are critical parameters that can influence their properties and performance (Mukherji et al., 2018). Techniques such as Transmission Electron Microscopy (TEM), Scanning Electron Microscopy (SEM), Atomic Force Microscopy (AFM), and Dynamic Light Scattering (DLS) are commonly used to measure the size and shape of nanoparticles (Kumar & Dixit 2017). These techniques provide information on the particle size distribution, shape, and morphology.

5.4.2 Composition Characterization

The composition of nanoparticles is another important parameter that can influence their properties and performance (Modena et al., 2019). Techniques such as Energy-Dispersive X-ray Spectroscopy (EDS) and X-Ray Diffraction (XRD) can be used to identify the chemical composition of nanoparticles. Fourier-Transform Infrared Spectroscopy (FTIR) and Raman spectroscopy can be used to determine the functional groups present on the surface of the nanoparticles (Kumar & Dixit 2017).

5.4.3 Surface Properties Characterization

The surface properties of nanoparticles can also be important for their behaviour and performance (Modena et al., 2019). Techniques such as zeta potential measurements, surface area analysis, and Surface-Enhanced Raman Scattering (SERS) can be used to evaluate the surface properties of nanoparticles (Kumar & Dixit 2017). These techniques provide information on the surface charge, surface area, and surface chemistry of the nanoparticles (Kumar & Dixit, 2017).

5.5 APPLICATION OF MICROBE-SYNTHESIZED NANOPARTICLES IN PLANT DISEASE MANAGEMENT

Microbe-synthesized NanoParticles (MNPs) have shown potential for use in plant disease management. Their antimicrobial properties make them effective against various plant pathogens, and their small size allows them to easily penetrate plant tissues (Singh et al., 2016; Worrall et al., 2018;). Gupta et al. (2018) showed that silver nanoparticles synthesised by microorganisms can be used as an antifungal agents. Microbes have also produced zinc oxide nanoparticles that have shown antifungal action against pathogenic fungus such *Aspergillus flavus, Aspergillus niger, Aspergillus fumigatus, Fusarium culmorum* and *Fusarium oxysporium* (Rajiv et al., 2013). In addition to their antimicrobial properties, MNPs can also enhance plant growth and stimulate plant defences against pathogens. For example, copper oxide nanoparticles synthesized by microbes have been shown to induce plant defence responses and reduce the severity of bacterial and fungal infections (Ashraf et al., 2021; Li et al., 2023). Overall, the use of MNPs with or without beneficial microbes in agriculture has the potential to provide an effective and sustainable method for managing plant diseases, while also promoting plant growth and health (Ansari et al., 2024; Ansari and Mahmood 2017b; Ansari et al., 2017; Ansari et al., 2023; Boyno et al., 2024).

5.5.1 Microbe-Synthesized Nanoparticles Mediated Plant Disease Management

MNPs have been shown to have potential as a tool for plant disease management, with promising results against a variety of plant pathogens. Some of the plant diseases that have been targeted using MNPs include:

5.5.1.1 Fungal Diseases

Fungal pathogens are one of the most significant challenges to crop production, and they can cause severe yield losses in a wide range of plants (Demirer Durak and Ok 2019; Çevik et al., 2022; Ansari and Mahmood 2017a; Ansari and Mahmood 2019a; Ansari and Mahmood 2019b; Ansari et al., 2020; Ansari et al., 2020; Ansari et al., 2019; Mahmood et al., 2019; Ansari and Khan 2012a). MNP usage has produced encouraging outcomes in the fight against fungi-caused illnesses. Numerous studies have shown that nanoparticles made by microbes, such as silver, zinc oxide, copper, and titanium dioxide nanoparticles, have antifungal properties (Gupta et al., 2018; Parveen & Siddiqui 2022; Li et al., 2023). For example, it has been demonstrated that silver nanoparticles produced by the bacteria *Bacillus subtilis* prevent the growth of the fungi *Fusarium oxysporum* and *Alternaria solani* (Sundararajan et al., 2016). Similarly, biosynthesised zinc oxide nanoparticles have the potential to control Fusarium wilt disease in eggplants (Abdelaziz et al., 2022). MNPs are considered to be able to penetrate the fungal cell wall and rupture the cell membrane, which results in cell death.

However, the exact mechanism of action of MNPs against fungal infections is not yet fully understood (Bera et al., 2014). Additionally, nanoparticles can induce oxidative stress and interfere with cellular processes, further contributing to their antifungal activity (Kumar et al., 2020). Overall, there is a lot of promise for MNPs in the treatment of fungal diseases, but more study is required to determine how effective and safe they are in real-world settings.

5.5.1.2 Bacterial Diseases

Bacterial pathogens can cause significant economic losses in agriculture by reducing crop yields and quality (Akköprü 2020). The use of MNPs has shown promise in controlling bacterial diseases in plants. For example, silver nanoparticles synthesised by bacteria such as *Pseudomonas stutzeri* and *Bacillus licheniformis* (Shanthi et al., 2016; Rajora et al., 2016), zinc oxide nanoparticles synthesised by *Bacillus cereus* (Ahmed et al., 2021), and copper nanoparticles synthesized by the bacterium *Agrobacterium* sp. (Bajaj et al., 2012) have been shown to have antibacterial activity against some plant bacterial diseases. According to these results, MNPs may be a useful substitute for traditional chemical pesticides in the treatment of bacterial plant diseases.

5.5.1.3 Viral Diseases

Plant viruses can cause severe diseases that reduce crop yields and quality (Rao and Reddy 2020). Unlike bacterial and fungal pathogens, there are limited options for controlling viral diseases in plants. However, some studies have investigated the potential of MNPs in combating viral diseases in plants. One study showed that silver nanoparticles synthesized by *Bacillus spp.* had antiviral activity against bean yellow mosaic virus, a common plant virus that infects a wide range of crops (Elbeshehy et al., 2015). The silver nanoparticles inhibited the replication of the virus in infected plants, resulting in reduced disease symptoms and increased plant growth (Elbeshehy et al., 2015). Another study found that Cai et al. (2019) reported on the antiviral activity of Zinc oxide nanoparticles against Tobacco Mosaic Virus (TMV) in vitro. While these studies show promising results, further research is needed to determine the effectiveness of MNPs against a broader range of plant viruses and to evaluate their safety for use in agriculture.

5.5.1.4 Nematodal Diseases

Nematodal diseases, also known as nematode diseases, are plant diseases caused by plant-parasitic nematodes (Ansari and Khan 2012b; Ibrahim et al., 2020; Devi et al., 2020; Ibrahim et al., 2022). Nematodes are microscopic roundworms that can infect various parts of plants, including roots, stems, and leaves (Ansari & Khan 2012). These pests can cause significant damage to crops and plants, leading to reduced yield and compromised plant health. Recently, nanoparticles synthesized by microorganisms are being investigated as a potential tool to control and manage nematode diseases in plants (Fabiyi et al., 2020).

Microbe-synthesized nanoparticles have shown nematicidal activity against plant-parasitic nematodes (Fabiyi et al., 2020). This activity is attributed to their ability to interact with nematodes, disrupt their physiology, damage their cuticles, and interfere with vital processes (Thakur & Shirkot 2017). These effects collectively result in reduced nematode viability, reproduction, and population size (Thakur & Shirkot 2017).

Microbe-synthesized nanoparticles, such as silver nanoparticles or gold nanoparticles, can interact with nematodes upon contact. For example, silver nanoparticles (Mahmoud et al., 2016) and gold nanoparticles (Thakur et al., 2018) were found to be lethal to *Meloidogyne incognita*. The nanoparticles adhere to the surface of nematodes, facilitating their entry into the nematode's body or interfering with the nematode's cuticle (outer protective layer) (Thakur & Shirkot 2017). Moreover, the nanoparticles may interfere with key enzymes or metabolic processes essential for the nematode's survival and reproduction (Mohamed et al., 2021). This disruption can negatively impact the nematode's overall health and lead to reduced viability. However, the precise mechanisms and modes of action of microbe-synthesized nanoparticles against nematodes may vary depending on the nanoparticle properties, nematode species, and specific interactions involved (Fabiyi et al., 2020).

5.5.1.5 Phytoplasma Diseases

The utilization of nanoparticles synthesized by microorganisms for the management of phytoplasma diseases is an area of ongoing research, but specific studies and reports on this topic are limited. Phytoplasmas are unique bacteria-like microorganisms that cause diseases in a wide range of plants, including various crops and ornamental plants (Noorizadeh et al., 2022). Currently, the management of phytoplasma diseases primarily involves cultural practices, vector control, and, in some cases, the use of antibiotics or plant resistance (Weintraub 2007; Tanno et al., 2018). However, the potential use of nanoparticles synthesized by microorganisms for managing phytoplasma diseases holds promise (Huliaieva et al., 2021). According to Ahmed et al. (2022), treatment of phytoplasma-infected broad bean plants with various nanoparticles resulted in severe damage, distortion, malformation, cell membrane lysis, cytoplasmic leakage, and complete lysis of phytoplasma cells. In addition, nanoparticles could be engineered or functionalized to detect phytoplasmas in plants. For example, quantum dots, one of the types of microbial nanoparticles, have been used as biosensors to detect *Phytoplasma aurantifolia* in linden and lemon trees (Rad et al., 2012; Kumar et al., 2022).

5.5.2 Modes of Action of Microbe-Synthesized Nanoparticles

The modes of action of MNPs can vary depending on the type of nanoparticle and the pathogen being targeted. However, some general mechanisms of action have been proposed based on experimental studies.

5.5.2.1 Physical Damage

Physical damage is one of the modes of action of MNPs against plant pathogens. Nanoparticles can interact with the pathogen's cell membrane or cell wall, leading to physical damage and disruption of the membrane or wall (Bera et al., 2014). This can cause the leakage of cellular contents and eventually cell death of the pathogen. For example, silver nanoparticles have been shown to attach to the cell membrane of fungi and bacteria, leading to membrane disruption and cell death (Siddiqi et al., 2018). Similarly, copper nanoparticles have been shown to disrupt the cell wall of fungi and bacteria, causing leakage of cellular contents and cell death (Ouda 2014). Physical damage is a promising mode of action for MNPs, as it can lead to rapid and effective control of plant pathogens. However, it is important to note that the specificity of this mode of action may be limited, as it can also affect beneficial microorganisms in the soil. Therefore, it is essential to carefully evaluate the potential ecological impacts of using nanoparticles for plant disease management.

5.5.2.2 Reactive Oxygen Species

Reactive Oxygen Species (ROS) are chemically reactive molecules that contain oxygen, such as superoxide and hydroxyl radicals (Mhamdi & Van Breusegem, 2018. These molecules are generated naturally during cellular metabolism and play important roles in signalling and regulation of cellular processes (Mhamdi & Van Breusegem, 2018. However, an excess of ROS can cause oxidative stress, leading to damage to cellular components such as proteins, lipids, and nucleic acids (Mhamdi & Van Breusegem, 2018. MNPs can generate ROS through various mechanisms, such as the release of metal ions or photocatalysis (Bhattacharya et al., 2022). The ROS produced may harm the pathogen's cell wall, membrane, and other cellular components, ultimately resulting in cell death (Lemire et al., 2013). The exact mechanism by which ROS cause damage varies depending on the type of ROS generated and the target cell components (Lemire et al., 2013). For example, superoxide radicals can damage proteins by oxidizing amino acid residues, while hydroxyl radicals can cause DNA strand breaks and lipid peroxidation (Bhattacharya et al., 2022). MNPs have the ability to effectively stop the development and spread of plant diseases by producing oxidative stress. Many different kinds of nanoparticles, including metal and metal oxide nanoparticles, have an essential mode of action known as ROS production (Bhattacharya et al., 2022).

5.5.2.3 Disruption of Enzymatic Activity

Certain MNPs have the ability to disrupt enzymatic activity in plant pathogens (Bhattacharya et al., 2022). Enzymes are crucial for many processes in the pathogen's growth and virulence, such as cell wall synthesis, nutrient acquisition, and toxin production. By inhibiting the activity of these enzymes, nanoparticles can disrupt these processes and reduce the pathogen's ability to cause disease. For example, silver nanoparticles synthesized by the bacterium *Bacillus licheniformis* have been shown to inhibit the activity of enzymes involved in the synthesis of the bacterial cell wall, leading to reduced growth and virulence of the bacterial pathogen (Kalimuthu et al., 2008; Ahmed et al., 2020). Similarly, zinc oxide nanoparticles synthesized by the fungus *Trichoderma harzianum* have been shown to inhibit the activity of enzymes involved in the production of fungal toxins, leading to reduced virulence of the fungal pathogen (Zaki et al., 2021). Overall, disruption of enzymatic activity is another mechanism by which MNPs can exert their antifungal or antibacterial effects and help manage plant diseases.

5.5.2.4 Induction of Plant Defence Mechanisms

Induction of plant defence mechanisms is another mode of action of MNPs. When nanoparticles come into contact with plants, they can activate the plant's innate immune system, which triggers a series of defence responses to protect the plant from pathogen attack (Worrall et al., 2018;). This immune response includes the production of various defence-related molecules such as Pathogenesis-Related (PR) proteins, phytohormones, and ROS, which can inhibit pathogen growth and spread (Mhamdi & Van Breusegem 2018 Bhattacharya et al., 2022). Some nanoparticles, such as silver and copper nanoparticles, have been shown to induce the expression of defence-related genes and enhance the production of defence-related molecules in plants. For example, silver nanoparticles have been shown to induce the expression of genes involved in the biosynthesis of plant defence molecules, such as jasmonic acid and ethylene, leading to enhanced resistance against pathogens (Kaveh et al., 2013). Similarly, copper nanoparticles have been reported to induce the production of ROS and enhance the expression of PR proteins, resulting in increased resistance against bacterial and fungal pathogens (Hernández-Hernández et al., 2018). The induction of plant defence mechanisms by MNPs can also lead to Systemic Acquired Resistance (SAR), which is a long-lasting defence response that provides broad-spectrum protection against a wide range of pathogens (Prasad et al., 2017). Overall, the induction of plant defence mechanisms by MNPs is a promising mode of action that can enhance plant resistance against pathogens.

5.5.2.5 Interference with Signalling Pathways

Interference with signalling pathways is another mode of action by which MNPs can exhibit antifungal, antibacterial, or antiviral activity (Marslin et al., 2017). Pathogens rely on signalling pathways to sense their environment and communicate with other cells. When a nanoparticle interferes with these pathways, it can disrupt important cellular processes, leading to cell death. For example, some nanoparticles have been shown to inhibit the expression of virulence genes in pathogens, leading to reduced pathogenicity (Osonga et al., 2018). Similarly, some nanoparticles can interfere with Quorum Sensing (QS), a communication mechanism used by bacteria to coordinate group behaviour, leading to disruption of bacterial communities and reduced pathogenicity (Qin et al., 2022). For example, Gómez-Gómez et al. (2019) reported that zinc oxide nanoparticles, silver nanoparticles and titanium dioxide nanoparticles interfered with the QS of *Chromobacterium violaceum* bacteria and reduced its potential. In this context, nanoparticles can inhibit the growth and spread of pathogens such as bacteria by interfering with signalling pathways (Marslin et al.,

2017). Overall, the modes of action of MNPs are still being studied, and more research is needed to fully understand how these nanoparticles interact with plant pathogens and their host plants.

5.5.3 Efficacy of Microbe-Synthesized Nanoparticles in Plant Disease Management

The efficacy of MNPs in plant disease management varies depending on the type of nanoparticle, the target pathogen, and the mode of application. However, several studies have shown promising results in using these nanoparticles to control plant diseases. For example, in one study, silver nanoparticles synthesized by the fungus *Fusarium oxysporum* were found to be effective against various plant pathogens, including *Fusarium oxysporum, Fusarium solani,* and *Rhizoctonia solani* (Kamaluddin, 2018). Another study found that copper oxide nanoparticles synthesized by the bacterium *Pseudomonas stutzeri* showed significant antifungal activity against *Fusarium graminearum* and *Alternaria solani* (Bhuvaneshwari et al., 2022). Similarly, chitosan nanoparticles synthesized by the fungus *Trichoderma viride* have been shown to have antifungal activity against *Botrytis cinerea* and *Fusarium oxysporum.* Zinc oxide nanoparticles synthesized by the fungus *Aspergillus terreus* have been shown to have antifungal activity against *Alternaria alternata* and *Fusarium oxysporum* (Abd-Elsalam et al., 2018). In addition to their direct antifungal and antibacterial properties, MNPs have also been shown to enhance plant growth and induce systemic resistance against pathogens. For example, silver nanoparticles synthesized by the fungus *Aspergillus terreus* were found to enhance the growth of tomato plants and induce systemic resistance against *Botrytis cinerea.* (Rajeshkumar and Sivapriya 2020). Overall, while further research is needed to fully understand the efficacy and safety of MNPs in plant disease management, the current evidence suggests that they have great potential as a sustainable and effective alternative to traditional chemical pesticides.

5.5.4 Comparison of MNPs with Other Nanomaterials in Plant Disease Management

Compared to other nanomaterials, MNPs have several advantages in plant disease management (Singh et al., 2016; Worrall et al., 2018). Firstly, MNPs are eco-friendly and sustainable (Bhandari et al., 2023). They are synthesized using microbial cells, which are renewable and biodegradable resources, and do not generate toxic byproducts (Bhandari et al., 2023). In contrast, many other nanomaterials are synthesized using chemical methods that can be harmful to the environment and human health.

Secondly, MNPs have low toxicity and are generally considered safe for plants, animals, and humans (Buzea et al., 2007). This is because they are synthesized using natural and biocompatible materials, and the synthesis process does not involve the use of toxic chemicals. In contrast, some other nanomaterials have been shown to be toxic to plants and animals, and their long-term effects on the environment and human health are still not well understood (Buzea et al., 2007). Thirdly, MNPs can be synthesized using a variety of microorganisms, and the properties of the resulting nanoparticles can be tuned by adjusting the culture conditions and the type of microorganism used (Prasad et al., 2016). This makes them highly versatile and customizable for different applications in plant disease management. In contrast, some other nanomaterials have limited synthesis methods and properties, which may not be suitable for all applications (Prasad et al., 2016). Overall, the use of MNPs in plant disease management holds great promise, and their unique advantages make them an attractive alternative to other nanomaterials. However, further research is needed to fully understand their effectiveness and safety in different plant systems and environmental conditions.

5.5.5 Nanoparticle Delivery Systems for Plant Protection

Nanoparticle delivery systems have been developed to improve the efficacy and stability of MNPs in plant protection (Shang et al., 2019). These delivery systems can protect the nanoparticles from environmental stresses, such as UV radiation and oxidation, and enhance their adhesion to plant surfaces. Some common delivery systems include:

5.5.5.1 Polymeric Nanoparticles

These are composed of biodegradable polymers and can encapsulate MNPs for controlled release (Khodakovskaya and Marmiroli 2023). They can also protect the nanoparticles from degradation and enhance their uptake by plants (Khodakovskaya and Marmiroli 2023).

5.5.5.2 Liposomes

These are spherical structures composed of a lipid bilayer that can encapsulate MNPs (Date et al., 2007). They can protect the nanoparticles from environmental stresses and enhance their uptake by plant cells (Date et al., 2007).

5.5.5.3 Nanocapsules

These are composed of a polymeric shell that can encapsulate MNPs (Alomari et al., 2018). They can protect the nanoparticles from degradation and enhance their adhesion to plant surfaces (Alomari et al., 2018).

5.5.5.4 Nanofibers

These are composed of electrospun polymers that can be functionalized with MNPs (Meraz-Dávila et al., 2021). They can provide sustained release of the nanoparticles and enhance their adhesion to plant surfaces (Badgar et al., 2021). Overall, nanoparticle delivery systems can improve the efficacy and stability of MNPs in plant protection, but more research is needed to optimize their use in the field.

5.6 CHALLENGES AND FUTURE PERSPECTIVES

The potential of MNPs in sustainable agriculture is significant, as they offer a natural and eco-friendly alternative to chemical pesticides and fertilizers. They can help reduce the environmental impact of agriculture, by decreasing the use of harmful chemicals and improving soil health. Additionally, MNPs can be produced using low-cost and renewable resources, making them a viable option for farmers in developing countries. Moreover, the use of MNPs can also improve the nutritional quality of crops, as some nanoparticles have been shown to enhance the uptake of nutrients by plants (Gupta et al., 2018). This can lead to increased yields and better crop quality, ultimately benefiting both farmers and consumers. Despite the potential benefits of MNPs in plant disease management, there are also some challenges and limitations to their use:

5.6.1 Standardization of Nanoparticle Synthesis

The synthesis of nanoparticles by microorganisms can be affected by a range of factors, including nutrient availability, pH, temperature, and incubation time (Bogren et al., 2015). To ensure consistent and reliable synthesis, standardization of the synthesis process is required.

5.6.2 Characterization of Nanoparticles

Accurate characterization of the physical and chemical properties of nanoparticles is important for understanding their mode of action and potential toxicity (Shahbaz et al., 2022). However, the characterization of MNPs can be challenging due to their complex composition and small size.

5.6.3 Cost-Effectiveness

The cost of producing MNPs on a large scale may be a limiting factor in their widespread use in agriculture (Banerjee et al., 2021).

5.6.4 Regulatory Considerations

The regulatory requirements for the use of nanoparticles in agriculture are still being developed (Jain et al., 2018). There is a need for clear guidelines and standards for the production and use of MNPs.

5.6.5 Potential Toxicity

Although MNPs are generally considered to be safe and biocompatible, there is still a need for further research to assess their potential toxicity to plants and non-target organisms (Banerjee et al., 2021).

5.6.6 Field Efficacy

While laboratory studies have shown promising results, field studies are needed to assess the efficacy of MNPs in real-world agricultural settings (Elmer and White 2018). Addressing these challenges and limitations will be important for the development and adoption of MNPs as a sustainable tool for plant disease management. Some potential future directions for research and development of MNPs in plant disease management include:

5.6.7 Optimization of Synthesis Methods

There is still much to be learned about the optimal conditions for the synthesis of MNPs, including the selection of microorganisms, growth conditions, and other factors that can affect the properties of the resulting nanoparticles (Tariq et al., 2022).

5.6.8 Understanding the Mechanisms of Action and Efficient Delivery

While some modes of action of MNPs have been identified, there is still much to be learned about how they interact with plant pathogens and plant cells (Lum et al., 2021). Further research is needed to understand these interactions in more detail. The efficacy of MNPs can be influenced by the method of delivery, and more research is needed to develop efficient and effective delivery systems for these nanoparticles in agricultural settings (Mohamed et al., 2019).

5.6.9 Safety, Environmental Impact Assessment, Field Trial and Commercialization

As with any new technology, safety and environmental impact assessment is essential for the development and adoption of MNPs in agriculture (Zehra et al., 2021). More research is needed to fully understand the potential risks associated with the use of these nanoparticles, and to develop strategies to minimize any negative impacts on the environment or human health (Zehra et al., 2021). Plenty of research on field testing and commercialization is needed to demonstrate their effectiveness in real-world agricultural settings, and to make them available to farmers at an affordable cost (Elmer & White 2018).

5.7 CONCLUSIONS AND FUTURE PROSPECTS

Usage of MNPs in agriculture represents a promising alternative to traditional chemical pesticides for controlling plant diseases. Despite the potential advantages of MNPs, several challenges and limitations need to be addressed before their widespread adoption in agriculture. One of the major challenges is the lack of understanding of their long-term effects on the environment and human health. Further research is needed to assess the potential toxicity of these nanoparticles to plants and non-target organisms, as well as their impact on the environment. The production and commercialization of these nanoparticles face regulatory and economic barriers, limiting their widespread adoption. Optimization of synthesis methods, understanding the mechanisms of action, development of efficient delivery systems, and safety and environmental impact assessment are some of the key areas that require further investigation. If these challenges can be overcome, MNPs have the potential to revolutionize the agriculture industry and promote sustainable agricultural practices. The integration of MNPs with other approaches such as biological control, crop rotation, and cultural practices can enhance the effectiveness of plant disease management and reduce the reliance on chemical pesticides. In summary, the use of MNPs in agriculture is a promising field that offers many potential benefits, including cost-effectiveness, specificity, and environmental friendliness. However, further research and development are needed to address the challenges and limitations associated with their production and adoption. With continued research and development, MNPs could have a wide range of applications in various fields, providing a sustainable, effective, and environmentally friendly alternative to traditional approaches.

REFERENCES

Abdallah, Y., Liu, M., Ogunyemi, S. O., Ahmed, T., Fouad, H., Abdelazez, A., ... & Li, B. (2020). Bioinspired green synthesis of chitosan and zinc oxide nanoparticles with strong antibacterial activity against rice pathogen *Xanthomonas oryzae* pv. oryzae. *Molecules*, *25*(20), 4795.

Abdelaziz, A. M., Salem, S. S., Khalil, A. M., El-Wakil, D. A., Fouda, H. M., & Hashem, A. H. (2022). Potential of biosynthesized zinc oxide nanoparticles to control *Fusarium* wilt disease in eggplant (*Solanum melongena*) and promote plant growth. *BioMetals*, *35*(3), 601–616.

Abd-Ellatif, S., Ibrahim, A. A., Safhi, F. A., Abdel Razik, E. S., Kabeil, S. S., Aloufi, S., ... & Elshafie, H. S. (2022). Green synthesized of *Thymus vulgaris* chitosan nanoparticles induce relative WRKY-genes expression in *Solanum lycopersicum* against *Fusarium solani*, the causal agent of root rot disease. *Plants*, *11*(22), 3129.

Abd-Elsalam, K. A. (Ed.). (2019). *Carbon Nanomaterials for Agri-Food and Environmental Applications*. Elsevier.

Abd-Elsalam, K. A., Vasil'kov, A. Y., Said-Galiev, E. E., Rubina, M. S., Khokhlov, A. R., Naumkin, A. V., ... & Alghuthaymi, M. A. (2018). Bimetallic blends and chitosan nanocomposites: Novel antifungal agents against cotton seedling damping-off. *European Journal of Plant Pathology*, *151*, 57–72.

Abdo, A. M., Fouda, A., Eid, A. M., Fahmy, N. M., Elsayed, A. M., Khalil, A. M. A., ... & Soliman, A. M. (2021). Green synthesis of Zinc oxide nanoparticles (ZnO-NPs) by *Pseudomonas aeruginosa* and their activity against pathogenic microbes and common house mosquito, *Culex pipiens*. *Materials*, *14*(22), 6983.

Abdollahdokht, D., Gao, Y., Faramarz, S., Poustforoosh, A., Abbasi, M., Asadikaram, G., & Nematollahi, M. H. (2022). Conventional agrochemicals towards nano-biopesticides: An overview on recent advances. *Chemical and Biological Technologies in Agriculture*, *9*(1), 1–19.

Abdollahnia, M., Makhdoumi, A., Mashreghi, M., & Eshghi, H. (2020). Exploring the potentials of halophilic prokaryotes from a solar saltern for synthesizing nanoparticles: The case of silver and selenium. *PLoS One*, *15*(3), e0229886.

Adeel, M., Farooq, T., White, J. C., Hao, Y., He, Z., & Rui, Y. (2021). Carbon-based nanomaterials suppress tobacco mosaic virus (TMV) infection and induce resistance in Nicotiana benthamiana. *Journal of Hazardous Materials*, *404*, 124167.

Agarwal, H., Menon, S., Kumar, S. V., & Rajeshkumar, S. (2018). Mechanistic study on antibacterial action of zinc oxide nanoparticles synthesized using green route. *Chemico-Biological Interactions*, *286*, 60–70.

Ahmad, F., Ashraf, N., Ashraf, T., Zhou, R. B., & Yin, D. C. (2019). Biological synthesis of metallic nanoparticles (MNPs) by plants and microbes: Their cellular uptake, biocompatibility, and biomedical applications. *Applied Microbiology and Biotechnology*, *103*, 2913–2935.

Ahmed, E. A., Shoala, T., Abdelkhalik, A., El-Garhy, H. A., Ismail, I. A., & Farrag, A. A. (2022). Nanoinhibitory impacts of salicylic acid, glycyrrhizic acid ammonium salt, and boric acid nanoparticles against *phytoplasma* associated with faba bean. *Molecules*, *27*(5), 1467.

Ahmed, T., Shahid, M., Noman, M., Niazi, M. B. K., Mahmood, F., Manzoor, I., ... & Chen, J. (2020). Silver nanoparticles synthesized by using *Bacillus cereus* SZT1 ameliorated the damage of bacterial leaf blight pathogen in rice. *Pathogens*, *9*(3), 160.

Ahmed, T., Wu, Z., Jiang, H., Luo, J., Noman, M., Shahid, M., ... & Li, B. (2021). Bioinspired green synthesis of zinc oxide nanoparticles from a native *Bacillus cereus* strain RNT6: Characterization and antibacterial activity against rice panicle blight pathogens *Burkholderia glumae* and *B. gladioli*. *Nanomaterials*, *11*(4), 884.

Akköprü, A. (2020). Potential using of transgenerational resistance against common bacterial blight in *Phaseolus vulgaris*. *Crop Protection*, *127*, 104967.

Alabdallah, N. M., Irshad, M. A., Rizwan, M., Nawaz, R., Inam, A., Mohsin, M., ... & Ali, S. (2023). Synthesis, characterization and antifungal potential of titanium dioxide nanoparticles against fungal disease (*Ustilago tritici*) of wheat (*Triticum aestivum* L.). *Environmental Research*, *228*, 115852.

Alfy, H., Ghareeb, R. Y., Soltan, E. L., & Farag, D. A. (2020). Impact of chitosan nanoparticles as insecticide and nematicide against *Spodoptera littoralis*, *Locusta migratoria*, and *Meloidogyne incognita*. *Plant Cell Biotechnology Molecular Biology*, *21*, 126–140.

Ali, A., Shah, T., Ullah, R., Zhou, P., Guo, M., Ovais, M., ... & Rui, Y. (2021). Review on recent progress in magnetic nanoparticles: Synthesis, characterization, and diverse applications. *Frontiers in Chemistry*, *9*, 629054.

Alimardani, V., Abolmaali, S. S., & Borandeh, S. (2019). Antifungal and antibacterial properties of graphene-based nanomaterials: A mini-review. *Journal of Nanostructures*, *9*(3), 402–413.

Alomari, A. A., Kloub Fares, K. E., & Moustafa, N. E. (2018). Green synthesis of assembled silver nanoparticles in nano capsules of *Peganum harmala* L leaf extract. Antibacterial activity and conjugate investigation. *Cogent Chemistry*, *4*(1), 1532374.

Ansari, R. A., & Khan, T. A. (2012a). Parasitic association of root-knot nematode, *Meloidogyne incognita* on guava. *e-Journal of Science Technology*, *5*(12), 65–67.

Ansari, R. A., & Khan, T. A. (2012b). Diversity and community structure of phytonematodes associated with guava in and around Aligarh, Uttar Pradesh, India. *Trends in Bioscience*, *5*(3), 202–204.

Ansari, R. A., & Mahmood, I. (2017a). Determination of disease incidence caused by *Meloidogyne* spp. and or *Fusarium udum* on pigeonpea in Aligarh district: A survey. *Trends in Biosciences*, *10*(24), 5239–5243.

Ansari, R. A., & Mahmood, I. (2017b). Optimization of organic and bio-organic fertilizers on soil properties and growth of pigeon pea. *Scientia Horticulturae*, *226*, 1–9.

Ansari, R. A., & Mahmood, I. (2019a). *Plant Health Under Biotic Stress – Volume II, Microbial Interactions*, Springer Nature, Singapore. ISBN: 978-981-13-6040-4.

Ansari, R. A., & Mahmood, I. (2019b). *Plant Health Under Biotic Stress – Volume I, Organic Strategies*, Springer Nature, Singapore. ISBN: 978-981-13-6043-5.

Ansari, R. A., Mahmood, I., Rizvi, R., Sumbul, A., & Safiuddin. (2017). Siderophores: Augmentation of soil health and crop productivity. *Probiotics in Agroecosystem*, 291–312.

Ansari, R. A., Rizvi, R., & Mahmood, I. (2020). *Management of Phytonematodes: Management of Phytonematodes: Recent Advances and Future Challenges*, Springer Nature, Singapore. ISBN: 978-981-15-4087-5.

Ansari, R. A., Rizvi, R., & Mahmood, I. (2023). *Mycorrhizal Symbiosis and Agroecosystem Restoration*, Springer.

Ansari, R. A., Rizvi, R., Sumbul, A., & Mahmood, I. (2020). Plant-growth-promoting rhizobacteria (PGPR)-based sustainable management of phytoparasitic nematodes: Current understandings and future challenges. In *Management of Phytonematodes: Recent Advances and Future Challenges* (pp. 51–71).

Ansari, R. A., Sumbul, A., Rizvi, R., & Mahmood, I. (2019). Organic soil amendments: Potential tool for soil and plant health management. In *Plant Health Under Biotic Stress: Volume 1: Organic Strategies* (pp. 1–35).

Ansari, R. A., Sumbul, A., Rizvi, R., & Mahmood, I. (2024). Impact of environmental gases on mycorrhizal symbiosis and its influence on ecosystem functioning under the current climate change scenario. In *Mycorrhizal Symbiosis and Agroecosystem Restoration* (pp. 51–76). Singapore: Springer Nature Singapore.

Ashraf, H., Anjum, T., Riaz, S., Ahmad, I. S., Irudayaraj, J., Javed, S., ... & Naseem, S. (2021). Inhibition mechanism of green-synthesized copper oxide nanoparticles from *Cassia fistula* towards *Fusarium oxysporum* by boosting growth and defence response in tomatoes. *Environmental Science: Nano*, *8*(6), 1729–1748.

Badgar, K., Prokisch, J., & El-Ramady, H. (2021). Nanofibers for sustainable agriculture: A short communication. *Egyptian Journal of Soil Science*, *61*(3), 373–380.

Bajaj, M., Schmidt, S., & Winter, J. (2012). Formation of Se (0) nanoparticles by *Duganella* sp. and *Agrobacterium* sp. isolated from Se-laden soil of North-East Punjab, India. *Microbial Cell Factories*, *11*, 1–14.

Balasubramanian, S., Kala, S. M. J., & Pushparaj, T. L. (2020). Biogenic synthesis of gold nanoparticles using *Jasminum auriculatum* leaf extract and their catalytic, antimicrobial and anticancer activities. *Journal of Drug Delivery Science and Technology*, *57*, 101620.

Banerjee, A., Sarkar, A., Acharya, K., & Chakraborty, N. (2021). Nanotechnology: An emerging hope in crop improvement. *Letters in Applied NanoBioScience*, *10*(4), 2784–2803.

Banik, A., Vadivel, M., Mondal, M., & Sakthivel, N. (2022). Molecular mechanisms that mediate microbial synthesis of metal nanoparticles. In *Microbial Metabolism of Metals and Metalloids* (pp. 135–166). Springer International Publishing, Cham.

Barapatre, A., Aadil, K. R., & Jha, H. (2016). Synergistic antibacterial and antibiofilm activity of silver nanoparticles biosynthesized by lignin-degrading fungus. *Bioresources and Bioprocessing*, *3*, 1–13.

Beatty, P. H., & Lewis, J. D. (2019). Cowpea Mosaic virus nanoparticles for cancer imaging and therapy. *Advanced Drug Delivery Reviews*, *145*, 130–144.

Benzerara, K., Miot, J., Morin, G., Ona-Nguema, G., Skouri-Panet, F., & Ferard, C. (2011). Significance, mechanisms and environmental implications of microbial biomineralization. *Comptes Rendus Geoscience*, *343*(2–3), 160–167.

Bera, R. K., Mandal, S. M., & Raj, C. R. (2014). Antimicrobial activity of fluorescent Ag nanoparticles. *Letters İn Applied Microbiology*, *58*(6), 520–526.

Bhandari, G., Dhasmana, A., Chaudhary, P., Gupta, S., Gangola, S., Gupta, A., ... & Slama, P. (2023). A perspective review on green nanotechnology in agro-ecosystems: Opportunities for sustainable agricultural practices & environmental remediation. *Agriculture*, *13*(3), 668.

Bhardwaj, B., Singh, P., Kumar, A., Kumar, S., & Budhwar, V. (2020). Eco-friendly greener synthesis of nanoparticles. *Advanced Pharmaceutical Bulletin*, *10*(4), 566.

Bhattacharya, J., Nitnavare, R., Shankhapal, A., & Ghosh, S. (2022). Microbially synthesized nanoparticles: Aspect in plant disease management. In *Biocontrol Mechanisms of Endophytic Microorganisms* (pp. 303–325). Academic Press.

Bhuvaneshwari, V., Ramasamy, N. K., Kumar, S. I., Kalaivani, S., Vaidehi, D., & Kumar, D. K. (2022). Antimicrobial activity of copper nanomaterials: Current status and future perspectives. In *Copper Nanostructures: Next-Generation of Agrochemicals for Sustainable Agroecosystems* (pp. 453–475).

Bogren, S., Fornara, A., Ludwig, F., del Puerto Morales, M., Steinhoff, U., Fougt Hansen, M., ... & Johansson, C. (2015). Classification of magnetic nanoparticle systems—synthesis, standardization and analysis methods in the nanomag project. *International Journal of Molecular Sciences*, *16*(9), 20308–20325.

Boroumand Moghaddam, A., Namvar, F., Moniri, M., Md. Tahir, P., Azizi, S., & Mohamad, R. (2015). Nanoparticles biosynthesized by fungi and yeast: A review of their preparation, properties, and medical applications. *Molecules*, *20*(9), 16540–16565.

Borovaya, M. N., Burlaka, O. M., Yemets, A. I., & Blume, Y. B. (2015). Biosynthesis of quantum dots and their potential applications in biology and biomedicine. In *Nanoplasmonics, Nano-Optics, Nanocomposites, and Surface Studies: Selected Proceedings of the Second FP7 Conference and the Third International Summer School Nanotechnology: From Fundamental Research to Innovations, August 23–30, 2014, Yaremche-Lviv, Ukraine* (pp. 339–362). Springer International Publishing.

Boyno, E. (2022). *Yoğurt üretiminde peyniraltı suyu tozu içeren zein nanolifinin yoğurdun kalite özelliklerine etkisi* (master's thesis). Fen Bilimleri Enstitüsü, Van.

Boyno, G., Ansari, R. A., Durak, E. D., Güneş, H., Çevik, R., & Demir, S. (2024). Arbuscular mycorrhizal technology in sustainable agriculture: Current knowledge and challenges in agroforestry. *Mycorrhizal Symbiosis and Agroecosystem Restoration*, 173–195.

Boyno, G., Demir, S., & Danesh, Y. R. (2022). Effects of some biological agents on the growth and biochemical parameters of tomato plants infected with *Alternaria solani* (Ellis & Martin) Sorauer. *European Journal of Plant Pathology*, *162*(1), 19–29.

Buzea, C., Pacheco, I. I., & Robbie, K. (2007). Nanomaterials and nanoparticles: Sources and toxicity. *Biointerphases*, *2*(4), MR17–MR71.

Cabuzu, D., Cirja, A., Puiu, R., & Mihai Grumezescu, A. (2015). Biomedical applications of gold nanoparticles. *Current Topics in Medicinal Chemistry*, *15*(16), 1605–1613.

Cai, L., Liu, C., Fan, G., Liu, C., & Sun, X. (2019). Preventing viral disease by ZnONPs through directly deactivating TMV and activating plant immunity in *Nicotiana benthamiana*. *Environmental Science: Nano*, *6*(12), 3653–3669.

Carabineiro, S. A. C. (2017). Applications of gold nanoparticles in nanomedicine: Recent advances in vaccines. *Molecules*, *22*(5), 857.

Carratalá, J. V., Brouillette, E., Serna, N., Sánchez-Chardi, A., Sánchez, J. M., Villaverde, A., ... & Malouin, F. (2020). In vivo bactericidal efficacy of GWH1 antimicrobial peptide displayed on protein nanoparticles, a potential alternative to antibiotics. *Pharmaceutics*, *12*(12), 1217.

Çeşmeli, S., & Biray Avci, C. (2019). Application of titanium dioxide (TiO2) nanoparticles in cancer therapies. *Journal of Drug Targeting*, *27*(7), 762–766.

Çevik, R., Demir, S., Türkölmez, Ş., & Boyno, G. (2022). The effect of *Clonostachys rosea* (sch.) schroers and samuels against *verticillium* wilt (*Verticillium dahliae* Kleb.) and early blight [*Alternaria solani* (Ell. and G. Martin) Sor.] diseases in tomato plants. *Yuzuncu Yıl University Journal of Agricultural Sciences*, *32*(2), 372–382.

Chandrasekaran, M., Kim, K. D., & Chun, S. C. (2020). Antibacterial activity of chitosan nanoparticles: A review. *Processes*, *8*(9), 1173.

Chen, M. C., Koh, P. W., Ponnusamy, V. K., & Lee, S. L. (2022). Titanium dioxide and other nanomaterials based antimicrobial additives in functional paints and coatings. *Progress in Organic Coatings*, *163*, 106660.

Chinni, S. V., Sanniraj, A., Koduru, J. R., & Lebaka, V. R. (2023). Nanobioremediation—An emerging eco-friendly approach for a sustainable environment. In *Hybrid Nanomaterials for Sustainable Applications* (pp. 199–219). Elsevier.

Chmykhalo, V., Belanova, A., Belousova, M., Butova, V., Makarenko, Y., Khrenkova, V., ... & Zolotukhin, P. (2021). Microbial-based magnetic nanoparticles production: A mini-review. *Integrative Biology*, *13*(4), 98–107.

Chung, Y. H., Cai, H., & Steinmetz, N. F. (2020). Viral nanoparticles for drug delivery, imaging, immunotherapy, and theranostic applications. *Advanced Drug Delivery Reviews*, *156*, 214–235.

Corredor, E., Testillano, P. S., Coronado, M. J., González-Melendi, P., Fernández-Pacheco, R., Marquina, C., ... & Risueño, M. C. (2009). Nanoparticle penetration and transport in living pumpkin plants: In situ subcellular identification. *BMC Plant Biology*, *9*, 1–11.

Cristea, C., Tertis, M., & Galatus, R. (2017). Magnetic nanoparticles for antibiotics detection. *Nanomaterials*, *7*(6), 119.

Date, A. A., Joshi, M. D., & Patravale, V. B. (2007). Parasitic diseases: Liposomes and polymeric nanoparticles versus lipid nanoparticles. *Advanced Drug Delivery Reviews*, *59*(6), 505–521.

Delgado-García, M., Rodríguez, J. A., Mateos-Díaz, J. C., Aguilar, C. N., Rodríguez-Herrera, R., & Camacho-Ruíz, R. M. (2018). Halophilic archaeal lipases and esterases: Activity, stability, and food applications. *Enzymes in Food Technology: Improvements and Innovations*, 243–262.

Demirer Durak, E., & Ok, F. (2019). Determination of anastomosis groups and pathogenicity of *Rhizoctonia solani* kühn isolates from tomato (*Solanum lycopersicum*) in lake van basin, Turkey. *Journal of the Institute of Science and Technology*, *9*(4), 1908–1915.

Devi, L. S., & Joshi, S. R. (2015). Ultrastructures of silver nanoparticles biosynthesized using endophytic fungi. *Journal of Microscopy and Ultrastructure*, *3*(1), 29–37.

Devi, T. S., Das, D., Ansari, R. A., Rizvi, R., Sumbul, A., & Mahmood, I. (2020). Role of organic additives in the sustainable management of phytoparasitic nematodes. In *Management of Phytonematodes: Recent Advances and Future Challenges* (pp. 279–295).

Di, X., Gomila, J. O., & Takken, F. L. (2017). Involvement of salicylic acid, ethylene and jasmonic acid signalling pathways in the susceptibility of tomato to *Fusarium oxysporum*. *Molecular Plant Pathology*, *18*(7), 1024–1035.

Díez-Pascual, A. M. (2021). State of the art in the antibacterial and antiviral applications of carbon-based polymeric nanocomposites. *International Journal of Molecular Sciences*, *22*(19), 10511.

Dikshit, P. K., Kumar, J., Das, A. K., Sadhu, S., Sharma, S., Singh, S., ... & Kim, B. S. (2021). Green synthesis of metallic nanoparticles: Applications and limitations. *Catalysts*, *11*(8), 902.

Din, M. I., & Rehan, R. (2017). Synthesis, characterization, and applications of copper nanoparticles. *Analytical Letters*, *50*(1), 50–62.

Dizaj, S. M., Mennati, A., Jafari, S., Khezri, K., & Adibkia, K. (2015). Antimicrobial activity of carbon-based nanoparticles. *Advanced Pharmaceutical Bulletin*, *5*(1), 19.

Djiwanti, S. R., & Kaushik, S. (2019). Nanopesticide: Future application of nanomaterials in plant protection. In *Plant Nanobionics: Volume 2, Approaches in Nanoparticles, Biosynthesis, and Toxicity* (pp. 255–298).

Dong, S., Qu, M., Rui, Q., & Wang, D. (2018). Combinational effect of titanium dioxide nanoparticles and nanopolystyrene particles at environmentally relevant concentrations on nematode Caenorhabditis elegans. *Ecotoxicology and Environmental Safety*, *161*, 444–450.

Duan, H., Wang, D., & Li, Y. (2015). Green chemistry for nanoparticle synthesis. *Chemical Society Reviews*, *44*(16), 5778–5792.

Durán, N., Durán, M., De Jesus, M. B., Seabra, A. B., Fávaro, W. J., & Nakazato, G. (2016). Silver nanoparticles: A new view on mechanistic aspects on antimicrobial activity. *Nanomedicine: Nanotechnology, Biology and Medicine*, *12*(3), 789–799.

Durán, N., Marcato, P. D., Durán, M., Yadav, A., Gade, A., & Rai, M. (2011). Mechanistic aspects in the biogenic synthesis of extracellular metal nanoparticles by peptides, bacteria, fungi, and plants. *Applied Microbiology and Biotechnology*, *90*, 1609–1624.

Ealia, S. A. M., & Saravanakumar, M. P. (2017, November). A review on the classification, characterisation, synthesis of nanoparticles and their application. In *IOP Conference Series: Materials Science and Engineering* (Vol. 263, No. 3, p. 032019). IOP Publishing.

Elangovan, R., Philip, L., & Chandraraj, K. (2010). Hexavalent chromium reduction by free and immobilized cell-free extract of *Arthrobacter rhombi*-RE. *Applied Biochemistry and Biotechnology*, *160*, 81–97.

Elbeshehy, E. K., Elazzazy, A. M., & Aggelis, G. (2015). Silver nanoparticles synthesis mediated by new isolates of *Bacillus* spp., nanoparticle characterization and their activity against Bean Yellow Mosaic Virus and human pathogens. *Frontiers in Microbiology*, *6*, 453.

Elmer, W., & White, J. C. (2018). The future of nanotechnology in plant pathology. *Annual Review of Phytopathology*, *56*, 111–133.

El-Naggar, N. E. A., Saber, W. I., Zweil, A. M., & Bashir, S. I. (2022). An innovative green synthesis approach of chitosan nanoparticles and their inhibitory activity against phytopathogenic *Botrytis cinerea* on strawberry leaves. *Scientific Reports*, *12*(1), 3515.

Erkoc, P., & Ulucan-Karnak, F. (2021). Nanotechnology-based antimicrobial and antiviral surface coating strategies. *Prosthesis*, *3*(1), 25–52.

Estelrich, J., Escribano, E., Queralt, J., & Busquets, M. A. (2015). Iron oxide nanoparticles for magnetically-guided and magnetically-responsive drug delivery. *International Journal of Molecular Sciences*, *16*(4), 8070–8101.

Fabiyi, O. A., Alabi, R. O., & Ansari, R. A. (2020). Nanoparticles' synthesis and their application in the management of phytonematodes: An overview. In *Management of Phytonematodes: Recent Advances and Future Challenges* (pp. 125–140).

Fabiyi, O. A., Ansari, R. A., Rizvi, R., Sumbul, A., & Mahmood, I. (2024). Influence of arbuscular mycorrhizal fungi on soil health amelioration and plant fitness under hostile environment. In *Mycorrhizal Symbiosis and Agroecosystem Restoration* (pp. 227–248). Singapore, Springer Nature Singapore.

Fahmy, S. A., Preis, E., Bakowsky, U., & Azzazy, H. M. E. S. (2020). Platinum nanoparticles: Green synthesis and biomedical applications. *Molecules*, *25*(21), 4981.

Falcaro, P., Ricco, R., Yazdi, A., Imaz, I., Furukawa, S., Maspoch, D., ... & Doonan, C. J. (2016). Application of metal and metal oxide nanoparticles@ MOFs. *Coordination Chemistry Reviews*, *307*, 237–254.

Foresti, M. L., Vázquez, A., & Boury, B. (2017). Applications of bacterial cellulose as precursor of carbon and composites with metal oxide, metal sulfide and metal nanoparticles: A review of recent advances. *Carbohydrate Polymers*, *157*, 447–467.

Gadirova, E. M. (2019). Photochemical degradation of phenol in the presence of titanium dioxide nanoparticles. *Известия Вузов. Прикладная Химия И Биотехнология*, *9*(2 (29)), 176–182.

Gal, A., Weiner, S., & Addadi, L. (2015). A perspective on underlying crystal growth mechanisms in biomineralization: Solution mediated growth versus nanosphere particle accretion. *CrystEngComm*, *17*(13), 2606–2615.

Garg, A., Sharma, G. S., Goyal, A. K., Ghosh, G., Si, S. C., & Rath, G. (2020). Recent advances in topical carriers of anti-fungal agents. *Heliyon*, *6*(8). e04663.

Gemishev, O., Panayotova, M., Gicheva, G., & Mintcheva, N. (2022). Green synthesis of stable spherical monodisperse silver nanoparticles using a cell-free extract of *Trichoderma reesei*. *Materials*, *15*(2), 481.

Gholami-Shabani, M., Shams-Ghahfarokhi, M., Gholami-Shabani, Z., Akbarzadeh, A., Riazi, G., Ajdari, S., ... & Razzaghi-Abyaneh, M. (2015). Enzymatic synthesis of gold nanoparticles using sulfite reductase purified from *Escherichia coli*: A green eco-friendly approach. *Process Biochemistry*, *50*(7), 1076–1085.

Goddard, Z. R., Marín, M. J., Russell, D. A., & Searcey, M. (2020). Active targeting of gold nanoparticles as cancer therapeutics. *Chemical Society Reviews*, *49*(23), 8774–8789.

Gómez-Gómez, B., Arregui, L., Serrano, S., Santos, A., Pérez-Corona, T., & Madrid, Y. (2019). Unravelling mechanisms of bacterial quorum sensing disruption by metal-based nanoparticles. *Science of The Total Environment*, *696*, 133869.

González-García, Y., Cadenas-Pliego, G., Alpuche-Solís, Á. G., Cabrera, R. I., & Juárez-Maldonado, A. (2022). Effect of carbon-based nanomaterials on *Fusarium* wilt in tomato. *Scientia Horticulturae*, *291*, 110586.

Guilger-Casagrande, M., & Lima, R. D. (2019). Synthesis of silver nanoparticles mediated by fungi: A review. *Frontiers in Bioengineering and Biotechnology*, *7*, 287.

Gupta, N., Upadhyaya, C. P., Singh, A., Abd-Elsalam, K. A., & Prasad, R. (2018). Applications of silver nanoparticles in plant protection. *Nanobiotechnology Applications in Plant Protection*, 247–265.

Gupta, T. K., Budarapu, P. R., Chappidi, S. R., YB, S. S., Paggi, M., & Bordas, S. P. (2019). Advances in carbon based nanomaterials for bio-medical applications. *Current Medicinal Chemistry*, *26*(38), 6851–6877.

Han, Y., Han, Y., Du, G., Zhang, T., Guo, Q., Yang, H., ... & Xu, Y. (2021). Physiological effect of colloidal carbon quantum dots on *Bursaphelenchus xylophilus*. *RSC Advances*, *11*(11), 6212–6220.

Hebbalalu, D., Lalley, J., Nadagouda, M. N., & Varma, R. S. (2013). Greener techniques for the synthesis of silver nanoparticles using plant extracts, enzymes, bacteria, biodegradable polymers, and microwaves. *ACS Sustainable Chemistry & Engineering*, *1*(7), 703–712.

Hernández-Hernández, H., Juárez-Maldonado, A., Benavides-Mendoza, A., Ortega-Ortiz, H., Cadenas-Pliego, G., Sánchez-Aspeytia, D., & González-Morales, S. (2018). Chitosan-PVA and copper nanoparticles improve growth and overexpress the SOD and JA genes in tomato plants under salt stress. *Agronomy*, *8*(9), 175.

Hu, C. C., Wu, G. H., Hua, T. E., Wagner, O. I., & Yen, T. J. (2018). Uptake of TiO_2 nanoparticles into C. elegans neurons negatively affects axonal growth and worm locomotion behaviour. *ACS Applied Materials & Interfaces*, *10*(10), 8485–8495.

Huliaieva, H., Pasichnyk, L., Tokovenko, I., Kalinichenko, A., Patyka, V., Bohdan, M., ... & Maksin, V. (2021). Influence of citrate nanoparticles on photochemical activity, resistance to pathogens and productivity of wheat. *Agriculture and Forestry= Poljoprivreda i šumarstvo*, *67*(4), 15–33.

Hulkoti, N. I., & Taranath, T. C. (2014). Biosynthesis of nanoparticles using microbes—A review. *Colloids and Surfaces B: Biointerfaces*, *121*, 474–483.

Husen, A. (2020). Carbon-based nanomaterials and their interactions with agricultural crops. In *Nanomaterials for Agriculture and Forestry Applications* (pp. 199–218). Elsevier.

Ibrahim, D. S. S., Elderiny, M. M., Ansari, R. A., Rizvi, R., Sumbul, A., & Mahmood, I. (2020). Role of *Trichoderma* spp. in the management of plant-parasitic nematodes infesting important crops. In *Management of Phytonematodes: Recent Advances and Future Challenges* (pp. 259–278).

Ibrahim, D. S. S., Mostafa, M., Abdellatif, A. A., & Ansari, R. A. (2022). Sustainable management of soil-borne plant pathogens. In *Earth Systems Protection and Sustainability: Volume 1* (pp. 35–56). Cham: Springer International Publishing.

Imada, K., Sakai, S., Kajihara, H., Tanaka, S., & Ito, S. (2016). Magnesium oxide nanoparticles induce systemic resistance in tomato against bacterial wilt disease. *Plant Pathology*, *65*(4), 551–560.

Iravani, S., & Varma, R. S. (2020). Bacteria in heavy metal remediation and nanoparticle biosynthesis. *ACS Sustainable Chemistry & Engineering*, *8*(14), 5395–5409.

Jacob, J. J., & Suthindhiran, K. (2016). Magnetotactic bacteria and magnetosomes–Scope and challenges. *Materials Science and Engineering: C, 68*, 919–928.

Jain, A., Ranjan, S., Dasgupta, N., & Ramalingam, C. (2018). Nanomaterials in food and agriculture: An overview on their safety concerns and regulatory issues. *Critical Reviews in Food Science and Nutrition, 58*(2), 297–317.

Jameel, M. S., Aziz, A. A., & Dheyab, M. A. (2020). Green synthesis: Proposed mechanism and factors influencing the synthesis of platinum nanoparticles. *Green Processing and Synthesis, 9*(1), 386–398.

Jeevanandam, J., Chan, Y. S., & Danquah, M. K. (2016). Biosynthesis of metal and metal oxide nanoparticles. *ChemBioEng Reviews, 3*(2), 55–67.

John, M. S., Nagoth, J. A., Zannotti, M., Giovannetti, R., Mancini, A., Ramasamy, K. P., ... & Pucciarelli, S. (2021). Biogenic synthesis of copper nanoparticles using bacterial strains isolated from an *Antarctic consortium* associated to a psychrophilic marine ciliate: Characterization and potential application as antimicrobial agents. *Marine Drugs, 19*(5), 263.

Jones, S. R., Wilson, T. D., Brown, M. E., Rahn-Lee, L., Yu, Y., Fredriksen, L. L., ... & Chang, M. C. (2015). Genetic and biochemical investigations of the role of MamP in redox control of iron biomineralization in *Magnetospirillum magneticum. Proceedings of the National Academy of Sciences, 112*(13), 3904–3909.

Jordan, J. T., Singh, K. P., & Cañas-Carrell, J. E. (2018). Carbon-based nanomaterials elicit changes in physiology, gene expression, and epigenetics in exposed plants: A review. *Current Opinion in Environmental Science & Health, 6*, 29–35.

Kalabegishvili, T. L., Murusidze, I. G., Prangishvili, D. A., Kvachadze, L. I., Kirkesali, E. I., Rcheulishvili, A. N., ... & Pavlov, S. S. (2015). Silver nanoparticles in *Sulfolobus islandicus* biomass for technological applications. *Advanced Science, Engineering and Medicine, 7*(9), 797–804.

Kalimuthu, K., Babu, R. S., Venkataraman, D., Bilal, M., & Gurunathan, S. (2008). Biosynthesis of silver nanocrystals by *Bacillus licheniformis. Colloids And Surfaces B: Biointerfaces, 65*(1), 150–153.

Kamaluddin, Z. N. (2018). Study influence of silver nanoparticles on growth rate of plant pathogens: *Rhizoctonia solani, Alternaria alternata, Fusarium solani, Fusarium oxysporum, Aspergillus terreus, Aspergillus parasiticus*, and Bio agent *Trichoderma harzianum. Journal of University of Babylon for Pure and Applied Sciences, 26*(5), 26–35.

Kannan, M., Bojan, N., Swaminathan, J., Zicarelli, G., Hemalatha, D., Zhang, Y., ... & Faggio, C. (2023). Nanopesticides in agricultural pest management and their environmental risks: A review. *International Journal of Environmental Science and Technology*, 1–26.

Kashyap, P. L., Kumar, S., Jasrotia, P., Singh, D. P., & Singh, G. P. (2019). Nanosensors for plant disease diagnosis: Current understanding and future perspectives. *Nanoscience for Sustainable Agriculture*, 189–205.

Kaur, P., Thakur, R., & Chaudhury, A. (2016). Biogenesis of copper nanoparticles using peel extract of *Punica granatum* and their antimicrobial activity against opportunistic pathogens. *Green Chemistry Letters and Reviews, 9*(1), 33–38.

Kaveh, R., Li, Y. S., Ranjbar, S., Tehrani, R., Brueck, C. L., & Van Aken, B. (2013). Changes in *Arabidopsis thaliana* gene expression in response to silver nanoparticles and silver ions. *Environmental Science & Technology, 47*(18), 10637–10644.

Keshari, N., Kranti, K. V. V. S. K., Gunda, N. K., & Ansari, R. A. (2024). Arbuscular mycorrhizal fungi: A potential agent for phytonematodes management in diverse agro-climatic zones. In *Mycorrhizal Symbiosis and Agroecosystem Restoration* (pp. 147–169). Singapore: Springer Nature Singapore.

Khalil, A. T., Ovais, M., Iqbal, J., Ali, A., Ayaz, M., Abbas, M., ... & Devkota, H. P. (2022). Microbes-mediated synthesis strategies of metal nanoparticles and their potential role in cancer therapeutics. In *Seminars İn Cancer Biology* 86(3), 693–705.

Khalil, M. S., Abd El-Aziz, M. H., & Selim, R. E. S. (2022). Physiological and morphological response of tomato plants to nano-chitosan used against bio-stress induced by root-knot nematode (*Meloidogyne incognita*) and Tobacco mosaic tobamovirus (TMV). *European Journal of Plant Pathology, 163*(4), 799–812.

Khan, M. A. U., Arshad, H., & Majid, A. (2022). QDs, plant diseases and potential risks. In *Quantum Dots for Plant Systems* (pp. 161–190). Cham: Springer International Publishing.

Khan, M. R., & Rizvi, T. F. (2014). Nanotechnology: Scope and application in plant disease management. *Plant Pathology Journal, 13*(3), 214–231.

Khan, M., Siddiqui, Z. A., Parveen, A., Khan, A. A., Moon, I. S., & Alam, M. (2022). Elucidating the role of silicon dioxide and titanium dioxide nanoparticles in mitigating the disease of the eggplant caused by *Phomopsis vexans*, *Ralstonia solanacearum*, and root-knot nematode *Meloidogyne incognita*. *Nanotechnology Reviews*, *11*(1), 1606–1619.

Khan, S. A., & Lee, C. S. (2020). Green biological synthesis of nanoparticles and their biomedical applications. *Applications of Nanotechnology for Green Synthesis*, 247–280.

Khanna, K., Sharma, N., Ohri, P., & Bhardwaj, R. (2022). Emerging trends of nanoparticles in sustainable agriculture: Current and future perspectives. *Plant and Nanoparticles*, 1–52.

Khodakovskaya, M., & Marmiroli, M. (2023). Polymeric nanoparticles for sustainable plant agriculture and food industry. *Frontiers in Plant Science*, *14*. 1183938.

Kitching, M., Ramani, M., & Marsili, E. (2015). Fungal biosynthesis of gold nanoparticles: Mechanism and scale up. *Microbial Biotechnology*, *8*(6), 904–917.

Kotliarevski, L., Cohen, R., Ramakrishnan, J., Wu, S., Mani, K. A., Amar-Feldbaum, R., ... & Mechrez, G. (2022). Individual coating of entomopathogenic nematodes with titania (TiO_2) nanoparticles based on oil-in-water Pickering emulsion: A new formulation for biopesticides. *Journal of Agricultural and Food Chemistry*, *70*(42), 13518–13527.

Kouzai, Y., Kimura, M., Yamanaka, Y., Watanabe, M., Matsui, H., Yamamoto, M., ... & Noutoshi, Y. (2016). Expression profiling of marker genes responsive to the defence-associated phytohormones salicylic acid, jasmonic acid and ethylene in *Brachypodium distachyon*. *BMC Plant Biology*, *16*, 1–11.

Kumar, A., & Dixit, C. K. (2017). Methods for characterization of nanoparticles. In *Advances in Nanomedicine for The Delivery of Therapeutic Nucleic Acids* (pp. 43–58). Woodhead Publishing.

Kumar, A., Choudhary, A., Kaur, H., Guha, S., Mehta, S., & Husen, A. (2022). Potential applications of engineered nanoparticles in plant disease management: A critical update. *Chemosphere*, *295*, 133798.

Kumar, H., Bhardwaj, K., Nepovimova, E., Kuča, K., Singh Dhanjal, D., Bhardwaj, S., ... & Kumar, D. (2020). Antioxidant functionalized nanoparticles: A combat against oxidative stress. *Nanomaterials*, *10*(7), 1334.

Kuroda, K., & Ueda, M. (2010). Engineering of microorganisms towards recovery of rare metal ions. *Applied Microbiology and Biotechnology*, *87*, 53–60.

Lahiri, D., Nag, M., Sheikh, H. I., Sarkar, T., Edinur, H. A., Pati, S., & Ray, R. R. (2021). Microbiologically-synthesized nanoparticles and their role in silencing the biofilm signalling cascade. *Frontiers in Microbiology*, *12*, 636588.

Lai, F., Wissing, S. A., Müller, R. H., & Fadda, A. M. (2006). *Artemisia arborescens* L essential oil-loaded solid lipid nanoparticles for potential agricultural application: Preparation and characterization. *Aaps Pharmscitech*, *7*, E10–E18.

Lallo da Silva, B., Abuçafy, M. P., Berbel Manaia, E., Oshiro Junior, J. A., Chiari-Andréo, B. G., Pietro, R. C. R., & Chiavacci, L. A. (2019). Relationship between structure and antimicrobial activity of zinc oxide nanoparticles: An overview. *International Journal of Nanomedicine*, *14*, 9395–9410.

Lateef, A., Darwesh, O. M., & Matter, I. A. (2021). Microbial nanobiotechnology: The melting pot of microbiology, microbial technology and nanotechnology. In *Microbial Nanobiotechnology: Principles and Applications* (pp. 1–19).

Laurent, S., Forge, D., Port, M., Roch, A., Robic, C., Vander Elst, L., & Muller, R. N. (2008). Magnetic iron oxide nanoparticles: Synthesis, stabilization, vectorization, physicochemical characterizations, and biological applications. *Chemical Reviews*, *108*(6), 2064–2110.

Lemire, J. A., Harrison, J. J., & Turner, R. J. (2013). Antimicrobial activity of metals: Mechanisms, molecular targets and applications. *Nature Reviews Microbiology*, *11*(6), 371–384.

Li, M., Noriega-Trevino, M. E., Nino-Martinez, N., Marambio-Jones, C., Wang, J., Damoiseaux, R., ... & Hoek, E. M. (2011). Synergistic bactericidal activity of Ag-TiO_2 nanoparticles in both light and dark conditions. *Environmental Science & Technology*, *45*(20), 8989–8995.

Li, X., Xu, H., Chen, Z. S., & Chen, G. (2011). Biosynthesis of nanoparticles by microorganisms and their applications. *Journal of Nanomaterials*, *2011*, 1–16.

Li, Y., Zhang, P., Li, M., Shakoor, N., Adeel, M., Zhou, P., ... & Rui, Y. (2023). Application and mechanisms of metal-based nanoparticles in the control of bacterial and fungal crop diseases. *Pest Management Science*, *79*(1), 21–36.

Liu, Z., Zhang, M., Han, X., Xu, H., Zhang, B., Yu, Q., & Li, M. (2016). TiO_2 nanoparticles cause cell damage independent of apoptosis and autophagy by impairing the ROS-scavenging system in Pichia pastoris. *Chemico-Biological Interactions*, *252*, 9–18.

López-Sagaseta, J., Malito, E., Rappuoli, R., & Bottomley, M. J. (2016). Self-assembling protein nanoparticles in the design of vaccines. *Computational and Structural Biotechnology Journal*, *14*, 58–68.

Lum, P. T., Sekar, M., Gan, S. H., Bonam, S. R., & Shaikh, M. F. (2021). Protective effect of natural products against Huntington's disease: An overview of scientific evidence and understanding their mechanism of action. *ACS Chemical Neuroscience*, *12*(3), 391–418.

Ma, H., Brennan, A., & Diamond, S. A. (2012). Photocatalytic reactive oxygen species production and phototoxicity of titanium dioxide nanoparticles are dependent on the solar ultraviolet radiation spectrum. *Environmental Toxicology and Chemistry*, *31*(9), 2099–2107.

Ma, J. L., Yin, B. C., Wu, X., & Ye, B. C. (2017). Simple and cost-effective glucose detection based on carbon nanodots supported on silver nanoparticles. *Analytical Chemistry*, *89*(2), 1323–1328.

Madalosso, R. G., Verdi, C. M., Souza, M. E., Cunha, J. A., Santos, P. S., Balardin, R. S., ... & Santos, R. C. V. (2017). In vitro and in vivo effects of *Melaleuca alternifolia* nanoparticles in *Meloidogyne* sp. *Journal of Asia-Pacific Entomology*, *20*(4), 1344–1349.

Mahmood, I., Rizvi, R., Sumbul, A., & Ansari, R. A. (2019). Potential role of plant growth promoting rhizobacteria in alleviation of biotic stress. In *Plant Health Under Biotic Stress: Volume 2: Microbial Interactions* (pp. 177–188).

Mahmoud, W. M., Abdelmoneim, T. S., & Elazzazy, A. M. (2016). The impact of silver nanoparticles produced by *Bacillus pumilus* as antimicrobial and nematicide. *Frontiers in Microbiology*, *7*, 1746.

Maity, D., Gupta, U., & Saha, S. (2022). Biosynthesized metal oxide nanoparticles for sustainable agriculture: Next-generation nanotechnology for crop production, protection and management. *Nanoscale*, *14*(38), 13950–13989.

Malandrakis, A. A., Kavroulakis, N., & Chrysikopoulos, C. V. (2022). Metal nanoparticles against fungicide resistance: Alternatives or partners? *Pest Management Science*, *78*(10), 3953–3956.

Marslin, G., Sheeba, C. J., & Franklin, G. (2017). Nanoparticles alter secondary metabolism in plants via ROS burst. *Frontiers in Plant Science*, *8*, 832.

McDaniel, A., Tonyali, B., Yucel, U., & Trinetta, V. (2019). Formulation and development of lipid nanoparticle antifungal packaging films to control postharvest disease. *Journal of Agriculture and Food Research*, *1*, 100013.

Mehta, M. R., Biradar, S. P., Mahajan, H. P., Bankhele, R. R., & Hivrale, A. U. (2022). Chitosan and chitosan-based nanoparticles in horticulture: Past, present and future prospects. In *Role of Chitosan and Chitosan-Based Nanomaterials in Plant Sciences* (pp. 453–474).

Meraz-Dávila, S., Pérez-García, C. E., & Feregrino-Perez, A. A. (2021). Challenges and advantages of electrospun nanofibers in agriculture: A review. *Materials Research Express*, *8*(4), 042001.

Mhamdi, A., & Van Breusegem, F. (2018). Reactive oxygen species in plant development. *Development*, *145*(15), dev164376.

Modena, M. M., Rühle, B., Burg, T. P., & Wuttke, S. (2019). Nanoparticle characterization: What to measure? *Advanced Materials*, *31*(32), 1901556.

Moghimipour, E., Abedishirehjin, S., Baghbadorani, M. A., & Handali, S. (2021). Bacteria and Archaea: A new era of cancer therapy. *Journal of Controlled Release*, *338*, 1–7.

Mohamed, A. A., Abu-Elghait, M., Ahmed, N. E., & Salem, S. S. (2021). Eco-friendly mycogenic synthesis of ZnO and CuO nanoparticles for in vitro antibacterial, antibiofilm, and antifungal applications. *Biological Trace Element Research*, *199*, 2788–2799.

Mohamed, M. A., & Abd-Elsalam, K. A. (2019). Magnetic nanoparticles: A unique gene delivery system in plant science. *Magnetic Nanostructures: Environmental and Agricultural Applications*, 95–108.

Mohamed, M. A., Hashim, A. F., Alghuthaymi, M. A., & Abd-Elsalam, K. A. (2018). Nano-carbon: Plant growth promotion and protection. *Nanobiotechnology Applications in Plant Protection*, 155–188.

Mohamed, M. A., Mohamed, A. E. M. A., & Abd-Elsalam, K. A. (2019). Magnetic nanoparticles in plant protection: Promises and risks. *Magnetic Nanostructures: Environmental and Agricultural Applications*, 225–246.

Mohamed, Y. M. A., Osman, S. A., Elshahawy, I. E., Soliman, G. M., & Ahmed, A. (2021). Charcoal rot and root-knot nematode control on faba bean by photosynthesized colloidal silver nanoparticles using bioactive compounds from *Moringa oleifera* leaf extract. *Journal of Plant Protection Research, 61*, 414–429.

Mouniga, R., Anita, B., Shanthi, A., Lakshmanan, A., & Karthikeyan, G. (2022). Phenol and antioxidant enzymatic activity in root knot nematode, *Meloidogyne incognita* infected tomato plants treated with chitosan nanoparticles. *The Pharma Innovation Journal, 11*(4), 241–245.

Mukherjee, P., Ahmad, A., Mandal, D., Senapati, S., Sainkar, S. R., Khan, M. I., ... & Sastry, M. (2001). Fungus-mediated synthesis of silver nanoparticles and their immobilization in the mycelial matrix: A novel biological approach to nanoparticle synthesis. *Nano Letters, 1*(10), 515–519.

Mukherji, S., Bharti, S., Shukla, G., & Mukherji, S. (2018). Synthesis and characterization of size-and shape-controlled silver nanoparticles. *Physical Sciences Reviews, 4*(1), 20170082.

Narendrakumar, G., Namasivayam, S. K. R., Manikanta, M., Saha, M., Dasgupta, T., Divyasri, N., ... & Preethi, T. V. (2018). Enhancement of biocontrol potential of biocompatible bovine serum albumin (BSA) based protein nanoparticles loaded bacterial chitinase against major plant pathogenic fungi *Alternaria alternata*. *Biocatalysis and Agricultural Biotechnology, 15*, 219–228.

Niehl, A., Appaix, F., Boscá, S., Van der Sanden, B., Nicoud, J. F., Bolze, F., & Heinlein, M. (2016). Fluorescent tobacco mosaic virus-derived bio-nanoparticles for intravital two-photon imaging. *Frontiers in Plant Science, 6*, 1244.

Noorizadeh, S., Golmohammadi, M., Bagheri, A., & Bertaccini, A. (2022). Citrus industry: Phytoplasma-associated diseases and related challenges for Asia, America and Africa. *Crop Protection, 151*, 105822.

Nurdin, M., Watoni, A. H., Yanti, N. A., Lilis, L., Maulidiyah, M., Irwan, I., ... & Arham, Z. (2023). Antifungal properties of TiO_2 nanoparticle as prevention wilt disease *Fusarium oxysporum* under UV light. In *AIP Conference Proceedings, 2683*(1), 0300035.

Oh, J. W., Chun, S. C., & Chandrasekaran, M. (2019). Preparation and in vitro characterization of chitosan nanoparticles and their broad-spectrum antifungal action compared to antibacterial activities against phytopathogens of tomato. *Agronomy, 9*(1), 21.

Ojha, S., Singh, D., Sett, A., Chetia, H., Kabiraj, D., & Bora, U. (2018). Nanotechnology in crop protection. In *Nanomaterials in Plants, Algae, and Microorganisms* (pp. 345–391). Academic Press.

Osonga, F. J., Akgul, A., Yazgan, I., Akgul, A., Ontman, R., Kariuki, V. M., ... & Sadik, O. A. (2018). Flavonoid-derived anisotropic silver nanoparticles inhibit growth and change the expression of virulence genes in *Escherichia coli* SM10. *RSC Advances, 8*(9), 4649–4661.

Ouda, S. M. (2014). Antifungal activity of silver and copper nanoparticles on two plant pathogens, *Alternaria alternata* and *Botrytis cinerea*. *Research Journal of Microbiology, 9*(1), 34.

Ovais, M., Khalil, A. T., Ayaz, M., Ahmad, I., Nethi, S. K., & Mukherjee, S. (2018). Biosynthesis of metal nanoparticles via microbial enzymes: A mechanistic approach. *International Journal of Molecular Sciences, 19*(12), 4100.

Owolade, O., & Ogunleti, D. (2008). Effects of titanium dioxide on the diseases, development and yield of edible cowpea. *Journal of Plant Protection Research, 48*(3), 329–336.

Panáček, A., Kvítek, L., Smékalová, M., Večeřová, R., Kolář, M., Röderová, M., ... & Zbořil, R. (2018). Bacterial resistance to silver nanoparticles and how to overcome it. *Nature Nanotechnology, 13*(1), 65–71.

Pardo, J., Peng, Z., & Leblanc, R. M. (2018). Cancer targeting and drug delivery using carbon-based quantum dots and nanotubes. *Molecules, 23*(2), 378.

Park, J. A., Kim, S. B., Lee, C. G., Lee, S. H., & Choi, J. W. (2014). Adsorption of bacteriophage MS2 to magnetic iron oxide nanoparticles in aqueous solutions. *Journal of Environmental Science and Health, Part A, 49*(10), 1116–1124.

Park, T. J., Lee, K. G., & Lee, S. Y. (2016). Advances in microbial biosynthesis of metal nanoparticles. *Applied Microbiology and Biotechnology, 100*, 521–534.

Parveen, A., & Siddiqui, Z. A. (2022). Foliar spray and seed priming of titanium dioxide nanoparticles and their impact on the growth of tomato, defence enzymes and some bacterial and fungal diseases. *Archives of Phytopathology and Plant Protection, 55*(5), 527–548.

Parveen, K., Banse, V., & Ledwani, L. (2016, April). Green synthesis of nanoparticles: Their advantages and disadvantages. In *AIP Conference Proceedings* (Vol. 1724, No. 1, p. 020048). AIP Publishing LLC.

Patel, A., Tiwari, S., Parihar, P., Singh, R., & Prasad, S. M. (2019). Carbon nanotubes as plant growth regulators: Impacts on growth, reproductive system, and soil microbial community. In *Nanomaterials in Plants, Algae and Microorganisms* (pp. 23–42). Elsevier Inc. Academic Press.

Patil, S., & Chandrasekaran, R. (2020). Biogenic nanoparticles: A comprehensive perspective in synthesis, characterization, application and its challenges. *Journal of Genetic Engineering and Biotechnology*, *18*, 1–23.

Patil, S., Ansari, R. A., Gurikar, C., Bhavan, S., & Hariprasad, K. (2024). Propagation of mycorrhizal fungi and their strength in plant health ameliorations. In *Mycorrhizal Symbiosis and Agroecosystem Restoration* (pp. 77–98). Singapore: Springer Nature Singapore.

Pedone, D., Moglianetti, M., De Luca, E., Bardi, G., & Pompa, P. P. (2017). Platinum nanoparticles in nanobiomedicine. *Chemical Society Reviews*, *46*(16), 4951–4975.

Pietroiusti, A., Stockmann-Juvala, H., Lucaroni, F., & Savolainen, K. (2018). Nanomaterial exposure, toxicity, and impact on human health. *Wiley Interdisciplinary Reviews: Nanomedicine and Nanobiotechnology*, *10*(5), e1513.

Pinilla, C. M. B., Lopes, N. A., & Brandelli, A. (2021). Lipid-based nanostructures for the delivery of natural antimicrobials. *Molecules*, *26*(12), 3587.

Ponnuvel, S., Subramanian, B., & Ponnuraj, K. (2015). Conformational change results in loss of enzymatic activity of jack bean urease on its interaction with silver nanoparticle. *The Protein Journal*, *34*, 329–337.

Prakash, A., Sharma, S., Ahmad, N., Ghosh, A., & Sinha, P. (2010). Bacteria mediated extracellular synthesis of metallic nanoparticles. *International Research Journal of the Biotechnology*, *1*(5), 071–079.

Prasad, R., Gupta, N., Kumar, M., Kumar, V., Wang, S., & Abd-Elsalam, K. A. (2017). Nanomaterials act as plant defence mechanism. *Nanotechnology: Food and Environmental Paradigm*, 253–269.

Prasad, R., Kumar, V., & Prasad, K. S. (2014). Nanotechnology in sustainable agriculture: Present concerns and future aspects. *African Journal of Biotechnology*, *13*(6), 705–713.

Prasad, R., Pandey, R., & Barman, I. (2016). Engineering tailored nanoparticles with microbes: quo vadis? *Wiley Interdisciplinary Reviews: Nanomedicine and Nanobiotechnology*, *8*(2), 316–330.

Purohit, J., Chattopadhyay, A., & Singh, N. K. (2019). Green synthesis of microbial nanoparticle: Approaches to application. In *Microbial Nanobionics: Volume 2, Basic Research and Applications* (pp. 35–60).

Qin, S., Xiao, W., Zhou, C., Pu, Q., Deng, X., Lan, L., ... & Wu, M. (2022). *Pseudomonas aeruginosa*: Pathogenesis, virulence factors, antibiotic resistance, interaction with host, technology advances and emerging therapeutics. *Signal Transduction and Targeted Therapy*, *7*(1), 199.

Qin, X., Xiang, X., Sun, X., Ni, H., & Li, L. (2016). Preparation of nanoscale *Bacillus thuringiensis* chitinases using silica nanoparticles for nematicide delivery. *International Journal of Biological Macromolecules*, *82*, 13–21.

Rad, F., Mohsenifar, A., Tabatabaei, M., Safarnejad, M. R., Shahryari, F., Safarpour, H., ... & Fotokian, M. (2012). Detection of *Candidatus Phytoplasma* aurantifolia with a quantum dots fret-based biosensor. *Journal of Plant Pathology*, *94*(3), 525–534.

Rahman, M. A., Abul Barkat, H., Harwansh, R. K., & Deshmukh, R. (2022). Carbon-based nanomaterials: Carbon nanotubes, graphene, and fullerenes for the control of burn infections and wound healing. *Current Pharmaceutical Biotechnology*, *23*(12), 1483–1496.

Rai, M., Ingle, A. P., Pandit, R., Paralikar, P., Shende, S., Gupta, I., ... & da Silva, S. S. (2018). Copper and copper nanoparticles: Role in management of insect-pests and pathogenic microbes. *Nanotechnology Reviews*, *7*(4), 303–315.

Rajeshkumar, S., & Naik, P. (2018). Synthesis and biomedical applications of cerium oxide nanoparticles–A review. *Biotechnology Reports*, *17*, 1–5.

Rajeshkumar, S., & Sivapriya, D. (2020). Fungus-mediated nanoparticles: Characterization and biomedical advances. *Nanoparticles in Medicine*, 185–199.

Rajiv, P., Rajeshwari, S., & Venckatesh, R. (2013). Bio-Fabrication of zinc oxide nanoparticles using leaf extract of *Parthenium hysterophorus* L. and its size-dependent antifungal activity against plant fungal pathogens. *Spectrochimica Acta Part A: Molecular and Biomolecular Spectroscopy*, *112*, 384–387.

Rajora, N., Kaushik, S., Jyoti, A., & Kothari, S. L. (2016). Rapid synthesis of silver nanoparticles by *Pseudomonas stutzeri* isolated from textile soil under optimised conditions and evaluation of their anti-microbial and cytotoxicity properties. *IET Nanobiotechnology*, *10*(6), 367–373.

Rao, G. P., & Reddy, M. G. (2020). Overview of yield losses due to plant viruses. In *Applied Plant Virology* (pp. 531–562). Academic Press.

Ravindran, A., Chandran, P., & Khan, S. S. (2013). Biofunctionalized silver nanoparticles: Advances and prospects. *Colloids and Surfaces B: Biointerfaces*, *105*, 342–352.

Resham, S., Khalid, M., & Kazi, A. G. (2015). Nanobiotechnology in agricultural development. In *PlantOmics: The Omics of Plant Science* (pp. 683–698).

Roh, J. Y., Park, Y. K., Park, K., & Choi, J. (2010). Ecotoxicological investigation of CeO_2 and TiO_2 nanoparticles on the soil nematode *Caenorhabditis elegans* using gene expression, growth, fertility, and survival as endpoints. *Environmental Toxicology and Pharmacology*, *29*(2), 167–172.

Rojas, J., Bedoya, M., & Ciro, Y. (2015). Current trends in the production of cellulose nanoparticles and nanocomposites for biomedical applications. *Cellulose-Fundamental Aspects and Current Trends*, *32*, 193–228.

Saallah, S., & Lenggoro, I. W. (2018). Nanoparticles carrying biological molecules: Recent advances and applications. *KONA Powder and Particle Journal*, *35*, 89–111.

Safarnejad, M. R., Samiee, F., Tabatabie, M., & Mohsenifar, A. (2017). Development of quantum dot-based nanobiosensors against citrus tristeza virus (CTV). *Sensors & Transducers*, *213*(6), 54.

Saleh, T. A. (2020). Nanomaterials: Classification, properties, and environmental toxicities. *Environmental Technology & Innovation*, *20*, 101067.

Sanghi, R., & Verma, P. (2009). Biomimetic synthesis and characterisation of protein capped silver nanoparticles. *Bioresource Technology*, *100*(1), 501–504.

Saravanan, A., Kumar, P. S., Karishma, S., Vo, D. V. N., Jeevanantham, S., Yaashikaa, P. R., & George, C. S. (2021). A review on biosynthesis of metal nanoparticles and its environmental applications. *Chemosphere*, *264*, 128580.

Sarsar, V., Selwal, M. K., & Selwal, K. K. (2016). Biogenic synthesis, optimisation and antibacterial efficacy of extracellular silver nanoparticles using novel fungal isolate *Aspergillus fumigatus* MA. *IET Nanobiotechnology*, *10*(4), 215–221.

Shafey, A. M. E. (2020). Green synthesis of metal and metal oxide nanoparticles from plant leaf extracts and their applications: A review. *Green Processing and Synthesis*, *9*(1), 304–339.

Shah, R., Eldridge, D., Palombo, E., & Harding, I. (2015). *Lipid Nanoparticles: Production, Characterization and Stability* (Vol. 1, pp. 23–43). Springer International Publishing, New York, NY, USA.

Shahbaz, M., Akram, A., Raja, N. I., Mukhtar, T., Mashwani, Z. U., Mehak, A., ... & Yousaf, T. (2022). Green synthesis and characterization of selenium nanoparticles and its application in plant disease management: A review. *Pakistan Journal of Phytopathology*, *34*(1), 189–102.

Shang, Y., Hasan, M. K., Ahammed, G. J., Li, M., Yin, H., & Zhou, J. (2019). Applications of nanotechnology in plant growth and crop protection: A review. *Molecules*, *24*(14), 2558.

Shanthi, S., Jayaseelan, B. D., Velusamy, P., Vijayakumar, S., Chih, C. T., & Vaseeharan, B. (2016). Biosynthesis of silver nanoparticles using a probiotic *Bacillus licheniformis* Dahb1 and their antibiofilm activity and toxicity effects in *Ceriodaphnia cornuta. Microbial Pathogenesis*, *93*, 70–77.

Shivaji, K., Mani, S., Ponmurugan, P., De Castro, C. S., Lloyd Davies, M., Balasubramanian, M. G., & Pitchaimuthu, S. (2018). Green-synthesis-derived CdS quantum dots using tea leaf extract: Antimicrobial, bioimaging, and therapeutic applications in lung cancer cells. *ACS Applied Nano Materials*, *1*(4), 1683–1693.

Shoala, T. (2018). Positive impacts of nanoparticles in plant resistance against different stimuli. In *Nanobiotechnology Applications in Plant Protection* (pp. 267–279).

Shoala, T. (2020). Carbon nanostructures: Detection, controlling plant diseases and mycotoxins. In *Carbon Nanomaterials for Agri-Food and Environmental Applications* (pp. 261–277). Elsevier.

Siddiqi, K. S., Husen, A., & Rao, R. A. (2018). A review on biosynthesis of silver nanoparticles and their biocidal properties. *Journal of Nanobiotechnology*, *16*(1), 1–28.

Singh, A., Gautam, P. K., Verma, A., Singh, V., Shivapriya, P. M., Shivalkar, S., ... & Samanta, S. K. (2020). Green synthesis of metallic nanoparticles as effective alternatives to treat antibiotics resistant bacterial infections: A review. *Biotechnology Reports*, *25*, e00427.

Singh, J., Dutta, T., Kim, K. H., Rawat, M., Samddar, P., & Kumar, P. (2018). 'Green'synthesis of metals and their oxide nanoparticles: Applications for environmental remediation. *Journal of Nanobiotechnology*, *16*(1), 1–24.

Singh, P., Kim, Y. J., Zhang, D., Yang, D. C. (2016). Biological synthesis of nanoparticles from plants and microorganisms. *Trends in Biotechnology*, *34*(7), 588–599.

Solanki, M. K., Kashyap, P. L., Kumari, B., Ansari, R. A., Sumbul, A., Rizvi, R., & Mahmood, I. (2021). Mycorrhizal fungi and its importance in plant health amelioration. In *Microbiomes and Plant Health* (pp. 205–223). Academic Press.

Sreelatha, S., Kumar, N., & Rajani, S. (2022b). Biological effects of Thymol loaded chitosan nanoparticles (TCNPs) on bacterial plant pathogen *Xanthomonas campestris* pv. *campestris*. *Frontiers in Microbiology*, *13*, 1085113.

Sreelatha, S., Kumar, N., Yin, T. S., & Rajani, S. (2022a). Evaluating the antibacterial activity and mode of action of thymol-loaded chitosan nanoparticles against plant bacterial pathogen *xanthomonas campestris* pv. *campestris*. *Frontiers in Microbiology*, *12*, 792737.

Stark, W. J., Stoessel, P. R., Wohlleben, W., & Hafner, A. J. C. S. R. (2015). Industrial applications of nanoparticles. *Chemical Society Reviews*, *44*(16), 5793–5805.

Subhapriya, S., & Gomathipriya, P. (2018). Green synthesis of titanium dioxide (TiO2) nanoparticles by *Trigonella foenum-graecum* extract and its antimicrobial properties. *Microbial Pathogenesis*, *116*, 215–220.

Sundararajan, B., Mahendran, G., Thamaraiselvi, R., & Ranjitha Kumari, B. D. (2016). Biological activities of synthesized silver nanoparticles from *Cardiospermum halicacabum* L. *Bulletin of Materials Science*, *39*, 423–431.

Suresh, A. K., Pelletier, D. A., Wang, W., Broich, M. L., Moon, J. W., Gu, B., ... & Doktycz, M. J. (2011). Biofabrication of discrete spherical gold nanoparticles using the metal-reducing bacterium *Shewanella oneidensis*. *Acta Biomaterialia*, *7*(5), 2148–2152.

Tang, S., & Zheng, J. (2018). Antibacterial activity of silver nanoparticles: Structural effects. *Advanced Healthcare Materials*, *7*(13), 1701503.

Tanno, K., Maejima, K., Miyazaki, A., Koinuma, H., Iwabuchi, N., Kitazawa, Y., ... & Namba, S. (2018). Comprehensive screening of antimicrobials to control phytoplasma diseases using an in vitro plant–phytoplasma co-culture system. *Microbiology*, *164*(8), 1048–1058.

Tariq, F., Ahmed, N., Afzal, M., Khan, M. A. U., & Zeshan, B. (2020). Synthesis, characterization and antimicrobial activity of *Bacillus subtilis*-derived silver nanoparticles against multidrug-resistant bacteria. *Jundishapur Journal of Microbiology*, *13*(5), e91934

Tariq, M., Mohammad, K. N., Ahmed, B., Siddiqui, M. A., & Lee, J. (2022). Biological synthesis of silver nanoparticles and prospects in plant disease management. *Molecules*, *27*(15), 4754.

Teniz, N. (2020). *Vanda yetiştirilen domates, biber ve kavun bitkilerinden izole edilen Fusarium spp. ve Rhizoctonia spp.nin teşhisi ve patojeniteleri etkisi* (master's thesis). Fen Bilimleri Enstitüsü, Van.

Thakkar, K. N., Mhatre, S. S., & Parikh, R. Y. (2010). Biological synthesis of metallic nanoparticles. *Nanomedicine: Nanotechnology, Biology and Medicine*, *6*(2), 257–262.

Thakur, R. K., & Shirkot, P. (2017). Potential of biogold nanoparticles to control plant pathogenic nematodes. *Journal of Bioanalysis & Biomedicine*, *9*(4), 220–222.

Thakur, R. K., Dhirta, B., & Shirkot, P. (2018). Studies on effect of gold nanoparticles on *Meloidogyne incognita* and tomato plants growth and development. *BioRxiv*, *26*, 428144.

Thiyagarajan, K., Bharti, V. K., Tyagi, S., Tyagi, P. K., Ahuja, A., Kumar, K., ... & Kumar, B. (2018). Retracted article: Synthesis of non-toxic, biocompatible, and colloidal stable silver nanoparticle using egg-white protein as capping and reducing agents for sustainable antibacterial application. *RSC Advances*, *8*(41), 23213–23229.

Tortella, G. R., Rubilar, O., Durán, N., Diez, M. C., Martínez, M., Parada, J., & Seabra, A. B. (2020). Silver nanoparticles: Toxicity in model organisms as an overview of its hazard for human health and the environment. *Journal of Hazardous Materials*, *390*, 121974.

Tripathi, D., Singh, M., & Pandey-Rai, S. (2022). Crosstalk of nanoparticles and phytohormones regulate plant growth and metabolism under abiotic and biotic stress. *Plant Stress*, *6*, 100107.

Tsekhmistrenko, S. I., Bityutskyy, V. S., Tsekhmistrenko, O. S., Horalskyi, L. P., Tymoshok, N. O., & Spivak, M. Y. (2020). Bacterial synthesis of nanoparticles: A green approach. *Biosystems Diversity*, *28*(1), 9–17.

Vallabani, N. S., & Singh, S. (2018). Recent advances and future prospects of iron oxide nanoparticles in biomedicine and diagnostics. *3 Biotech*, *8*(6), 279.

Vinardell, M. P., & Mitjans, M. (2015). Antitumor activities of metal oxide nanoparticles. *Nanomaterials*, *5*(2), 1004–1021.

Wang, R., Li, X., Nie, Z., Zhao, Y., & Wang, H. (2021). Metal/metal oxide nanoparticles-composited porous carbon for high-performance supercapacitors. *Journal of Energy Storage*, *38*, 102479.

Wang, T., Wen, X., Hu, Y., Zhang, X., Wang, D., & Yin, S. (2019). Copper nanoparticles induced oxidation stress, cell apoptosis and immune response in the liver of juvenile *Takifugu fasciatus*. *Fish & Shellfish İmmunology*, *84*, 648–655.

Weintraub, P. G. (2007). Insect vectors of phytoplasmas and their control-an update. *Bulletin of Insectology*, *60*(2), 169.

Worrall, E. A., Hamid, A., Mody, K. T., Mitter, N., & Pappu, H. R. (2018). Nanotechnology for plant disease management. *Agronomy*, *8*(12), 285.

Xu, G. C., Zhang, L. L., & Ni, Y. (2016). Enzymatic preparation of D-phenyllactic acid at high space-time yield with a novel phenylpyruvate reductase identified from *Lactobacillus* sp. CGMCC 9967. *Journal of Biotechnology*, *222*, 29–37.

Yan, L., Gu, Z., & Zhao, Y. (2013). Chemical mechanisms of the toxicological properties of nanomaterials: Generation of intracellular reactive oxygen species. *Chemistry–An Asian Journal*, *8*(10), 2342–2353.

Yao, S., Jin, B., Liu, Z., Shao, C., Zhao, R., Wang, X., & Tang, R. (2017). Biomineralization: From material tactics to biological strategy. *Advanced Materials*, *29*(14), 1605903.

Yin, C., Zhu, H., Lao, Y., Jiang, Y., & Gong, L. (2023). MicroRNAs in the exosome-like nanoparticles from orange *juice inhibit Citrus* blue mold caused by *Penicillium italicum*. *LWT*, *182*, 114781.

Yin, Y., Yang, X., Hu, L., Tan, Z., Zhao, L., Zhang, Z., ... & Jiang, G. (2016). Superoxide-mediated extracellular biosynthesis of silver nanoparticles by the fungus *Fusarium oxysporum*. *Environmental Science & Technology Letters*, *3*(4), 160–165.

Younis, A. B., Haddad, Y., Kosaristanova, L., & Smerkova, K. (2023). Titanium dioxide nanoparticles: Recent progress in antimicrobial applications. *Wiley Interdisciplinary Reviews: Nanomedicine and Nanobiotechnology*, *15*(3), e1860.

Zaki, S. A., Ouf, S. A., Albarakaty, F. M., Habeb, M. M., Aly, A. A., & Abd-Elsalam, K. A. (2021). *Trichoderma harzianum*-mediated ZnO nanoparticles: A green tool for controlling soil-borne pathogens in cotton. *Journal of Fungi*, *7*(11), 952.

Zehra, A., Rai, A., Singh, S. K., Aamir, M., Ansari, W. A., & Upadhyay, R. S. (2021). An overview of nanotechnology in plant disease management, food safety, and sustainable agriculture. *Food Security and Plant Disease Management*, 193–219.

Zhang, C., Li, X., Lyu, J., & Li, F. (2020). Comparison of carbonate precipitation induced by *Curvibacter* sp. HJ-1 and *Arthrobacter* sp. MF-2: Further insight into the biomineralization process. *Journal of Structural Biology*, *212*(2), 107609.

Zhu, L., Chen, L., Gu, J., Ma, H., & Wu, H. (2022). Carbon-based nanomaterials for sustainable agriculture: Their application as light converters, nanosensors, and delivery tools. *Plants*, *11*(4), 511.

Zong, J., Cobb, S. L., & Cameron, N. R. (2017). Peptide-functionalized gold nanoparticles: Versatile biomaterials for diagnostic and therapeutic applications. *Biomaterials Science*, *5*(5), 872–886.

6 Bio-Nanomaterials
The Next-Generation Weapon against Plant-Parasitic Nematodes in Commercial Crops

Gökhan Boyno, Mehmet Ramazan Rişvanlı, Sozan E. El-Abeid, Mohamed A. Mosa, and Dina S. S. Ibrahim

6.1 INTRODUCTION

Bio-nanomaterials have been increasingly gaining attention in the agricultural industry as a result of their unique properties and potential applications. The use of bio-nanomaterials for managing plant-parasitic nematodes, which are among the most dangerous agricultural pests, has been of significant interest. These nematodes can cause substantial yield losses in a wide range of crops, making them a major constraint in the production of vegetables and other crops (Ansari and Mahmood 2017a; Ansari and Mahmood 2019a; Ansari and Mahmood 2019b). Root-knot nematodes, for example, are endoparasites of most cultivated crops and are responsible for significant economic losses worldwide (Sasser, 1980). Several hundred ectoparasitic nematodes might feed on a plant without seriously affecting production while in other groups only single endo-parasitic nematodes can kill a plant or reduce its productivity (Ansari et al., 2020; Ansari et al., 2020; Ansari et al., 2019; Mahmood et al., 2019;). Endo-parasitic nematodes especially the root-knot nematodes (*Meloidongyne* spp.) have been found as one of the major constraints in the production of vegetables and other crops (Ansari and Khan 2012; Ibrahim et al., 2020; Devi et al., 2020; Ibrahim et al., 2022; Keshari et al., 2024). The infective second-stage juveniles penetrate the plant roots at the elongation zone and then migrate inter-cellularly, establishing feeding sites in the developing vascular tissues. Scientists have been working tirelessly to protect crops from root-knot nematode infection. However, the lack of plant resistance to most species of the root-knot nematode, coupled with the environmental restrictions on the usage of chemical nematocides for controlling plant parasitic nematodes, has made it necessary to look for more sustainable and eco-friendly disease control measures (Prasad et al., 2020; Solanki et al., 2020). On the other hand, worldwide climate changes are further enhancing the attacks of these parasites and then imposing the need for effective control strategies (Ansari et al., 2024; Ansari and Mahmood 2017b; Ansari et al., 2017; Ansari et al., 2023). The high cost and environmental impacts of nematocidal treatments with synthetic nematocides have raised increasing interest in searching for more effective, economic and safe alternatives, such as plant-derived compounds (Boyno et al., 2024; Fabiyi et al., 2024; Fabiyi et al., 2020; Patil et al., 2024; Solanki et al., 2021; Sumbul et al., 2020).

In recent years, bio-nanomaterials have emerged as an effective, economical, and safe alternative to traditional chemical nematocides. The secondary metabolism of many plants can produce nematocidal compounds that belong to several chemical classes mainly involved in defence mechanisms against predators, pests, and pathogens (Chitwood, 2002). The synthesis of nanoparticles through

 DOI: 10.1201/9781003256762-6

environmentally benign biological methods has offered a consistent, nontoxic, and eco-friendly approach for plant pathogen management due to their strong antimicrobial properties. Moreover, Nano-Materials (NMs) are carriers of agrochemicals, which facilitate site-targeted delivery of various nutrients needed for better growth and high productivity of plants (Singh et al., 2021).

Several routes have been developed for the biological or biogenic synthesis of nanoparticles from salts of the corresponding metals (Dhillon et al., 2012; Gan and Li, 2012). Microorganisms, whole plants, plant tissue and fruits, plant extracts, and marine algae have been used to produce nanoparticles (Rajesh et al., 2012). Biogenic synthesis is useful not only because of its reduced environmental impact compared with some of the physicochemical production methods (Shankar et al., 2004; Dahl et al., 2007), but also because it can produce large quantities of nanoparticles that are free of contamination and have a well-defined size and morphology (Hutchison, 2008). Biosynthetic routes can provide nanoparticles of a better-defined size and morphology than some of the physicochemical production methods (Raveendran et al., 2003). Due to the presence of a myriad of biomolecules in plant metabolites possessing bio reduction and stabilization ability, the exploration of such molecules could facilitate control over the size and morphology of nanoparticles. Biological synthesis is preferred nowadays because these methods are safe, cheap, eco-friendly, and do not involve any toxic substrate or by-product. Metals such as silver, copper, and zinc oxide can mainly be produced in their nano form Hence, the green synthesis of nanoparticles has offered a consistent, nontoxic, and eco-friendly approach for plant pathogen management due to their strong antimicrobial properties. A huge number of plants from all parts of the world are reported to facilitate the synthesis of NPs.

The use of bio-nanomaterials for nematode management represents a promising approach to improve the productivity of commercial crops while reducing the environmental impact of traditional chemical nematocides (Singh et al., 2021). The potential applications of bio-nanomaterials in crop protection and genetic manipulation of plants through nano-biosensors and nanodevices are also exciting research areas. Overall, the use of bio-nanomaterials in agriculture could lead to sustainable development and a better future for food security. The aim of this chapter is to discuss the use of bio-nanomaterials for nematode management in agriculture. It provides an overview of the potential applications of bio-nanomaterials in crop protection and genetic manipulation of plants. The chapter also describes the types of bio-nanomaterials that show promise for nematode management, including metallic nanoparticles, metal oxide nanoparticles, carbon-based nanoparticles, lipid-based nanoparticles, and polymer-based nanoparticles. The mechanisms of action of bio-nanomaterials against nematodes are also discussed. Overall, the chapter aims to provide insights into bio-nanomaterials' potential for sustainable and eco-friendly nematode control in agricultural systems.

6.2 TYPES OF BIO-NANOMATERIALS FOR NEMATODE MANAGEMENT

The field of bio-nanomaterials explores the use of nanoparticles derived from biological sources or synthesized through biological methods for various applications, including nematode management. Nanoparticles are tiny particles ranging from 1 to 100 nm in size, which gives them unique properties and behaviour compared to bulk materials (Ealia and Saravanakumar 2017). Bio-nanomaterials for nematode management involve utilizing nanoparticles with specific characteristics and properties to target and control nematodes, which are microscopic worms that can cause significant damage to plants and crops. The field of bio-nanomaterials offers various types of nanoparticles that show potential for nematode management (Figure 6.1). Here are some examples:

6.2.1 Metallic Nanoparticles

Metallic nanoparticles include silver, copper, and zinc which have been found to be promising nematocidal agents (Bhau et al., 2016; Sharmin et al., 2021). These nanoparticles possess inherent properties that effectively inhibit nematode growth and reproduction (Ardakani 2013). The nematocidal activity of metallic nanoparticles can be attributed to their antimicrobial properties (Lalitha et al., 2020). They have the ability to disrupt the integrity of the nematode's cell membranes, leading to cellular damage and eventual death (Lalitha et al., 2020). Additionally, metallic nanoparticles induce oxidative stress within nematodes, causing a disruption in their cellular functions and ultimately inhibiting their growth and reproduction (Huang et al., 2019).

Silver nanoparticles, in particular, have demonstrated strong nematocidal effects against various nematode species (Abbassy et al., 2017; Kalaiselvi et al., 2019). They have been shown to impair nematodes' movement and feeding behaviour, leading to reduced damage to plant roots and decreased reproduction rates (Kalaiselvi et al., 2019). Copper nanoparticles have also exhibited nematocidal activity by interfering with nematode physiology (Kausar 2022). They can disrupt the nematode's sensory perception, affecting their ability to locate and penetrate plant roots, thus reducing their pathogenicity (Khan et al., 2021). Similarly, zinc nanoparticles have been found to be effective in inhibiting nematode development and reproduction (Polak et al., 2014). They can interfere with nematode enzymatic processes, disrupt their cell signalling pathways, and induce oxidative stress, thereby impeding their growth and survival (Chang et al., 2012).

The antimicrobial and oxidative stress-inducing capabilities of metallic nanoparticles offer a promising avenue for nematode management (Bhau et al., 2016; Sharmin et al., 2021). Furthermore, by targeting key physiological processes and mechanisms within nematodes, metallic nanoparticles provide a potential alternative to conventional chemical nematocides, offering environmentally friendly and sustainable options for controlling nematode populations in agricultural settings.

6.2.2 Metal Oxide Nanoparticles

Metal oxide nanoparticles include titanium dioxide (TiO_2) and zinc oxide (ZnO) nanoparticles, which have emerged as promising agents for nematode management (Elmer et al., 2018; Thounaojam et al., 2021). These nanoparticles possess unique properties that enable them to effectively disrupt nematode physiology and impede their development and reproduction (Gonzalez-Moragas et al., 2017).

TiO_2 nanoparticles have exhibited nematocidal activity against various nematode species (Ardakani 2013). They can interfere with the nematode's sensory perception, affecting their ability to locate host plants and initiate feeding (Ardakani 2013). TiO_2 nanoparticles also possess photocatalytic properties, which, when activated by ultraviolet (UV) light, generate Reactive Oxygen Species (ROS) that induce oxidative stress within nematodes (Ardakani 2013; Dong et al., 2018). This oxidative stress can lead to cellular damage, impairing nematode growth, development, and reproductive capabilities (Ardakani 2013; Dong et al., 2018). ZnO nanoparticles have also demonstrated nematocidal effects against nematodes. These nanoparticles can disrupt nematode physiology through multiple mechanisms (Kalaba et al., 2021). ZnO nanoparticles can penetrate the nematode cuticle and interfere with their cellular processes, causing structural damage and disruption of essential physiological functions (Kalaba et al., 2021; El-Ansary et al., 2022). Additionally, ZnO nanoparticles can induce ROS production within nematodes, leading to oxidative stress and subsequent detrimental effects on their growth and reproduction (Huang et al., 2019). The unique physicochemical properties of metal oxide nanoparticles contribute to their nematocidal activity (Elmer et al., 2018; Thounaojam et al., 2021). Their small size and large surface area enables increased contact and interaction with nematodes, enhancing their efficacy (Kalaba et al., 2021). Furthermore, metal oxide nanoparticles can be functionalized or modified to improve their stability, dispersibility, and targeting specificity, making them more effective in nematode management (Kalaba et al., 2021). The use of metal oxide nanoparticles as nematocidal agents presents several advantages. They offer a potentially eco-friendly and sustainable alternative to conventional chemical nematocides, reducing the environmental impact associated with traditional pest control methods (Zikalala et al., 2018). Additionally, the specific targeting of nematodes while minimizing harm to beneficial organisms is a crucial aspect of sustainable agriculture, and metal oxide nanoparticles have shown promise in achieving this selectivity (Singh et al., 2021). Further research and development are necessary to optimize the formulation, application methods, and dosage of metal oxide nanoparticles for effective nematode management. However, their potential in disrupting nematode physiology and inhibiting their development and reproduction make metal oxide nanoparticles a promising tool for sustainable nematode control in agricultural systems (Elmer et al., 2018; Thounaojam et al., 2021).

6.2.3 Carbon-Based Nanoparticles

Carbon driven nanoparticles such as graphene and carbon nanotubes, have emerged as promising candidates for nematode management due to their unique properties and nematocidal effects. These nanoparticles offer potential solutions for controlling nematode populations by disrupting their physiology and inhibiting their survival and reproduction (Kralova and Jampilek 2023).

Graphene, a two-dimensional carbon nano-material, has shown nematocidal activity against various nematode species (Gareeb et al., 2019). Its high surface area and unique structure enable effective interaction with nematodes. Graphene nanoparticles can penetrate the nematode cuticle, the protective outer layer, and disrupt cellular processes (Ding et al., 2018;). By interacting with nematode cell membranes and altering their integrity, graphene nanoparticles can cause structural damage, leading to cell death and reduced nematode survival rates (Ding et al., 2018). On the other hand, Carbon NanoTubes (CNTs) possess a tubular structure composed of rolled graphene sheets (Das 2013). CNTs have demonstrated nematocidal effects by disrupting nematode physiology (Wu et al., 2013). They can penetrate the nematode cuticle and interact with cellular components, leading to the disruption of essential processes. CNTs have been shown to interfere with nematode locomotion, feeding behaviour, and reproduction, ultimately reducing nematode populations (Wu et al., 2013). Carbon-based nanoparticles' small size and unique physicochemical properties contribute to their nematocidal activity (Kralova and Jampilek, 2023). In addition, their high aspect ratio, large surface area, and strong interaction with biological molecules enable efficient interaction with nematodes, enhancing their effectiveness as nematocidal agents. Furthermore, carbon-based nanoparticles can be functionalized or combined with other compounds to improve their stability, dispersibility, and targeting specificity, enhancing their potential as targeted delivery systems for nematode control (Kralova & Jampilek, 2023). The use of carbon-based nanoparticles for nematode management offers several advantages. They provide an environmentally friendly alternative to conventional chemical nematocides, reducing the environmental impact associated with traditional pest control methods (Singh et al., 2021). Carbon-based nanoparticles also hold promise for selective targeting, minimizing harm to beneficial organisms and maintaining ecological balance in agricultural systems. However, further research is needed to optimize carbon-based nanoparticle application methods, dosage, and long-term effects for nematode management (Ding et al., 2018; Singh et al., 2021). Understanding their interactions with nematodes and their fate in the environment will contribute to their safe and effective agricultural implementation. Overall, carbon-based nanoparticles show great potential as innovative tools for sustainable nematode control and contribute to the development of environmentally friendly agricultural practices.

6.2.4 Lipid-Based Nanoparticles

Lipid driven nanomaterials particularly liposomes, have garnered attention for their potential in nematode control and management. These nanoparticles, composed of lipid bilayers, offer unique advantages for encapsulating and delivering nematocidal compounds, leading to enhanced effectiveness in combating nematode infestations (Colmenares et al., 2016). Liposomes, spherical vesicles with a lipid membrane structure, can encapsulate a variety of nematocidal compounds, including natural bioactive molecules or synthetic pesticides (Nuruzzaman et al., 2019). The encapsulation process protects the nematocidal agents from degradation and environmental factors and allows for controlled and targeted release at the desired site of action. The ability of liposomes to encapsulate hydrophilic and hydrophobic compounds makes them versatile carriers for a wide range of nematocidal agents (Fan et al., 2023). Hydrophilic compounds can be encapsulated within the aqueous core of liposomes, while hydrophobic compounds can be incorporated into the lipid bilayer (Gonzalez Gomez et al., 2019). This flexibility enables the delivery of diverse nematocidal compounds with different chemical properties, expanding the options for nematode control. Encapsulating nematocidal compounds within liposomes offers several advantages (Piao et al., 2020). Firstly, it improves the stability and solubility of the active ingredients, enhancing their shelf life and ensuring consistent efficacy (Nuruzzaman et al., 2016). Secondly, liposomes enable targeted delivery of the encapsulated compounds to specific nematode-infested areas, increasing the concentration and bioavailability at the target site while minimizing off-target effects (Nuruzzaman et al., 2016). This targeted delivery approach reduces the required dosage of nematocidal compounds, reducing environmental impact and potential harm to non-target organisms. Furthermore, the biocompatibility and biodegradability of liposomes contribute to their eco-friendly nature (de Oliveira et al., 2014). Liposomes can be designed using naturally derived lipids, making them compatible with environmentally sustainable agricultural practices (de Oliveira et al., 2014). However, further research is necessary to optimize lipid-based nanoparticle formulation, stability, and release kinetics for nematode management. Factors such as size, charge, and lipid composition can influence the efficacy and performance of liposomes as nematocidal carriers. In summary, lipid-based nanoparticles, particularly liposomes, offer a promising approach for nematode control. Their ability to encapsulate nematocidal compounds and facilitate targeted delivery enhances the effectiveness of these agents, while their eco-friendly nature aligns with sustainable agricultural practices. Continued research and development in lipid-based nanoparticle formulations can contribute to the development of efficient and environmentally friendly strategies for nematode management in crop production.

6.2.5 Polymer-Based Nanoparticles

Polymer driven nanomaterials particularly chitosan nanoparticles, have demonstrated promising nematocidal activity, providing a potential solution for nematode management in agricultural systems (Alfy et al., 2020; Mouniga et al., 2022). Chitosan, derived from chitin, the second most abundant biopolymer in nature, possesses unique properties that enable it to disrupt nematode physiology and impede their growth and reproduction (Alfy et al., 2020; Mouniga et al., 2022).

Chitosan nanoparticles exhibit nematocidal effects through multiple mechanisms. Firstly, chitosan can interact with the nematode cuticle, disrupting its integrity and leading to increased permeability (Lichtenberg et al., 2019). This disruption compromises the nematode's protective barrier, making it more susceptible to environmental stressors and reducing its survival rate. Additionally, chitosan nanoparticles can interfere with nematode sensory perception, impairing their ability to locate and infect host plants (Hidangmayum and Dwivedi 2022). Furthermore, chitosan has been found to possess antimicrobial properties, including antifungal and antibacterial activities (Hosseinnejad and Jafari 2016). These antimicrobial effects can indirectly contribute to nematode control by targeting microbial symbionts or pathogens associated with nematodes, disrupting their interactions and overall fitness (Khalil end Badawy 2012). The biocompatibility and biodegradability of chitosan nanoparticles make them particularly attractive for sustainable nematode management (Lichtenberg et al., 2019). Chitosan is derived from renewable sources, such as crustacean shells, and can be produced through eco-friendly processes (Alfy et al., 2020). It poses minimal risks to the environment and non-target organisms, aligning with sustainable agriculture principles. In addition to its direct nematocidal effects, chitosan nanoparticles can serve as carriers for other nematocidal compounds or bioactive molecules, enhancing their stability, solubility, and targeted delivery (Alfy et al., 2020; Mouniga et al., 2022). By encapsulating or

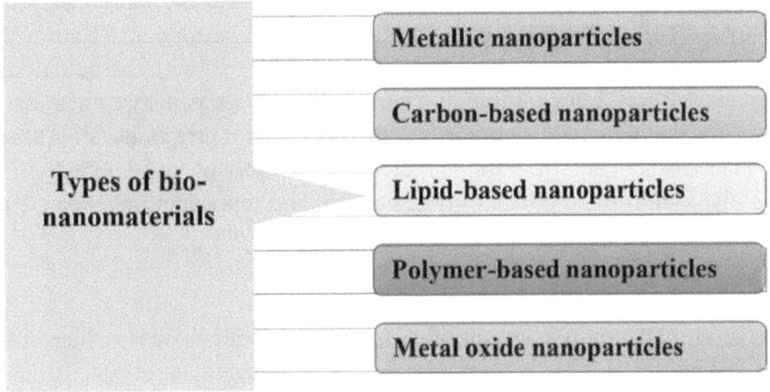

FIGURE 6.1 Types of bio-nanomaterials against nematodes: Metallic nanoparticles are nanoparticles made of metals, such as silver, gold, and copper. Metal oxide nanoparticles are nanoparticles made of metal oxides, such as zinc oxide and titanium dioxide. Carbon-based nanoparticles are nanoparticles made of carbon, such as fullerenes and carbon nanotubes. Lipid-based nanoparticles are nanoparticles made of lipids, such as liposomes and solid lipid nanoparticles. Polymer-based nanoparticles are nanoparticles made of polymers, such as poly(lactic-co-glycolic acid) (PLGA) and chitosan.

conjugating with nematocidal agents, chitosan nanoparticles can improve their efficacy and reduce their potential side effects, making them more suitable for practical applications. However, further research is needed to optimize chitosan nanoparticle formulation, dosage, and application methods for effective nematode management. Factors such as nanoparticle size, charge, and concentration can influence their interactions with nematodes and their overall performance (Mouniga et al., 2022). Polymer-based nanoparticles, particularly chitosan nanoparticles offer a promising avenue for sustainable nematode control in agriculture. Their ability to disrupt nematode physiology and inhibit their growth and reproduction, combined with their biocompatible and biodegradable nature, make them attractive alternatives to conventional chemical nematocides (Mouniga et al., 2022). Continued research and development in polymer-based nanoparticles can contribute to developing eco-friendly and effective strategies for nematode management, promoting sustainable crop production practices (Lichtenberg et al., 2019).

6.3 MECHANISMS OF BIO-NANOMATERIALS AGAINST NEMATODES

Bio-nanomaterials, due to their unique properties and characteristics, offer diverse mechanisms of action that contribute to their effectiveness in nematode management (Figure 6.2).

6.3.1 Physicochemical Interactions with Nematodes

Physicochemical interactions play a crucial role in the mechanisms of action exhibited by bio-nanomaterials against nematodes (Prabha et al., 2020). These interactions involve physical and chemical processes, allowing the nano-materials to interact effectively with and impact the nematode's biology (Gangwar et al., 2023). One aspect of these interactions is the adhesion of bio-nanomaterials to the nematode cuticle. Due to their nanoscale size and unique surface properties, these materials can adhere to the nematode's outer surface (Fakhrullin et al., 2021; Jin et al., 2021). This adhesion creates physical barriers that disrupt the nematode's normal functioning, impeding its movement and feeding activities. By forming a coating or layer on the cuticle, the bio-nanomaterials interfere with the nematode's ability to navigate through soil particles and reach host plants, limiting their mobility and access to nutrients (Kitherian 2017). Furthermore, the chemical properties of bio-nanomaterials play a significant role in their interactions with nematodes (Kausar 2022). The surface chemistry of the nanoparticles allows for specific interactions with the nematode's physiology, leading to detrimental effects. For example, the nano-materials may release chemical compounds or ions that can penetrate the nematode's cuticle and disrupt its vital processes (El-Ansary et al., 2022). These chemical interactions can affect the nematode's metabolism, enzyme activities, and cellular functions, ultimately compromising its survival, growth, and reproduction (Sabry 2019).

6.3.2 Disruption of Nematode Cell Membranes

Disruption of nematode cell membranes is another significant mechanism by which bio-nanomaterials exert their effects (Wang et al., 2018; Morsy et al., 2019). These nanomaterials have the ability to penetrate the nematode cuticle, which is the protective outer layer of the nematode's body and subsequently interact with the cell membranes (Wang et al., 2018; Morsy et al., 2019).

Upon penetration, the bio-nanomaterials can cause structural damage to the cell membranes, leading to the disruption of their integrity (Wang et al., 2018; Morsy et al., 2019). This disruption has severe consequences for the nematode's physiology and vital processes (Rajasekharan and Lee 2020). One important aspect affected by membrane disruption is osmoregulation, which is the maintenance of water and ion balance within the nematode's body (Lapresta-Fernández et al., 2012). The loss of membrane integrity caused by the nano-materials can disturb osmoregulation, resulting in imbalances in cellular water and ion concentrations. This disruption of osmoregulation can lead to cellular dehydration or swelling, impairing the nematode's normal functioning and survival (Lamitina et al., 2004) Additionally, the disruption of nematode cell membranes affects nutrient absorption. Nematodes rely on their cell membranes to facilitate the uptake of essential nutrients from the surrounding environment. The structural damage caused by the bio-nanomaterials can compromise the integrity of the nutrient transport systems in the nematode's cell membranes (Lapresta-Fernández et al., 2012). As a result, the nematode's ability to absorb and utilize nutrients efficiently is hindered leading to nutritional deficiencies that can impact its growth, development and reproductive capacity.

6.3.3 Inhibition of Nematode Enzymes and Metabolic Pathways

Inhibition of nematode enzymes and metabolic pathways represents another important mechanism through which bio-nanomaterials exert their effects on nematodes (Li et al., 2020). These nanomaterials have the ability to interfere with specific enzymes and disrupt key metabolic pathways within the nematode's body (Nikam et al., 2022). By targeting nematode enzymes, bio-nanomaterials can disrupt the catalytic activity necessary for various biochemical reactions (Li et al., 2020). These enzymes play crucial roles in processes such as digestion, energy production, and reproductive functions. Certain bio-nanomaterials can inhibit the activity of digestive enzymes responsible for breaking down nutrients, thereby impairing the nematode's ability to extract essential molecules for growth and development efficiently (Li et al., 2021). This inhibition can lead to nutritional deficiencies and energy depletion, negatively impacting the nematode's overall fitness and survival. For example, Mohammed & Reda, (2021) revealed that the use of bio-nanomaterial inhibited the growth of *Meloidogyne* spp. by reducing its feeding. In addition, nanomaterials caused the death of Caenorhabditis elegans by affecting its digestive enzymes (Li et al., 2021). Wu et al. (2016) found that nanomaterials exposure could affect the defecation cycle of nematodes and increase fat storage in the intestinal tract, which led to the death of nematodes. Furthermore, bio-nanomaterials can interfere with metabolic pathways, which are complex networks of biochemical reactions responsible for the synthesis and breakdown of molecules within the nematode's cells (Gonzalez-Moragas et al., 2017). By disrupting these pathways, the nano-materials can impede the production of vital metabolites, including those involved in energy generation and reproduction. For example, certain nano-materials can disrupt the electron transport chain, a crucial metabolic pathway involved in energy production through oxidative phosphorylation (Ahn et al., 2014; Chatterjee et al., 2014). This disruption can lead to a decrease in Adenosine TriPhosphate (ATP) synthesis, which is the main energy currency in living organisms (Ahn et al., 2014; Chatterjee et al., 2014). As a result, the nematode's energy reserves are depleted, compromising its physiological processes and overall fitness.

The inhibition of nematode enzymes and metabolic pathways by bio-nanomaterials can have profound effects on nematode physiology, ultimately leading to mortality or reduced fertility (Li et al., 2020; Nikam et al., 2022). By targeting specific enzymes and pathways essential for nematode survival and reproduction, bio-nanomaterials offer a precise and effective approach to nematode management (Li et al., 2021). Furthermore, the ability of these nano-materials to disrupt vital biochemical processes highlights their potential as powerful tools in sustainable agriculture, providing eco-friendly alternatives to conventional chemical-based approaches (Li et al., 2021).

6.3.4 Induction of Plant Defence Mechanisms

Induction of plant defence mechanisms represents a powerful mechanism of action exhibited by bio-nanomaterials against nematodes (Rajwade et al., 2020). When bio-nanomaterials are applied to plants, they can stimulate the plant's innate immune system, activating various defence mechanisms that help combat nematode infestations (Rajwade et al., 2020).

Upon exposure to bio-nanomaterials, plants can release signalling molecules such as hormones, phytoalexins, and Reactive Oxygen Species (ROS), which act as molecular messengers in the plant's defence signalling pathways (Zhang et al., 2020; Mouniga et al., 2022). These signalling molecules serve to activate specific defence responses against nematodes. For instance, the synthesis and release of defence compounds, such as terpenoids, phenolics, and alkaloids, can be stimulated by bio-nanomaterials (Zhang et al., 2020; Mouniga et al., 2022). These compounds have toxic or repellent effects on nematodes, hindering their development, movement, and reproduction (Zhang et al., 2020; Mouniga et al., 2022). Furthermore, bio-nanomaterials can enhance the plant's physical barriers against nematode attack. They can trigger the reinforcement of cell walls, making them more resistant to nematode penetration (Yu et al., 2022). This reinforcement involves the deposition of additional layers of structural components, such as lignin and cellulose, which create physical barriers that restrict nematode movement and feeding (Feofilova & Mysyakina, 2016). Moreover, the application of bio-nanomaterials can activate systemic resistance in plants (Alamri et al., 2022). Systemic resistance is a defence mechanism in which the plant's immune response is triggered locally at the site of application and in distal parts of the plant. This systemic response involves the production of defence-related proteins and signalling molecules that provide long-lasting protection against nematodes. By inducing systemic resistance, bio-nanomaterials can confer protection to the entire plant, making it more resilient to nematode infestations (Alamri et al., 2022).

The induction of plant defence mechanisms by bio-nanomaterials represents a sustainable approach to nematode management in agriculture (Mouniga et al., 2022). Bio-nanomaterials offer an eco-friendly alternative to chemical-based pesticides by harnessing the plant's natural defence systems. They promote plant health and resilience while minimizing the risks associated with conventional pesticide use, such as environmental pollution and negative impacts on beneficial organisms.

6.4 FACTORS INFLUENCING THE EFFECTIVENESS OF BIO-NANOMATERIALS

The effectiveness of bio-nanomaterials for nematode management is influenced by various factors that can affect their interaction with nematodes and their performance in the agricultural system (Li et al., 2021). Therefore, understanding these factors is crucial for optimizing the use of bio-nanomaterials and maximizing their efficacy.

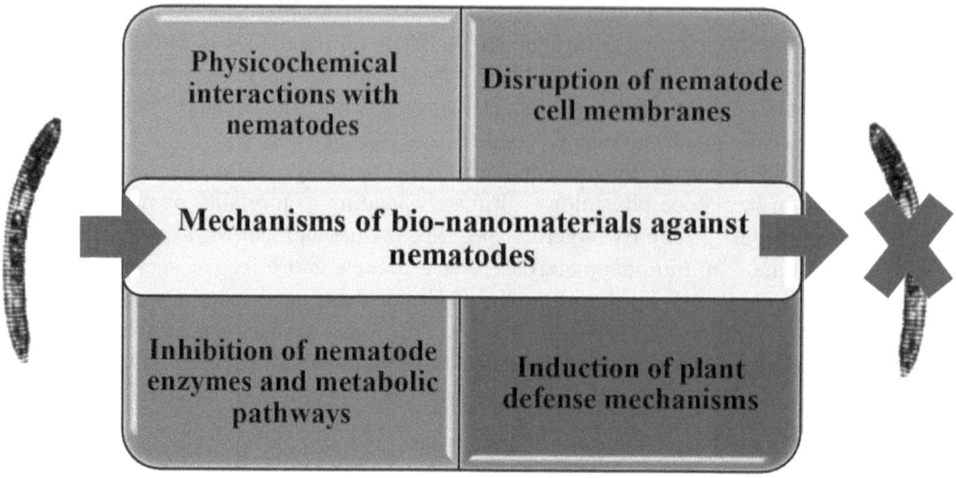

FIGURE 6.2 Mechanisms of bio-nanomaterials against nematodes: Physicochemical interactions involve the binding of nanoparticles to nematode surfaces, which can lead to dehydration and death of the nematodes. Disruption of nematode cell membranes can also be caused by the binding of nanoparticles to the membrane, which can lead to cellular leakage and death. Inhibition of nematode enzymes and metabolic pathways can occur through the disruption of normal cellular processes, leading to reduced growth and reproduction of nematodes. Induction of plant defence mechanisms involves the activation of plant immune responses, which can lead to increased resistance to nematode infestation.

6.4.1 Particle Size and Shape

The size and shape of bio-nanomaterials play a critical role in their interaction with nematodes (Thakur and Shirkot 2017). Nanoparticles with smaller sizes and specific shapes have been shown to have enhanced penetration and adhesion properties, allowing them to effectively interact with nematodes and disrupt their physiological processes (Thakur and Shirkot 2017; Moon et al., 2019). Smaller nanoparticles can also penetrate the nematode cuticle more easily, leading to increased internalization and subsequent toxic effects (Thakur and Shirkot 2017). For example, nanomaterials with dimensions in the range of 1 to 10 nanometres (nm) have been shown to penetrate the cuticle of nematodes faster, causing toxic effects on the nematode and leading to death (Wu et al., 2016; Li et al., 2021).

6.4.2 Surface Charge

The surface charge of bio-nanomaterials affects their stability, dispersibility, and interaction with nematodes (Stavitskaya et al., 2020; Schultz et al., 2021). Nanoparticles with positive or negative surface charges can exhibit different levels of attraction or repulsion toward nematodes, influencing their adhesion, uptake, and overall toxicity (Schultz et al., 2021). The surface charge of nanoparticles can be modified through functionalization or coating techniques to optimize their interaction with nematodes (Stavitskaya et al., 2020; Schultz et al., 2021).

6.4.3 Concentration and Dose

The concentration and dose of bio-nanomaterials applied to the agricultural system are important factors in determining their effectiveness (Sitia et al., 2022). The concentration of nanoparticles should be carefully optimized to ensure sufficient exposure and toxicity to nematodes without causing harm to plants or the environment (Tsai et al., 2021). Higher concentrations may lead to improved efficacy but can also increase the risk of potential adverse effects (Tsai et al., 2021).

6.4.4 Application Method

The application method plays a crucial role in the distribution and efficacy of bio-nanomaterials in nematode management (Sonawane et al., 2022). Different application methods, such as foliar spray, soil drench, seed treatment, or incorporation into irrigation systems, can influence the delivery, dispersion, and targeting of bio-nanomaterials (Kumar et al., 2022; Mallaiah and Muthamillan 2022). The selection of an appropriate application method depends on the specific crop, nematode species, and desired outcomes.

6.4.5 Soil Properties

Soil properties, including texture, pH, organic matter content, and microbial activity, can impact the behaviour and effectiveness of bio-nanomaterials in nematode management (Javed et al., 2019; Perea Velez et al., 2021). Soil characteristics can influence nanoparticle stability, mobility, and bioavailability, affecting their distribution and interaction with nematodes. Understanding the soil properties and adapting the application strategies accordingly can help optimize the effectiveness of bio-nanomaterials.

It is important to note that the interactions between bio-nanomaterials and nematodes are complex and multifaceted (Hou et al., 2022a). The effectiveness of bio-nanomaterials can also be influenced by other factors such as nematode species, life stages, and the presence of other environmental stressors (Hou et al., 2022b). Therefore, comprehensive research and field trials are necessary to evaluate the effectiveness of bio-nanomaterials under specific conditions and to develop tailored strategies for nematode management in diverse agricultural systems.

6.5 APPLICATIONS OF BIO-NANOMATERIALS IN NEMATODE MANAGEMENT IN COMMERCIAL CROPS

Bio-nanomaterials have the potential to revolutionize nematode management in commercial crops. Here are some potential applications of bio-nanomaterials for nematode management:

6.5.1 Nematocidal Nanoparticles

Nanoparticles such as metal oxide nanoparticles (for example, copper oxide, zinc oxide) or carbon-based nanoparticles, can be engineered to possess nematocidal properties (Fabiyi et al., 2020; Akhter et al., 2020). These nanoparticles can be incorporated into soil amendments or applied as seed treatments to control nematode populations (Fouda et al., 2020). The nanoparticles can disrupt nematode physiology, inhibit their reproduction, or induce toxicity, effectively reducing nematode populations in the soil (Akhter et al., 2020).

6.5.2 Nanoemulsion Formulations

Nanoemulsions are stable suspensions of nanosized oil or water droplets in another liquid (Gupta et al., 2019). They can be formulated with nematocidal compounds, such as essential oils or plant extracts with proven nematocidal activity (Akhter et al., 2020). Nanoemulsions can enhance the delivery and efficacy of these compounds, enabling targeted application to nematode-infested soil (Baronia et al., 2020). They can penetrate the nematode cuticle and disrupt their physiological processes, leading to nematode mortality or reduced reproduction (Sonawane et al., 2022).

6.5.3 Nanofertilizers with Nematocidal Properties

Nanostructured fertilizers can be designed to simultaneously provide essential nutrients to plants and exhibit nematocidal activity (Bratovcic et al., 2021; Tuncsoy 2021). These nanofertilizers can be functionalized with nematocidal compounds, such as chitosan or plant-derived bioactive molecules, which can inhibit nematode feeding or disrupt their life cycle (Bratovcic et al., 2021; Yu et al., 2021). By incorporating nematocidal properties into fertilizers, it is possible to promote plant growth while suppressing nematode populations in the rhizosphere (Bratovcic et al., 2021; Tuncsoy 2021).

6.5.4 Nanosensors for Nematode Detection

Nanotechnology-based sensors can be developed to detect and monitor nematode populations in commercial crops (Dubey and Mailapalli 2016; Boddula et al., 2019). These nanosensors can utilize specific recognition elements, such as aptamers or antibodies, to selectively bind to nematode biomarkers (Boddula et al., 2019). The presence of nematodes can be detected through signal transduction mechanisms, enabling early detection and precise monitoring of nematode infestations (Boddula et al., 2019). This information can help farmers implement timely nematode management strategies.

6.5.5 Nanocarriers for Nematode-Targeted Delivery

Nanoparticles and nanocarriers can serve as effective delivery vehicles for nematocidal compounds or biological control agents, such as beneficial nematodes or entomopathogenic fungi (Chariou and Steinmetz 2017; Zhang et al., 2020). The nanocarriers can protect the bioactive agents from degradation, enhance their stability, and facilitate targeted delivery to nematode-infested areas (Zhang et al., 2020). This approach improves the efficiency of biological control methods and reduces the need for excessive pesticide application. It is important to note that while bio-nanomaterials hold promise for nematode management, their practical application and commercialization require further research, including safety assessments and optimization of formulation and delivery methods (Chariou and Steinmetz 2017). Additionally, regulatory considerations regarding the use of nanomaterials in agriculture need to be addressed for wider adoption in commercial crop production.

6.6 USE OF BIO-NANOMATERIALS IN THE MANAGEMENT OF NEMATODES OF COMMERCIAL CROPS

Nematodes pose a significant challenge to commercial crop production worldwide. Their infestations can lead to severe damage to plant roots, resulting in stunted growth, reduced yields, and even complete crop failure. Conventional nematode management strategies, such as the use of chemical pesticides, have demonstrated adverse effects on the environment and human health. Consequently, there is a growing interest in exploring novel and sustainable approaches to address nematode populations. Amongst the promising avenues of research in nematode management, bio-nanomaterials have garnered attention.

In particular, bio-nanomaterials have been investigated for nematode management in various commercial crops including cotton, jute, sugarcane, tobacco and sugar beet. Some potential applications of bio-nanomaterials in these crops are:

6.6.1 Cotton

Cotton (*Gossypium hirsutum* L.) is a widely cultivated commercial crop susceptible to nematode damage, particularly from root-knot nematodes (Kumar et al., 2021; Rişvanlı, 2022). There are several types of nematodes that can infect cotton plants, but some of the most common ones include: Root-knot nematodes (*Meloidogyne* spp.), Reniform nematodes (*Rotylenchulus reniformis*), Columbia lance nematodes (*Hoplolaimus columbus*), Sting nematodes (*Belonolaimus* spp.) and Lesion nematodes (*Pratylenchus* spp.) (Machado et al., 2012; Pan et al., 2019). Nematode control strategies typically involve a combination of cultural, chemical, and biological methods (Starr et al., 2007). In recent years, there has been a growing interest in using bio-nanomaterials as a potential eco-friendly and sustainable approach to manage nematode diseases in cotton (Husen & Iqbal 2019). Bio-nanomaterials such as silver nanoparticles and chitosan nanoparticles have been investigated for nematode control in cotton fields (Tuncsoy, 2021; Biradar et al., 2022). These nanoparticles can be applied as seed treatments, soil amendments, or incorporated into foliar sprays to reduce nematode populations and improve cotton yield (Gad 2019;).

6.6.2 Jute

Jute (*Corchorus* spp.) is a group of plants belonging to the *Corchorus* genus. These plants are primarily cultivated for their fibres, which are used to produce various products, particularly sacks, bags, ropes, and textiles (Haque et al., 2008; Srivastava et al., 2010).

The nematodes that commonly affect jute plants include root-knot nematodes (*Meloidogyne* spp.) (Daramola et al., 2015. Bio-nanomaterials offer a promising approach for nematode management in jute crops due to their potential eco-friendly nature and low toxicity to non-target organisms. For example, Azeez et al. (2019) reported that silver nanoparticles were effective against nematodes in the *Corchorus olitorius* plant. These nanoparticles can be applied as soil amendments to inhibit nematode development and protect jute plants (Shah et al., 2021; Khader et al., 2023).

It is important to note that the field of bio-nanomaterials for nematode management in jute is still in the early stages of research and development (Shah et al., 2021). While the potential benefits are promising, further studies are needed to assess the efficacy, safety, and environmental impact of these bio-nanomaterials in practical agricultural settings.

6.6.3 Sugarcane

Sugarcane crops are susceptible to various nematode species, including root-knot nematodes and lesion nematodes (Bhuiyan et al., 2019; Manteghi et al., 2022). Bio-nanomaterials, such as titanium dioxide nanoparticles, have been explored for nematode management in sugarcane fields (Kumar et al., 2021). These nanoparticles can be applied as seed treatments or incorporated into soil amendments to suppress nematode populations and enhance sugarcane productivity (Khan et al., 2019).

6.6.4 Tobacco

Tobacco (*Nicotiana tabacum* L.) is a significant commercial crop used for the production of cigarettes, cigars, snuff, and other tobacco products. Nematode infestations, particularly root-knot nematodes (*Meloidogyne* spp.) and cyst nematodes (*Heterodera* spp.), can lead to reduced tobacco yields and quality (Xing et al., 2017; Li et al., 2018).

By incorporating bio-nanomaterials into the development of commercial crops such as tobacco, it may be possible to reduce nematode-induced losses, increase tobacco yield and quality, and promote more sustainable and environmentally friendly agricultural practices in the tobacco industry (Kalita et al., 2021). Many bio-nanomaterials, such as gold and silver nanoparticles, are important to investigate for their potential in nematode management in tobacco cultivation, which is important for the sustainability of agriculture in the future (Luo et al., 2016; Kalita et al., 2021). These nanoparticles can be applied as soil amendments or incorporated into foliar sprays to reduce nematode infestation and protect tobacco plants (Kalita et al., 2021; Kralova & Jampilek 2023).

6.6.5 Sugar Beet

Sugar beet (*Beta vulgaris* L.) is a biennial plant belonging to the Amaranthaceae family (Cooke & Scott 2012). It is a commercial crop grown primarily for its high sugar content, which is extracted from the roots and used for a variety of purposes, including sugar production, biofuel production and animal feed (Biancardi et al., 2010). Sugar beet crops are vulnerable to nematode attacks, particularly from cyst nematodes (*Heterodera schachtii*) and root-knot nematodes (*Meloidogyne* spp.) (Sönmez et al., 2014; Gohar et al., 2023). Silver and Zinc oxide nanoparticles have shown promise in the management of nematodes in sugar beet (Fadei Tehrani & Fati, 2020). These nanoparticles have been shown to protect sugar beet plants as they disrupt nematode physiology and inhibit their growth and reproduction (Fadei Tehrani & Fati 2020). Furthermore, Hallmann et al. (2019) showed that nanotechnological approaches significantly improved the biocontrol efficiency of *H. schachtii* in sugar beet.

6.7 ADVANTAGES OF BIO-NANOMATERIALS IN COMMERCIAL CROPS

The advantages of using bio-nanomaterials for nematode management in commercial crops include (Bhau et al., 2016):

6.7.1 Enhanced Efficacy

Bio-nanomaterials can exhibit higher efficacy in controlling nematodes than conventional methods (Cao et al., 2022. Their small size allows for better penetration and interaction with nematodes, leading to improved targeting and efficacy (Mohamed et al., 2019).

6.7.2 Reduced Environmental Impact

Bio-nanomaterials are generally considered more environmentally friendly compared to chemical nematocides (Khare et al., 2011). They have lower toxicity and minimal residual effects, minimizing the risk of harming non-target organisms, soil, water, and the overall ecosystem (Ma et al., 2009; Boros and Ostafe 2020).

6.7.3 Sustainable and Renewable

Many bio-nanomaterials can be derived from renewable sources such as plant extracts or microbial synthesis (Kumari and Singh 2020; An et al., 2022). This makes them more sustainable and aligns with the principles of green chemistry and sustainable agriculture (Kumari and Singh 2020; An et al., 2022).

6.7.4 Potential for Biodegradability

Some bio-nanomaterials have the advantage of being biodegradable. They can break down into harmless byproducts over time, reducing concerns about long-term accumulation or persistence in the environment (Djiwanti and Kaushik 2019).

6.7.5 Compatibility with Integrated Pest Management (IPM)

Bio-nanomaterials can be integrated into IPM approaches, which aim to combine multiple pest management strategies for effective control (Khandelwal et al., 2016). They can complement other control methods, such as cultural practices, biological control agents, and resistant crop varieties, providing a holistic and sustainable approach to nematode management (Singh et al., 2019).

6.7.6 Reduced Resistance Development

Bio-nanomaterials often have novel modes of action, which can help mitigate the development of resistance in nematode populations (Udalova et al., 2018). Targeting multiple physiological processes makes it more difficult for nematodes to develop resistance compared to traditional chemical pesticides (Bernard et al., 2019).

6.7.7 Potential for Targeted Delivery

Bio-nanomaterials can be engineered to have specific properties such as controlled release or targeted delivery to the site of nematode infestation (Thorn et al., 2021). This increases their efficiency and reduces the amount of material required for effective nematode management.

6.8 LIMITATIONS OF USING BIO-NANOMATERIALS FOR NEMATODE MANAGEMENT

Limitations of using bio-nanomaterials for nematode management in commercial crops include (Wu et al., 2019; Figure 6.3).

6.8.1 Limited Knowledge and Understanding

The field of bio-nanomaterials is relatively new, and there is still much to learn about their long-term effects on the environment, non-target organisms, and crop productivity (Dwivedi et al., 2016). Further research is needed to fully understand their potential risks and benefits (Dwivedi et al., 2016).

6.8.2 Lack of Standardized Regulations

The regulatory frameworks for bio-nanomaterials vary across countries, and there are a lack of standardized guidelines for their use in agriculture (Hanna et al., 2016). This can create uncertainties and challenges for their commercial adoption (Hanna et al., 2016).

6.8.3 Cost Considerations

The production and application of bio-nanomaterials can be expensive compared to conventional nematode management methods (Camara et al., 2019). The high costs may limit their accessibility and practicality for small-scale farmers or regions with limited resources.

6.8.4 Stability and Shelf Life

Some bio-nanomaterials may have limited stability and shelf life, requiring specialized storage conditions and careful handling (Du et al., 2023). This can add complexity and logistical challenges to their use in commercial agriculture (Handy et al., 2012).

6.8.5 Potential Off-Target Effects

While bio-nanomaterials are generally considered to have lower environmental impact, unintended off-target effects are possible (Lichtenberg et al., 2019). The nanoparticles may interact with non-target organisms or disrupt beneficial soil microorganisms, affecting overall ecosystem health (Côa et al., 2020).

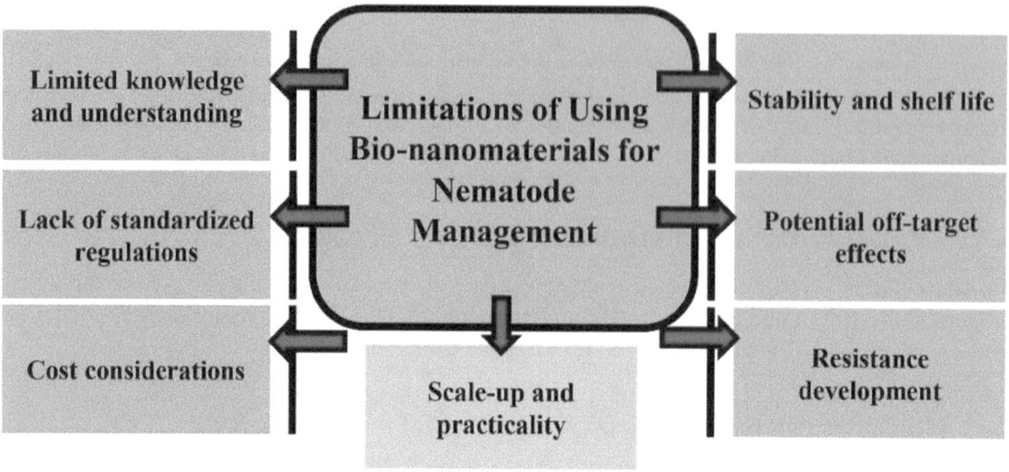

FIGURE 6.3 Limitations of using bio-nanomaterials for nematode management: Limited knowledge and understanding of the long-term effects of exposure to nanoparticles on human health and the environment. Another limitation is the lack of standardized regulations for the use of bio-nanomaterials in agriculture. Cost considerations, stability and shelf life, and potential off-target effects are also considerations when using bio-nanomaterials. In addition, there is the potential for the development of resistance in nematode populations, which could reduce the effectiveness of bio-nanomaterials over time. Finally, scaling up the production of bio-nanomaterials and their practicality in the field are also areas of concern.

6.8.6 Resistance Development

Although bio-nanomaterials have the potential to mitigate resistance development, there is still a possibility that nematodes can develop resistance to these materials over time (Tuncsoy 2021). Continuous and exclusive use of bio-nanomaterials without proper rotation or integration with other control strategies may contribute to resistance issues.

6.8.7 Scale-Up and Practicality

Scaling up bio-nanomaterial production to meet the demands of large-scale commercial agriculture can be challenging (McCourt et al., 2022). Manufacturing processes need to be optimized, and logistics for distribution and application need to be established for practical implementation (Handy et al., 2012).

It is important to address these limitations through further research, technological advancements, and regulatory frameworks to ensure bio-nanomaterials' safe and effective use in commercial crops' nematode management.

6.9 MAJOR CHALLENGES

The integration of bio-nanomaterials with other existing management strategies holds great potential for improving nematode control in commercial crops (Li et al., 2021). Several benefits can be achieved by combining bio-nanomaterials with traditional methods such as crop rotation, soil amendments, and cultural practices (Bhau et al., 2016). Firstly, the combined use of bio-nanomaterials with traditional strategies can provide a multi-dimensional approach to nematode management. Bio-nanomaterials can specifically target nematodes at the microscopic level, while traditional methods address broader factors such as crop diversity, soil health, and agronomic practices. This synergy between approaches can lead to enhanced overall effectiveness in controlling nematode populations (Bhau et al., 2016). Secondly, integrating bio-nanomaterials with other strategies can provide a more sustainable and environmentally friendly solution (Kumari and Singh 2020; An et al., 2022). Bio-nanomaterials can potentially reduce reliance on conventional chemical pesticides, which can negatively impact beneficial organisms and the environment (Ma et al., 2009; Boros and Ostafe 2020). By combining them with other management strategies, growers can adopt an Integrated Pest Management (IPM) approach focusing on long-term sustainability while effectively managing nematodes (Khandelwal et al., 2016). Moreover, the integration of bio-nanomaterials with other strategies can offer increased flexibility and adaptability. Different nematode species and populations may exhibit varying susceptibilities to different management methods. By combining bio-nanomaterials with other strategies, growers can tailor their nematode management approaches based on specific crop, soil, and nematode characteristics. This allows for a more customized and precise approach, optimizing the efficacy of nematode control (Djiwanti and Kaushik 2019).

One key aspect of environmental concern is the fate of bio-nanomaterials in the environment. It is crucial to understand how these nanoparticles behave once released into ecosystems. Factors such as their persistence, mobility, and potential for accumulation in the environment need to be evaluated. This information helps determine the potential for bio-nanomaterials to enter water bodies, soil, or other ecological compartments and their potential impacts on non-target organisms (Ma et al., 2009; Boros and Ostafe 2020). Assessing the potential ecological impacts of bio-nanomaterials is essential. While designed to target specific pests like nematodes, these materials may interact with other organisms in unintended ways. Research is needed to evaluate their effects on beneficial organisms, including pollinators, predators, and soil microorganisms. Understanding any potential disruptions to ecosystems and the broader implications for biodiversity is crucial for responsible use. Another important aspect is the potential impacts on human health and safety. Exposure to bio-nanomaterials can occur during their production, application, or through contact with treated crops (Shatkin and Kim 2015). Evaluating these materials' toxicity and health effects is essential to ensure worker safety, consumer safety, and the overall well-being of those involved in the agricultural value chain. Regulatory bodies play a crucial role in addressing these environmental and safety concerns. Developing and implementing appropriate regulations and guidelines can ensure the responsible use of bio-nanomaterials. These regulations should take into account factors such as labelling requirements, application guidelines, and risk assessment frameworks specific to bio-nanomaterials.

To address these concerns, comprehensive and independent studies are necessary to evaluate the potential risks and benefits associated with the use of bio-nanomaterials. Collaboration between researchers, regulatory agencies, industry stakeholders, and environmental organizations is vital to ensure a balanced and science-based approach to addressing environmental and safety considerations. Developing and using bio-nanomaterials for nematode management requires careful regulatory oversight to ensure their safe and responsible use (Kitherian 2017). Therefore, establishing regulatory frameworks is essential to address potential risks, evaluate efficacy, and provide guidelines for their application in commercial agriculture. One key aspect of regulatory considerations is the assessment of bio-nanomaterial efficacy (Faunce and Watal 2010). Regulatory authorities need to evaluate the effectiveness of these materials in controlling nematode infestations (Faunce and Watal 2010). This involves conducting rigorous scientific studies and field trials to determine their performance compared to traditional management strategies. Standardized protocols and evaluation criteria should be established to ensure consistent and reliable assessment of bio-nanomaterial efficacy. Another crucial regulatory consideration is assessing the potential risks associated with bio-nanomaterials (Zheng et al., 2017). Regulatory authorities must evaluate the environmental and human health impacts of these materials. This includes assessing their toxicity, potential for bioaccumulation, and any unintended ecological consequences. Risk assessment frameworks specific to bio-nanomaterials need to be developed to guide regulatory decisions and ensure their safe use. Regulatory guidelines are necessary to provide clear instructions on the application of bio-nanomaterials in commercial crops. These guidelines can include proper dosage, application methods, and timing recommendations to maximize effectiveness while minimizing potential risks (Sadrieh and Espandiari 2006). Standardized labelling requirements should also be established to ensure accurate product information and safe handling. Monitoring the environmental impact of bio-nanomaterials is an essential regulatory consideration. Regular surveillance and reporting systems can help track their potential effects on ecosystems, non-target organisms, and soil health (Côa et al., 2020). Long-term monitoring studies can provide valuable data on their environmental fate and any potential cumulative effects. This information can inform regulatory decisions and guide adjustments to guidelines and restrictions, if necessary. Addressing these future directions and challenges will require continued research, collaboration, and knowledge exchange among scientists, policymakers, growers, and other stakeholders. By addressing environmental and safety concerns and establishing regulatory frameworks, integrating bio-nanomaterials with other management strategies can be optimized to enhance nematode control while minimizing potential risks.

6.10 CONCLUSIONS AND FUTURE OUTLOOK

Bio-nanomaterials show promising prospects for managing plant-parasitic nematodes in commercial crops. Their unique properties and mechanisms of action make them practical tools for nematode control. They can specifically target nematodes, reduce populations, and enhance crop productivity. Integrating bio-nanomaterials with traditional management strategies enhances their overall effectiveness. However, challenges include environmental and safety concerns requiring thorough assessments and regulatory guidelines. Future directions involve enhancing mechanistic understanding, optimizing formulations, conducting field-scale evaluations, developing integrated management strategies, addressing risk assessments, and fostering collaboration. By addressing these aspects, bio-nanomaterials can contribute to sustainable nematode management in commercial crops, improving agricultural productivity and food security.

REFERENCES

Abbassy, M. A., Abdel-Rasoul, M. A., Nassar, A. M., & Soliman, B. S. (2017). Nematocidal activity of silver nanoparticles of botanical products against root-knot nematode, Meloidogyne incognita. *Archives of phytopathology Phytopathology and plant Plant Protection*, *50*(17–18), 909–926.

Ahn, J. M., Edom, H. J., Yang, X., Meyer, J. N., & Choi, J. (2014). Comparative toxicity of silver nanoparticles on oxidative stress and DNA damage in the nematode, Caenorhabditis elegans. *Chemosphere*, *108*, 343–352.

Akhter, G., Khan, A., Ali, S. G., Khan, T. A., Siddiqi, K. S., & Khan, H. M. (2020). Antibacterial and nematocidal properties of biosynthesized Cu nanoparticles using extract of holoparasitic plant. *SN Applied Sciences*, *2*, 1–6.

Alamri, S., Nafady, N. A., El-Sagheer, A. M., El-Aal, M. A., Mostafa, Y. S., Hashem, M., & Hassan, E. A. (2022). Current utility of arbuscular mycorrhizal fungi and hydroxyapatite nanoparticles in suppression of Tomato Root-Knot Nematode. *Agronomy*, *12*(3), 671.

Alfy, H., Ghareeb, R. Y., Soltan, E., & Farag, D. A. (2020). Impact of chitosan nanoparticles as insecticide and nematocide against *Spodoptera littoralis*, *Locusta migratoria*, and *Meloidogyne incognita*. *Plant Cell Biotechnology Molecular Biology*, *21*, 126–140.

An, C., Sun, C., Li, N., Huang, B., Jiang, J., Shen, Y., ... & Wang, Y. (2022). Nano-materials and nanotechnology for the delivery of agrochemicals: strategies towards sustainable agriculture. *Journal of Nanobiotechnology*, *20*(1), 1–19.

Ansari, R. A., & Khan, T. A. (2012). Parasitic association of root-knot nematode, *Meloidogyne incognita* on guava. *e e-Journal of Science Technology*, *5*(12), 65–67.

Ansari, R. A., & Mahmood, I, . (2019a). *Plant Health Under Biotic Stress – Volume II, Microbial Interactions*, Springer Nature, Singapore. ISBN: 978-981-13-6040-4; DOI: 10.1007/978-981-13-6040-4

Ansari, R. A., & Mahmood, I, . (2019b). *Plant Health Under Biotic Stress – Volume I, Organic Strategies*, Springer Nature, Singapore. ISBN: 978-981-13-6043-5; DOI: 10.1007/978-981-13-6043-5

Ansari, R. A., & Mahmood, I. (2017a). Determination of disease incidence caused by *Meloidogyne* spp. and or *Fusarium udum* on pigeonpea in Aligarh district: a survey. *Trends in Biosciences*, 10(24), 5239–5243.

Ansari, R. A., & Mahmood, I. (2017b). Optimization of organic and bio-organic fertilizers on soil properties and growth of pigeon pea. *Scientia Horticulturae*, *226*, 1–9.

Ansari, R. A., & TATa, K. (2012b). Diversity and community structure of phytonematodes associated with guava in and around Aligarh, Uttar Pradesh, India. *Trends in Biosciences, Trends Biosci*, *5*(3), 202–204.

Ansari, R. A., Mahmood, I., Rizvi, R., Sumbul, A., & Safiuddin. (2017). Siderophores: augmentation of soil health and crop productivity. *Probiotics in agroecosystem*, 291–312.

Ansari, R. A., Rizvi, R. & Mahmood, I. (2020). *Management of Phytonematodes: Management of Phytonematodes: Recent Advances and Future Challenges*, Springer Nature, Singapore. ISBN: 978-981-15-4087-5; DOI: 10.1007/978-981-15-4087-5

Ansari, R. A., Rizvi, R., & Mahmood, I. (Eds.). (2024). Mycorrhizal Symbiosis and Agroecosystem Restoration. Springer Nature Singapore, Imprint: Springer.

Ansari, R. A., Rizvi, R., Sumbul, A., & Mahmood, I. (2020). Plant-growth-promoting Rhizobacteria (PGPR)-based sustainable management of phytoparasitic nematodes: current understandings and future challenges. *Management of phytonematodes: Recent Advances and Future Challenges*, 51–71.

Ansari, R. A., Sumbul, A., Rizvi, R., & Mahmood, I. (2019). Organic soil amendments: potential tool for soil and plant health management. *Plant Health Under Biotic Stress: Volume 1: Organic Strategies*, 1–35.

Ansari, R. A., Sumbul, A., Rizvi, R., & Mahmood, I. (2024). Impact of environmental gases on mycorrhizal symbiosis and its influence on ecosystem functioning under the current climate change scenario. In *Mycorrhizal Symbiosis and Agroecosystem Restoration* (pp. 51–76). Singapore: Springer Nature Singapore.

Aprista-Fernández, A., Fernández, A., & Blasco, J. (2012). Nano ecotoxicity effects of engineered silver and gold nanoparticles in aquatic organisms. *TrAC Trends in Analytical Chemistry*, *32*, 40–59.

Ardakani, A. S. (2013). Toxicity of silver, titanium and silicon nanoparticles on the root-knot nematode, *Meloidogyne incognita*, and growth parameters of tomato. *Nematology*, *15*(6), 671–677.

Azeez, L., Lateef, A., Wahab, A. A., Rufai, M. A., Salau, A. K., Ajayi, E. I., ... & Adebisi, B. (2019). Phytomodulatory effects of silver nanoparticles on *Corchorus olitorius*: its antiphytopathogenic and hepatoprotective potentials. *Plant Physiology and Biochemistry*, *136*, 109–117.

Baronia, R., Kumar, P., Singh, S. P., & Walia, R. K. (2020). Silver nanoparticles as a potential nematocide against. *Journal of Nematology*, *52*(1), 1–9.

Bernard, G. C., Fitch, J., Min, B., Shahi, N., Egnin, M., Ritte, I., ... & Bonsi, C. (2019). Potential nematocidal activity of silver nanoparticles against the root-knot nematode (Meloidogyne incognita). *Online Journal of Complementary & Alternative Medicine*, *2*, 000531.

Bhau, B. S., Phukon, P., Ahmed, R., Gogoi, B., Borah, B., Baruah, J. & Wann, S. B. (2016). A novel tool of nanotechnology: nanoparticle mediated control of nematode infection in plants. *In: Microbial Inoculants in Sustainable Agricultural Productivity: Vol. 2: Functional Applications*, 253–269.

Bhuiyan, S. A., Garlick, K., & Piperidis, G. (2019). *Saccharum spontaneum*, a novel source of resistance to root-lesion and root-knot nematodes in sugarcane. *Plant Disease*, *103*(9), 2288–2294.

Biancardi, E., McGrath, J. M., Panella, L.W., Lewellen, R. T. & Stevanato, P. (2010). Sugar Beet. In: Bradshaw, J. (eds) *Root and Tuber Crops. Handbook of Plant Breeding*, vol 7. 173–219, Springer, New York, NY.

Biradar, S. P., Mehta, M. R., Mahajan, H. P., Bankhele, R. R., & Hivrale, A. U. (2022). Biological activities of chitosan-based nanomaterials. In *Role of chitosan and chitosan-based nanomaterials in plant sciences* (pp. 119–142). Academic Press.

Boddula, R., Trivedi, U., Pothu, R., Rajput, M. S., & Saran, A. (2019). Nanopesticides and Nanosensors in Agriculture. *Plant Nanobionics: Volume 1, Advances in the Understanding of Nanomaterials Research and Applications*, 165–181.

Boros, B. V., & Ostafe, V. (2020). Evaluation of ecotoxicology assessment methods of nano-materials and their effects. *Nano-Materials*, *10*(4), 610.

Boyno, G., Ansari, R. A., Durak, E. D., Güneş, H., Çevik, R., & Demir, S. (2024). Arbuscular Mycorrhizal Technology in Sustainable Agriculture: Current Knowledge and Challenges in Agroforestry. *Mycorrhizal Symbiosis and Agroecosystem Restoration*, 173–195.

Bratovcic, A., Hikal, W. M., Said-Al Ahl, H. A., Tkachenko, K. G., Baeshen, R. S., Sabra, A. S., & Sany, H. (2021). Nanopesticides and nanofertilizers and agricultural development: Scopes, advances and applications. *Open Journal of Ecology*, *11*(4), 301–316.

Camara, M. C., Campos, E. V. R., Monteiro, R. A., do Espirito Santo Pereira, A., de Freitas Proença, P. L., & Fraceto, L. F. (2019). Development of stimuli-responsive nano-based pesticides: emerging opportunities for agriculture. *Journal of Nanobiotechnology*, *17*(1), 1–19.

Cao, J., Guenther, R. H., Sit, T. L., Lommel, S. A., Opperman, C. H., & Willoughby, J. A. (2015). Development of abamectin loaded plant virus nanoparticles for efficacious plant parasitic nematode control. *ACS Applied Materials & Interfaces*, *7*(18), 9546–9553.

Cao, Y., Qiao, Y., Cui, S., & Ge, J . (2022). Origin of metal cluster tuning enzyme activity at the bio-nano interface. *JACS Au*, *2*(4), 961–971.

Chang, Y. N., Zhang, M., Xia, L., Zhang, J., & Xing, G. (2012). The toxic effects and mechanisms of CuO and ZnO nanoparticles. *Materials*, *5*(12), 2850–2871.

Chariou, P. L., & Steinmetz, N. F. (2017). Delivery of pesticides to plant parasitic nematodes using tobacco mild green mosaic virus as a nanocarrier. *ACS Nano*, *11*(5), 4719–4730.

Chatterjee, N., Eom, H. J., & Choi, J. (2014). Effects of silver nanoparticles on oxidative DNA damage–repair as a function of p38 MAPK status: A a comparative approach using human Jurkat T cells and the nematode *Caenorhabditis elegans*. *Environmental and Molecular Mutagenesis*, *55*(2), 122–133.

Chitwood, D. J. (2002). Phytochemical based strategies for nematode control. *Annual review Review of Phytopathology*, *40*(1), pp. 221–249.

Côa, F., Bortolozzo, L. S., Petry, R., Da Silva, G. H., Martins, C. H., de Medeiros, A. M. & Martinez, D. S. T. (2020). Environmental toxicity of nanopesticides against non-target organisms: the state of the art. *Nanopesticides: From research and development to mechanisms of action and sustainable use in agriculture*, 227–279.

Colmenares, D., Sun, Q., Shen, P., Yue, Y., McClements, D. J., & Park, Y. (2016). Delivery of dietary triglycerides to *Caenorhabditis elegans* using lipid nanoparticles: Nanoemulsion-based delivery systems. *Food Chemistry*, *202*, 451–457.

Cooke, D. A., & Scott, J. E. (2012). The sugar beet crop. Springer Science & Business Media.

Dahl, J. A., Maddux, B. L., & Hutchison, J. E. (2007). Toward greener nanosynthesis. *Chemical Reviews*, *107*(6), 2228–2269.

Daramola, F. Y., Popoola, J. O., Eni, A. O., & Sulaiman, O. (2015). Characterization of root-knot nematodes (*Meloidogyne* spp.) associated *with Abelmoschus esculentus, Celosia argentea* and *Corchorus olitorius*. *Asian Journal of Biological Sciences*, *8*(1), 42–50.

Das, S. (2013). A review on Carbon nano-tubes-A new era of nanotechnology. *International Journal of Emerging Technology and Advanced Engineering*, *3*(3), 774–783.

de Oliveira, J. L., Campos, E. V. R., Bakshi, M., Abhilash, P. C., & Fraceto, L. F. (2014). Application of nanotechnology for the encapsulation of botanical insecticides for sustainable agriculture: prospects and promises. *Biotechnology Advances*, *32*(8), 1550–1561.

Devi, T. S., Das, D., Ansari, R. A., Rizvi, R., Sumbul, A., & Mahmood, I. (2020). Role of organic additives in the sustainable management of phytoparasitic nematodes. *Management of Phytonematodes: Recent Advances and Future Challenges*, 279–295.

Dhillon, G. S., Brar, S. K., Kaur, S., & Verma, M. (2012). Green approach for nanoparticle biosynthesis by fungi: current trends and applications. *Critical reviews Reviews in Biotechnology*, *32*(1), 49–73.

Ding, X., Rui, Q., Zhao, Y., Shao, H., Yin, Y., Wu, Q., & Wang, D. (2018). Toxicity of graphene oxide in nematodes with a deficit in the epidermal barrier caused by RNA interference knockdown of unc-52. *Environmental Science & Technology Letters*, *5*(11), 622–628.

Djiwanti, S. R., & Kaushik, S. (2019). Nanopesticide: future application of nanomaterials in plant protection. *Plant Nanobionics: Volume 2, Approaches in Nanoparticles, Biosynthesis, and Toxicity*, 255–298.

Dong, S., Qu, M., Rui, Q., & Wang, D. (2018). Combinational effect of titanium dioxide nanoparticles and nanopolystyrene particles at environmentally relevant concentrations on nematode Caenorhabditis elegans. *Ecotoxicology and environmental Environmental Safety*, *161*, 444–450.

Du, J., Wang, C., Liu, Y., Xue, C., Ge, J., Si, G. & Li, B. (2023). One-pot construction of epoxy resin nanocarrier delivering abamectin and its efficacy on plant root-knot nematodes. *Pest Management Science*, *79*(9), 3103–3113.

Du, J., Wang, C., Liu, Y., Xue, C., Ge, J., Si, G., ... & Li, B. (2023). One-pot construction of epoxy resin nanocarrier delivering abamectin and its efficacy on plant root-knot nematodes. *Pest Management Science*, 1–11.

Dubey, A., & Mailapalli, D. R. (2016). Nanofertilisers, nanopesticides, nanosensors of pest and nanotoxicity in agriculture. *Sustainable Agriculture Reviews: Volume 19*, 307–330.

Durán, N., & Seabra, A. B. (2012). Metallic oxide nanoparticles: state of the art in biogenic syntheses and their mechanisms. *Applied Microbiology and Biotechnology*, 95, 275–288.

Dwivedi, S., Saquib, Q., Al-Khedhairy, A. A., & Musarrat, J. (2016). Understanding the role of nanomaterials in agriculture. *Microbial Inoculants in Sustainable Agricultural Productivity: Vol. 2: Functional Applications*, 271–288.

Ealia, S. A. M., & Saravanakumar, M. P. (2017, November). A review on the classification, characterisation, synthesis of nanoparticles and their application. In *IOP conference Conference Series: Materials Science and Engineering* (Vol. 263, No. 3, p. 032019). IOP Publishing.

El-Ansary, M. S. M., Hamouda, R. A., & Elshamy, M. M. (2022). Using biosynthesized zinc oxide nanoparticles as a pesticide to alleviate the toxicity on banana infested with parasitic-nematode. *Waste and biomass valorization*, 1–11.

Elmer, W., Ma, C., & White, J. (2018). Nanoparticles for plant disease management. *Current Opinion in Environmental Science & Health*, *6*, 66–70.

Fabiyi, O. A., Alabi, R. O., & Ansari, R. A. (2020). Nanoparticles' synthesis and their application in the management of phytonematodes: An overview. *Management of phytonematodes: recent advances and future challenges*, 125–140.

Fabiyi, O. A., Ansari, R. A., Rizvi, R., Sumbul, A., & Mahmood, I. (2024). Influence of arbuscular mycorrhizal fungi on soil health amelioration and plant fitness under hostile environment. In *Mycorrhizal Symbiosis and Agroecosystem Restoration* (pp. 227–248). Singapore: Springer Nature Singapore.

Fabiyi, O., Olatunji, G., Atolani, O., & Olawuyi, O. (2020). Preparation of bio-nematocidal nanoparticles of Eucalyptus officinalis for the control of cyst nematode (Heterodera sacchari). *Journal of Animal Plant Science*, *30*(5), 1172–1177.

Fadei Tehrani, A., & Fati, Z. (2020). Effect of Silver silver and Zinc zinc Oxide oxide Nanoparticles nanoparticles on Sugar sugar Beet beet Cyst Nematode nematode (Heterodera schachtii). *Journal of Applied Research in Plant Protection*, *8*(4), 47–59.

Fakhrullin, R., Nigamatzyanova, L., & Fakhrullina, G. (2021). Dark-field/hyperspectral microscopy for detecting nanoscale particles in environmental nanotoxicology research. *Science of the Total Environment*, *772*, 145478.

Fan, Z., Wang, L., Qin, Y., & Li, P. (2023). Activity of chitin/chitosan/chitosan oligosaccharide against plant pathogenic nematodes and potential modes of application in agriculture: A review. *Carbohydrate Polymers*, *306*, 120592.

Faunce, T., & Watal, A. (2010). Nanosilver and global public health: international regulatory issues. *Nanomedicine*, *5*(4), 617–632.

Feofilova, E. P., & Mysyakina, I. S. (2016). Lignin: Chemical chemical structure, biodegradation, and practical application (a review). *Applied Biochemistry and Microbiology*, *52*, 573–581.

Fouda, M. M., Abdelsalam, N. R., Gohar, I. M. A., Hanfy, A. E., Othman, S. I., Zaitoun, A. F., ... & El-Naggar, M. (2020). Utilization of High high throughput microcrystalline cellulose decorated silver nanoparticles as an eco-nematicidnematocide on root-knot nematodes. *Colloids and Surfaces B: Biointerfaces*, *188*, 110805.

Gad, S. B. (2019). Efficacy of soaking cotton seeds within salicylic acid and potassium silicate on reducing reniform nematode infection. *Archives of Phytopathology and Plant Protection*, *52*(15–16), 1149–1160.

Gan, P. P., & Li, S. F. Y. (2012). Potential of plant as a biological factory to synthesize gold and silver nanoparticles and their applications. *Reviews in Environmental Science and Bio/Technology, 11*, 169–206.

Gangwar, J., Kadanthottu Sebastian, J., Puthukulangara Jaison, J., & Kurian, J. T. (2023). Nano-technological interventions in crop production—A review. *Physiology and Molecular Biology of Plants*, *29*(1), 93–107.

Gareeb, R. Y., Elnouby, M. S., Hasan, M. A., Ticu, S. I. M. O. N. A., Popa, A., Bungau, S., & Hafez, E. E. (2019). New trend for using the reduced graphene oxide as effective and eco-friendly Nematicide. *Mater Plast*, *56*(1).

Gohar, I., Alyamani, A., Shafi, M. E., Mohamed, E. A., Ghareeb, R. Y., Desoky, E. M., ... & Elnahal, A. S. (2023). A quantitative and qualitative assessment of sugar beet genotype resistance to root-knot nematode, Meloidogyne incognita. *Frontiers in Plant Science*, *13*, 966377.

Gonzalez Gomez, A., Syed, S., Marshall, K., & Hosseinidoust, Z. (2019). Liposomal nanovesicles for efficient encapsulation of staphylococcal antibiotics. *ACS Omega*, *4*(6), 10866–10876.

Gonzalez-Moragas, L., Yu, S. M., Benseny-Cases, N., Stürzenbaum, S., Roig, A., & Laromaine, A. (2017). Toxicogenomics of iron oxide nanoparticles in the nematode C. elegans. *Nanotoxicology*, *11*(5), 647–657.

Gupta, P. K., Bhandari, N., Shah, H. N., Khanchandani, V., Keerthana, R., Nagarajan, V., & Hiremath, L. (2019). An update on nanoemulsions using nanosized liquid in liquid colloidal systems. İn *Nanoemulsions-Properties, Fabrications and Applications* (pp. 1–20).

Hallmann, J., Gutberlet, V., Jakobs-Schönwandt, D., Vorlop, K. D., Müller, J., & Patel, A. V. (2019). Effect of additives on the efficacy of microencapsulated Hirsutella rhossiliensis controlling Heterodera schachtii on sugar beets. *Biological Control*, *128*, 40–47.

Handy, R. D., van den Brink, N., Chappell, M., Mühling, M., Behra, R., Dušinská, M., ... & Riediker, M. (2012). Practical considerations for conducting ecotoxicity test methods with manufactured nanomaterials: what have we learnt so far?. *Ecotoxicology*, *21*, 933–972.

Hanna, S. K., Cooksey, G. A., Dong, S., Nelson, B. C., Mao, L., Elliott, J. T., & Petersen, E. J. (2016). Feasibility of using a standardized Caenorhabditis elegans toxicity test to assess nano-material toxicity. *Environmental Science: Nano*, *3*(5), 1080–1089.

Haque, S. M. A., Mosaddeque, H. Q. M., Sultana, K., Islam, M. N., & Rahman, M. L. (2008). Effect of different trap crops against root knot nematode disease of jute. *Journal of Innovation and Development Strategy*, *2*(3), 42–47.

Hidangmayum, A., & Dwivedi, P. (2022). Chitosan based nanoformulation for sustainable agriculture with special reference to abiotic stress: a review. *Journal of Polymers and the Environment*, *30*(4), 1264–1283.

Hosseinnejad, M., & Jafari, S. M. (2016). Evaluation of different factors affecting antimicrobial properties of chitosan. *International journal Journal of biological Biological Macromolecules*, *85*, 467–475.

Hou, J., Hu, C., Li, P., & Lin, D. (2022a). Multidimensional bioresponses in nematodes contribute to the antagonistic toxic interaction between pentachlorophenol and TiO2 nanoparticles in soil. *Journal of Hazardous Materials*, *424*, 127587.

Hou, J., Hu, C., Wang, Y., Zhang, J., White, J. C., Yang, K., & Lin, D. (2022b). Nano-bio interfacial interactions determined the contact toxicity of nTiO2 to nematodes in various soils. *Science of The Total Environment*, *835*, 155456.

Huang, C. W., Li, S. W., & Liao, V. H. C. (2019). Long-term sediment exposure to ZnO nanoparticles induces oxidative stress in *Caenorhabditis elegans*. *Environmental Science: Nano*, *6*(8), 2602–2614.

Husen, A., & Iqbal, M. (2019). *Nanomaterials and plant Plant potentialPotential: an An overview Overview* (pp. 3–29). Springer International Publishing.

Hutchison, J. E. (2008). Greener nanoscience: a proactive approach to advancing applications and reducing implications of nanotechnology. *ACS Nano*, 2(3), 395–402.

Ibrahim, D. S. S., Mostafa, M., Abdellatif, A. A., & Ansari, R. A. (2022). Sustainable management of soil-borne plant pathogens. In *Earth Systems Protection and Sustainability: Volume 1* (pp. 35–56). Cham: Springer International Publishing.

Ibrahim, D. S.S., Elderiny, M. M., Ansari, R. A., Rizvi, R., Sumbul, A., & Mahmood, I. (2020). Role of *Trichoderma* spp. in the management of plant-parasitic nematodes infesting important crops. *Management of Phytonematodes: Recent Advances and Future Challenges*, 259–278.

Javed, Z., Dashora, K., Mishra, M., Fasake, V. D., & Srivastva, A. (2019). Effect of accumulation of nanoparticles in soil health-a concern on future. *Frontiers in Nanoscience Nanotechnology*, *5*, 1–9.

Jin, T., Liu, T., Lam, E., & Moores, A. (2021). Chitin and chitosan on the nanoscale. *Nanoscale Horizons*, *6*(7), 505–542.

Kalaba, M. H., Moghannem, S. A., El-Hawary, A. S., Radwan, A. A., Sharaf, M. H., & Shaban, A. S. (2021). Green synthesized ZnO nanoparticles mediated by Streptomyces plicatus: Characterizations, antimicrobial and nematocidal activities and cytogenetic effects. *Plants*, *10*(9), 1760.

Kalaiselvi, D., Mohankumar, A., Shanmugam, G., Nivitha, S., & Sundararaj, P. (2019). Green synthesis of silver nanoparticles using latex extract of Euphorbia tirucalli: a novel approach for the management of root knot nematode, Meloidogyne incognita. *Crop Protection*, *117*, 108–114.

Kalita, R., Saha, O., Rahman, N., Tiwari, S., & Phukon, M. (2021). Nanotechnology in agriculture. *Nanobiotechnology: Mitigation of Abiotic Stress in Plants*, 101–116.

Kausar, S. (2022). Application of copper-based nano-materials against parasitic nematodes. In *Copper Nanostructures: Next-Generation of Agrochemicals for Sustainable Agroecosystems* (pp. 263–290). Elsevier.

Keshari, N., Kranti, K. V. V. S. K., Gunda, N. K., & Ansari, R. A. (2024). Arbuscular mycorrhizal fungi: A a potential agent for phytonematodes management in diverse agro-climatic zones. In *Mycorrhizal Symbiosis and Agroecosystem Restoration* (pp. 147–169). Singapore: Springer Nature Singapore.

Khader, A., Ibrahim, M., Alkhathami, F., Elsawy, H., El-Kader, N. A., Shaker, E., ... & Mahmoud, E. (2023). Properties of Nanonano-Amendments and Their Effect on Some Soil Properties and Root-Knot Nematode and Yield Attributes of Tomato Plant. *Agriculture*, *13*(2), 366.

Khalil, M. S. & Badawy, M.E.I (2012). Nematicidal activity of a biopolymer chitosan at different molecular weights against root-knot nematode, *Meloidogyne incognita*. Plant Protection Science, *Prague, 48*(4): 170–178.

Khalil, M. S., & Badawy, M. E. (2012). Nematocidal activity of a biopolymer chitosan at different molecular weights against root-knot nematode, Meloidogyne incognita. *Plant Protection Science*, *48*(4), 170–178.

Khan, F., Ansari, T., Shariq, M., & Siddiqui, M. A. (2021). Nanotechnology: A a New new Beginning beginning to Mitigate mitigate the Effect effect of Plant-Parasitic Nematodes. In *Innovative Approaches in Diagnosis and Management of Crop Diseases* (pp. 19–43). Apple Academic Press.

Khan, S. M., Ali, S., Nawaz, A., Bukhari, S. A. H., Ejaz, S., & Ahmad, S. (2019). Integrated pest and disease management for better agronomic crop production. *Agronomic Crops: Volume 2: Management Practices*, 385–428.

Khandelwal, N., Barbole, R. S., Banerjee, S. S., Chate, G. P., Biradar, A. V., Khandare, J. J., & Giri, A. P. (2016). Budding trends in integrated pest management using advanced micro-and nano-materials: Challenges challenges and perspectives. *Journal of Environmental Management*, *184*, 157–169.

Khare, P., Sonane, M., Pandey, R., Ali, S., Gupta, K. C., & Satish, A. (2011). Adverse effects of TiO_2 and ZnO nanoparticles in soil nematode, Caenorhabditis elegans. *Journal of Biomedical Nanotechnology*, *7*(1), 116–117.

Kitherian, S. (2017). Nano and bio-nanoparticles for insect control. *Research Journal of Nanoscience and Nanotechnology*, *7*(1), 1–9.

Kralova, K., & Jampilek, J. (2023). Applications of nanomaterials in plant disease management and protection. In *Nanotechnology in agriculture and agroecosystems* (pp. 239–296). Elsevier.

Kumar, A., Choudhary, A., Kaur, H., Guha, S., Mehta, S., & Husen, A. (2022). Potential applications of engineered nanoparticles in plant disease management: a critical update. *Chemosphere*, *295*, 133798.

Kumar, D., Kumar, V., Kumar, A., & Mann, S. S. (2021). Root-Knot nematode problem and their management in cotton crop. *ENTOMOLOGY*, 105.

Kumar, K., Sridhar, J., Choudhary, V. K., Singh, H. K., Parameshwari, B., Kumar, K. S., & Sivalingam, P. N. (2021). New innovations and approaches for biotic stress management of crops. In *Innovations in Agriculture for a Self-Reliant India* (pp. 265–292). CRC Press.

Kumari, R., & Singh, D. P. (2020). Nano-biofertilizer: an emerging eco-friendly approach for sustainable agriculture. *Proceedings of the National Academy of Sciences, India Section B: Biological Sciences*, *90*, 733–741.

Lalitha, K., Kalaimurgan, D., Nithya, K., Venkatesan, S., & Shivakumar, M. S. (2020). Antibacterial, antifungal and mosquitocidal efficacy of copper nanoparticles synthesized from entomopathogenic nematode: Insectinsect–host relationship of bacteria in secondary metabolites of Morganella morganii sp.(PMA1). *Arabian Journal for Science and Engineering*, *45*, 4489–4501.

Lamitina, S. T., Morrison, R., Gilbert, W. M. & Strange, K. (2004). Adaptation of the nematode *Caenorhabditis elegans* to extreme osmotic stress. *American J. Physiology-Cell Physiology*, *286*:4, C785–C791.

Li, X., Hu, J., Qiu, R., Zhang, X., Chen, Y., & He, D. (2020). Joint toxic effects of polystyrene nanoparticles and organochlorine pesticides (chlordane and hexachlorocyclohexane) on Caenorhabditis elegans. *Environmental Science: Nano*, *7*(10), 3062–3073.

Li, X., Xing, X., Xu, S., Zhang, M., Wang, Y., Wu, H., ... & Yang, T. (2018). Genome-wide identification and functional prediction of tobacco lncRNAs responsive to root-knot nematode stress. *PLoS One*, *13*(11), e0204506.

Li, Y., Zhong, L., Zhang, L., Shen, X., Kong, L., & Wu, T. (2021). Research advances on the adverse effects of nanomaterials in a model organism, *Caenorhabditis elegans*. *Environmental Toxicology and Chemistry*, *40*(9), 2406–2424.

Lichtenberg, S. S., Tsyusko, O. V., Palli, S. R., & Unrine, J. M. (2019). Uptake and bioactivity of chitosan/double-stranded RNA polyplex nanoparticles in Caenorhabditis elegans. *Environmental Science & Technology*, *53*(7), 3832–3840.

Luo, X., Xu, S., Yang, Y., Li, L., Chen, S., Xu, A., & Wu, L. (2016). Insights into the ecotoxicity of silver nanoparticles transferred from Escherichia coli to Caenorhabditis elegans. *Scientific Reports*, *6*(1), 36465.

Ma, H., Bertsch, P. M., Glenn, T. C., Kabengi, N. J., & Williams, P. L. (2009). Toxicity of manufactured zinc oxide nanoparticles in the nematode Caenorhabditis elegans. *Environmental Toxicology and Chemistry: An International Journal*, *28*(6), 1324–1330.

Machado, A. C. Z., Ferraz, L. C. C. B., & Inomoto, M. M. (2012). Pathogenicity of Pratylenchus brachyurus on cotton plants. *Journal of Cotton Science*, *16*(4), 268–271.

Mahmood, I., Rizvi, R., Sumbul, A., & Ansari, R. A. (2019). Potential role of plant growth promoting Rhizobacteria in alleviation of biotic stress. *Plant Health Under Biotic Stress: Volume 2: Microbial Interactions*, 177–188.

Mallaiah, B., & Muthamillan, M. (2022). Changes in Activity activity of Phenylalanine phenylalanine Ammonia Lyase (PAL), Phenols phenols and β 1, 3 Glucanase in Crossandra Plants affected by Fusarium incarnatum (Desm.) Sacc and treated with Bioagents, Organic organic Amendment Silver Nanoparticles and Fungicide in Pot Culture. *International Journal of Environment and Climate Change*, *12*(10), 1052–1059.

Manteghi, A., Danesh, Y. R., Hesar, A. M., Demir, S., Boyno, G., Catani, L., & Semprucci, F. (2022). Is the Arbuscular arbuscular Mycorrhizal mycorrhizal Fungus fungus *Funneliformis mosseae* a Suitable suitable Agent agent to Control control Criconematid Populations. *Diversity*, *14*(11), 898.

McCourt, K. M., Cochran, J., Abdelbasir, S. M., Carraway, E. R., Tzeng, T. R. J., Tsyusko, O. V., & Vanegas, D. C. (2022). Potential Environmental environmental and Health health Implications implications from the Scaled-Up up Production and Disposal of Nano-materials Used in Biosensors. *Biosensors*, *12*(12), 1082.

Mohamed, E. A., Elsherbiny, S. F., & Abdulsamad, D. (2019). Evaluation of in vitro nematocidal efficiency of copper nanoparticles against root-knot nematode Meloidogyne incognita. *South Asian Journal of Parasitology*, *2*(1), 1–6.

Mohammed, L. H., & Reda, I. M. A. (2021). Evaluation the efficiency of some tomato hybrids, biological and chemical factors against tomato root knot caused by *Meloidogyne* spp. *Journal of Kerbala for Agricultural Sciences*, *8*(2), 22–35.

Moon, J., Kwak, J. I., & An, Y. J. (2019). The effects of silver nano-material shape and size on toxicity to Caenorhabditis elegans in soil media. *Chemosphere*, *215*, 50–56.

Morsy, K., Fahmy, S., Mohamed, A., Ali, S., El–Garhy, M., & Shazly, M. (2019). Optimizing and evaluating the antihelminthic activity of the biocompatible zinc oxide nanoparticles against the ascaridid nematode, Parascaris equorum in vitro. *Acta Parasitologica*, *64*, 873–886.

Mouniga, R., Anita, B., Shanthi, A., Lakshmanan, A., & Karthikeyan, G. (2022). Phenol and antioxidant enzymatic activity in root knot nematode, *Meloidogyne incognita* infected tomato plants treated with chitosan nanoparticles. *The Pharma Innovation Journal*, *11*(4), 241–245.

Nikam, P. B., Salunkhe, J. D., Mohite, B. V., Chaudhari, R. S., & Patil, S. V. (2023). Biogenic synthesis of selenium nanomaterial and its application as anti-Nematode booster in Solanum lycopersicum (Tomato). *Vegetos*, *36*(4), 1458–1464.

Nuruzzaman, M. D., Rahman, M. M., Liu, Y., & Naidu, R. (2016). Nanoencapsulation, nano-guard for pesticides: a new window for safe application. *Journal of Agricultural and Food Chemistry*, *64*(7), 1447–1483.

Nuruzzaman, M., Liu, Y., Rahman, M. M., Dharmarajan, R., Duan, L., Uddin, A. F. M. J., & Naidu, R. (2019). Nanobiopesticides: Composition and preparation methods. In *Nano-biopesticides today and future perspectives* (pp. 69–131). Academic Press.

Pan, X., Nichols, R. L., Li, C., & Zhang, B. (2019). MicroRNA-target gene responses to root knot nematode (Meloidogyne incognita) infection in cotton (*Gossypium hirsutum* L.). *Genomics*, *111*(3), 383–390.

Patil, S., Ansari, R. A., Gurikar, C., Bhavan, S., & Hariprasad, K. (2024). Propagation of mycorrhizal fungi and their strength in plant health ameliorations. In *Mycorrhizal Symbiosis and Agroecosystem Restoration* (pp. 77–98). Singapore: Springer Nature Singapore.

Perea Velez, Y. S., Carrillo-González, R., & Gonzalez-Chavez, M. D. C. A. (2021). Interaction of metal nanoparticles–plants–microorganisms in agriculture and soil remediation. *Journal of Nanoparticle Research*, *23*(9), 206.

Piao, X., Zhang, L., Zhang, S., & Yi, F. (2020). Nematocidal action of microencapsulated essential oil of flesh fingered citron. *Journal of Chemistry*, *2020*, 1–7.

Polak, N., Read, D. S., Jurkschat, K., Matzke, M., Kelly, F. J., Spurgeon, D. J., & Stürzenbaum, S. R. (2014). Metalloproteins and phytochelatin synthase may confer protection against zinc oxide nanoparticle induced toxicity in *Caenorhabditis elegans*. *Comparative Biochemistry and Physiology Part C: Toxicology & Pharmacology*, *160*, 75–85.

Prabha, A. S., Dorothy, R., Jancirani, S., Rajendran, S., Singh, G., & Kumaran, S. S. (2020). Recent advances in the study of toxicity of polymer-based nanomaterials. *Nanotoxicity*, 143–165.

Prasad, B. N., Subramanyam, B., Lakshmipathi, R. N., Ansari, R. A., Rizvi, R., Sumbul, A., & Rachmi, C. M. (2020). Utilization of beneficial microorganisms in sustainable control of phytonematodes. *Management of phytonematodes: Recent advances and future challenges*, 317–337.

Rajasekharan, S. K., & Lee, J. (2020). Hydropic anthelmintics against parasitic nematodes. *PLoS Pathogens*, *16*(1), e1008202.

Rajesh, S., Raja, D. P., Rathi, J. M., & Sahayaraj, K. (2012). Biosynthesis of silver nanoparticles using *Ulva fasciata* (Delile) ethyl acetate extract and its activity against *Xanthomonas campestris* pv. *malvacearum*. *Journal of Biopesticides*, *5*, 119.

Rajwade, J. M., Chikte, R. G., & Paknikar, K. M. (2020). Nano-materials: new weapons in a crusade against phytopathogens. *Applied Microbiology and Biotechnology*, *104*, 1437–1461.

Raveendran, P., Fu, J. & Wallen, S. L. (2003). Completely "Green" Synthesis synthesis and Stabilization of Metal metal Nanoparticles. *Journal of American chemical society*, *125*, 13940–13941.

Rişvanlı, M. R. (2022). *Trichoderma harzianum uygulanmış patates ve pamuk bitkisi üzerinde, Spodoptera exigua'nın popülasyon performansı ve beslenme kapasitesinin belirlenmesi.* Van Yüzüncü Yıl Üniversitesi, Fen Bilimleri Enstitüsü, Van.

Sabry, A. K. H. (2019). Role of nanotechnology applications in plant-parasitic nematode control. *Nanobiotechnology Applications in Plant Protection*, *2*, 223–240.

Sadrieh, N., & Espandiari, P. (2006). Nanotechnology and the FDA: what are the scientific and regulatory considerations for products containing nano-materials. *Nanotech. L. & Bus.*, *3*, 339.

Sasser, J. N. (1980). Root knot nematodes: A a global menace to crop production. *Plant Disease*, *64*, 36–41.

Schultz, C. L., Bart, S., Lahive, E., & Spurgeon, D. J. (2021). What is on the outside matters—surface charge and dissolve organic matter association affect the toxicity and physiological mode of action of polystyrene nanoplastics to C. elegans. *Environmental Science & Technology*, *55*(9), 6065–6075.

Shah, S. S., Shaikh, M. N., Khan, M. Y., Alfasane, M. A., Rahman, M. M., & Aziz, M. A. (2021). Present status and future prospects of jute in nanotechnology: A a review. *The Chemical Record*, *21*(7), 1631–1665.

Shankar, S. S., Rai, A., Ankamwar, B., Singh, A., Ahmad, A., & Sastry, M. (2004). Biological synthesis of triangular gold nanoprisms. *Nature Materials*, *3*(7), 482–488.

Sharmin, S., Rahaman, M. M., Sarkar, C., Atolani, O., Islam, M. T., & Adeyemi, O. S. (2021). Nanoparticles as antimicrobial and antiviral agents: A a literature-based perspective study. *Heliyon*, *7*(3), e06456.

Shatkin, J. A., & Kim, B. (2015). Cellulose nano-materials: life cycle risk assessment, and environmental health and safety roadmap. *Environmental Science: Nano*, *2*(5), 477–499.

Siddiqui, Z. A., Khan, A., Khan, M. R., & Abd-Allah, E. F. (2018). Effects of zinc oxide nanoparticles (ZnO NPs) and some plant pathogens on the growth and nodulation of lentil (Lens culinaris Medik.). *Acta Phytopathologica et Entomologica Hungarica*, *53*(2), 195–211.

Singh, A., Bhardwaj, R., & Singh, I. K. (2019). Biocontrol agents: potential of biopesticides for integrated pest management. *Biofertilizers for sustainable agriculture and environment*, 413–433.

Singh, R. P., Handa, R., & Manchanda, G. (2021). Nanoparticles in sustainable agriculture: An an emerging opportunity. *Journal of Controlled Release*, *329*, 1234–1248.

Sitia, G., Fiordaliso, F., Violatto, M. B., Alarcon, J. F., Talamini, L., Corbelli, A., ... & Bigini, P. (2022). Food-Grade grade Titanium titanium Dioxide dioxide Induces induces Toxicity toxicity in the Nematode

nematode Caenorhabditis elegans and Acute acute Hepatic hepatic and Pulmonary pulmonary Responses responses in Mice. *Nano-materials, 12*(10), 1669.

Solanki, M. K., Kashyap, P. L., Kumari, B., Ansari, R. A., Sumbul, A., Rizvi, R., & Mahmood, I. (2021). Mycorrhizal fungi and its importance in plant health amelioration. In *Microbiomes and Plant Health* (pp. 205–223). Academic Press.

Solanki, M.K., Kashyap, P.L., Ansari, R.A. & Kumari, B. (2020). *Microbiomes and Plant Health: Panoply and Their Applications*, Academic Press, Elsevier, The Netherland.

Sonawane, H., Shelke, D., Camshare, M., Dixit, N., Math, S., Sen, S., ... & Sarma, H. (2022). Fungi-derived agriculturally important nanoparticles and their application in crop stress management–Prospects and environmental risks. *Environmental Research, 212*, 113543.

Sönmez, Ç., Telecoil, İ. H., Yücel, A. M., & Öktem, H. A. (2014). Transgenic Nicotiana tabacum cultivar Samsun plants carrying the wild sugar beet Hs1pro1 gene have resistance to root-knot nematodes. *Turkish Journal of Biology, 38*(2), 200–207.

Srivastava, R. K., Kumar, N., Singh, R. K., & Awasthi, L. P. (2010). Efficacy of Biobio-agents against root-knot nematode on Jute. *Annals of Plant Protection Sciences, 18*(2), 555–556.

Starr, J. L., Koenning, S. R., Kirkpatrick, T. L., Robinson, A. F., Roberts, P. A., & Nichols, R. L. (2007). The future of nematode management in cotton. *Journal of Nematology, 39*(4), 283.

Stavitskaya, A., Shakhbazova, C., Cherednichenko, Y., Nigamatzyanova, L., Fakhrullina, G., Khaertdinov, N., ... & Fakhrullin, R. (2020). Antibacterial properties and in vivo studies of tannic acid-stabilized silver–halloysite nano-materials. *Clay Minerals, 55*(2), 112–119.

Sumbul, A., Ansari, R. A., Rizvi, R., & Mahmood, I. (2020). Azotobacter: A a potential bio-fertilizer for soil and plant health management. *Saudi Journal of Biological Sciences, 27*(12), 3634–3640.

Tenchov, R., Bird, R., Curtze, A. E., & Zhou, Q. (2021). Lipid nanoparticles— from liposomes to mRNA vaccine delivery, a landscape of research diversity and advancement. *ACS Nano, 15*(11), 16982–17015.

Thakur, R. K., & Shirkot, P. (2017). Potential of biogold nanoparticles to control plant pathogenic nematodes. *Journal of Bioanalytical and Biomedicine, 9*(4), 220–222.

Thorn, C. R., Raju, D., Lacdao, I., Gilbert, S., Sivarajah, P., Howell, P. L., ... & Thomas, N. (2021). Protective liquid crystal nanoparticles for targeted delivery of PslG: a biofilm dispersing enzyme. *ACS Infectious Diseases, 7*(8), 2102–2115.

Thounaojam, T. C., Meetei, T. T., Devi, Y. B., Panda, S. K., & Upadhyaya, H. (2021). Zinc oxide nanoparticles (ZnO-NPs): a promising nanoparticle in renovating plant science. *Acta Physiologiae Plantarum, 43*, 1–21.

Tsai, M. H., Chao, H. R., Jiang, J. J., Su, Y. H., Mariene-syne, P. C., Tayo, L. L., ... & Hsu, W. L. (2021). Toxicity of Low-dose Graphene Oxide Nanoparticles in an in-vivo Wild Type of *Caenorhabditis elegans* Model. *Aerosol and Air Quality Research, 21*(5), 200559.

Tuncsoy, B. (2021). Nematocidal activity of silver nano-materials against plant-parasitic nematodes. In *Silver Nano-materials for Agri-Food Applications* (pp. 527–548). Elsevier.

Udalova, Z. V., Folmanis, G. E., Khasanov, F. K., & Zinovieva, S. V. (2018). Selenium nanoparticles—an inducer of tomato resistance to the root-knot nematode Meloidogyne incognita (Kofoid et white, 1919) Chitwood 1949. In *Doklady Biochemistry and Biophysics* (Vol. 482, pp. 264–267). Pleiades Publishing.

Wang, R., Song, B., Wu, J., Zhang, Y., Chen, A., & Shao, L. (2018). Potential adverse effects of nanoparticles on the reproductive system. *International Journal of Nanomedicine, 13*, 8487.

Wu, Q., Li, Y., Li, Y., Zhao, Y., Ge, L., Wang, H., & Wang, D. (2013). Crucial role of the biological barrier at the primary targeted organs in controlling the translocation and toxicity of multi-walled carbon nanotubes in the nematode *Caenorhabditis elegans. Nanoscale, 5*(22), 11166–11178.

Wu, Q., Zhi, L., Qu, Y., & Wang, D. (2016). Quantum dots increased fat storage in intestine of *Caenorhabditis elegans* by influencing molecular basis for fatty acid metabolism. *Nanomedicine: Nanotechnology, Biology and Medicine, 12*(5), 1175–1184.

Wu, T., Xu, H., Liang, X., & Tang, M. (2019). Caenorhabditis elegans as a complete model organism for biosafety assessments of nanoparticles. *Chemosphere, 221*, 708–726.

Xing, X., Li, X., Zhang, M., Wang, Y., Liu, B., Xi, Q., ... & Yang, T. (2017). Transcriptome analysis of resistant and susceptible tobacco (*Nicotiana tabacum*) in response to root-knot nematode *Meloidogyne incognita* infection. *Biochemical and biophysical Biophysical research Research Communications, 482*(4), 1114–1121.

Yu, J., Wang, D., Geetha, N., Khawar, K. M., Jogaiah, S., & Mujtaba, M. (2021). Current trends and challenges in the synthesis and applications of chitosan-based nanocomposites for plants: A a review. *Carbohydrate Polymers*, *261*, 117904.

Yu, J., Yu, X., Li, C., Ayaz, M., Abdulsalam, S., Peng, D., ... & Huang, W. (2022). Silicon Mediated mediated Plant plant Immunity immunity against Nematodes: Summarizing the Underline Defence Mechanisms in Plant Nematodes Interaction. *International Journal of Molecular Sciences*, *23*(22), 14026.

Zhang, D. X., Liu, G., Jing, T. F., Luo, J., Wei, G., Mu, W., & Liu, F. (2020). Lignin-modified electronegative epoxy resin nanocarriers effectively deliver pesticides against plant root-knot nematodes *(Meloidogyne incognita)*. *Journal of Agricultural and Food Chemistry*, *68*(47), 13562–13572.

Zhang, F., You, X., Zhu, T., Gao, S., Wang, Y., Wang, R., ... & Qian, B. (2020). Silica nanoparticles enhance germ cell apoptosis by inducing reactive oxygen species (ROS) formation in *Caenorhabditis elegans*. *The Journal of Toxicological Sciences*, *45*(3), 117–129.

Zheng, N., Sun, D. D., Zou, P., & Jiang, W. (2017). Scientific and regulatory considerations for generic complex drug products containing nano-materials. *The AAPS Journal*, *19*, 619–631.

Zikalala, N., Matshetshe, K., Parani, S., & Oluwafemi, O. S. (2018). Biosynthesis protocols for colloidal metal oxide nanoparticles. *Nano-Structures & Nano-Objects*, *16*, 288–299.

7 Maximizing the Potential of Nanoparticles for Sustainable Plant Pathogen Management

Addressing Phytotoxicity and Residues

Geeta Rautela, Aiman Zafar, Aisha Sumbul, Rizwan Ali Ansari, Irshad Mahmood, Arsh Rehan Siddiqui, and Rose Rizvi

7.1 INTRODUCTION

Nanotechnology has revolutionized many fields of research such as medical sciences, physical sciences, engineering, agriculture and allied sciences. NanoParticles (NPs) exhibit size-dependent qualities, being less than 100 nm in order to enhance their surface area (Akbari et al., 2011). They have an advanced role in improved agriculture, for example, seed germination, hormone delivery, transfer of target genes, nano-biosensors and nano-barcoding, and the like (Hayles et al., 2017). NPs are also employed in the management of plant pathogens such as bacteria and fungi (Worrall et al., 2018). Nanotechnology-driven metal oxide and metal NPs such as silver, zinc oxide, and titanium oxide demonstrate potent antiviral and antifungal properties (Kim et al., 2018). NPs have a great role in sustainable agriculture as plant growth regulators, nanobiosensors, nanofertilizers, in nutrient management, and in protection against phytopathogens and nanopesticides. There are many benefits of NPs but another aspect of this technology is that it might have a negative impact on the environment and this cannot be overlooked (Hong et al., 2014). NPs induce oxidative stress in organisms within the ecosystem. In agriculture, NPs can lead to phytotoxicity and genotoxicity in plants. According to Tolaymat et al., (2015), NPs, when interacting with plants, can affect the food chain and subsequently human health. Toxicity of NPs in plants alters biochemicals, inhibits root elongation, causes DNA mutation and reduces seed germination (Atha et al., 2012). NPs synthesized through physical processes, such as radiation and high temperature, may not align well with ecological considerations (Alzahrani et al., 2015). The chemical methods produce NPs and enable higher production levels within shorter timeframes (He et al., 2015). It is crucial to acknowledge that chemical-based synthesis methods are under scrutiny by researchers due to their potential impact on non-targeted microorganisms within the ecosystem. In contrast, biological synthesis of NPs through living organisms offers a simple, cost-effective, rapid, clean, non-toxic, and eco-friendly alternative (Sharma et al., 2019). Various NPs have been utilized to enhance crop yield and quality by modifying cellular metabolisms, improving nutrient uptake and enhancing enzymatic activity (Podder et al., 2018). The NPs, if combined with fungicides, herbicides, and pesticides exhibited a positive effect in the inhibition of plant pathogens. NPs in the form of biosensors are employed in the

DOI: 10.1201/9781003256762-7

detection of phytopathogens (Elmer et al., 2018). This chapter provides a thorough examination of how NPs are utilized in managing plant diseases caused by various pathogens. Additionally, it examines the increasing concern about how NPs influence plant growth and yield. It provides detailed information insights into addressing potential phytotoxicity concerns related to NPs application and presents viable solutions.

7.2 SYNTHESIS OF NPS

7.2.1 Physical Methods

Physical methods of NP synthesis include colloidal dispersal, spray pyrolysis, thermal evaporation, pulsed laser desorption, sputter deposition, diffusion flame, ball milling, and lithographic techniques (Joeroger et al., 2000). According to Vinayagam et al. (2020) many forces participate such as crystallization, physical fragmentation, and amorphous condensation in the formation of NP synthesis and this method produced uniform, pure and crystalline NPs.

7.2.2 Chemical Methods

Generally non-judicious use of chemicals is very dangerous for the environment and they are very difficult to manage. Examples are: spray pyrolysis, direct precipitation, microemulsion, and micro-assisted combustion. Benefit of these processes is the production of a huge number of NPs. However, they are not as eco-friendly as NPs synthesized through biological means (Nasrollahzadeh et al., 2020).

7.2.3 Biological Methods

The biological method for synthesizing NPs is characterized by its simplicity, lack of toxicity, and cost-effectiveness. Sadiq et al. (2015) demonstrate the utilization of both eukaryotic and prokaryotic organisms in the synthesis of metal and metal oxide NPs. In recent years NPs biofabricated through the bacteria, plant and fungi have exhibited better stability, polydispersity and dimension. The green synthesis of NPs using plant extracts demonstrates antimicrobial activity, eco-friendliness and simplicity. In the biosynthesis of NPs, various organisms such as fungi, algae and plants are utilized, leading to the significant production of NPs (Kumar et al., 2014). By the employment of living organisms, increased NPs production and biocompatibility have been observed (Alavi and Karimi, 2018). Biological systems such as plants, animal cell cultures, and microorganisms are other methods for the production of NPs. Polysaccharide, phenolics, proteins, nucleic acids and amino acids behave like capping and reducing agents when NPs are synthesized through the biological systems (see Table 7.1).

7.2.3.1 Biosynthesis of NPs by Microorganisms

Fungi, bacteria and yeast are utilized for this purpose which have great value in microbial synthesis of NPs. The presence of different biomolecules, enzymes and protein function as reducing agents in the production of NPs. During the incubation period, the microorganisms both intracellular and extracellular synthesize the NPs. Fungi could function as better agents for the synthesis of NPs because fungi generate extracellular enzymes that catalyse the heavy metal ions. The NPs synthesize intracellularly through the microbial cell metal ions, and transform into NPs due to intracellular enzymes in the cells. NPs synthesized from extracellular metal ions adhere to the surface of microbial cells. Cellular enzymes then catalyse the conversion of these metal ions into NPs (Kalabegishvili et al., 2012).

7.2.3.2 Bacterial Mediated NP Synthesis

A suitable cell can serve as a crucial cellular factory for the synthesis of NPs. The *Bacillus* spp. is utilized for the synthesis of AgNPs, size range 10–20 nm (Sunkar and Nhachiyar 2012). Selenium NPs (SeNPs) were synthesized during *Klebesiella pneumoniae* from selenium chloride solution (Fesharaki et al., 2010). CuNPs was fabricated from the *Thermoacetobacter* sp. X513size range from 5–150 and 5–10 nm and the confirmation was achieved by UV-Visible Spectroscopy and FTIR (Jang et al., 2015). *Bacillus licheniformis* extracellularly synthesized ZnONPs which are hexagonal in shape and with a size of 100 nm (Abinaya et al., 2018). According to Srinath et al., (2018) the *Bacillus subtilis* is employed in the fabrication of AgNPs 20–25 nm size and spherical in shape. *Streptomyces* spp. OSI1 and *Streptomyces* spp. OSNP14 strain synthesized AgNPs, exhibited antibacterial activity with size range of 8 nm (Bakhtiari-Sardari et al., 2020). The *Lactobacillus lantarum* TA4 synthesized ZnONPs using its cell biomass and supernatant. FTIR analysis identified the presence of hydroxyl and carboxyl groups on the surface of the synthesized ZnONPs, with these groups serving as stabilizing and reducing agents during the synthesis process (Yusof et al., 2020). MgO, ZnO and MnO_2NPs were synthesized using *Paenibacillus polymyxa* strain Sx3, and identification of NPs was conducted by different techniques such as SEM and TEM which determined the MgONPs (size 10–18 nm), ZnONPs (size 56–110 nm) and MnO_2NPs (size 19–63 nm). The NPs were characterized as spherical and cubic in shape demonstrating antimicrobial activity against *Xanthomonas oryzae* pv. *oryzae* (Ogunyemi et al., 2020).

7.2.3.3 Fungi Mediated Synthesis of NPs

Fungi are additional objects for the fabrication of NPs. According to Singh et al. (2016) fungi are the ideal objects for the synthesis of NPs and manufacture huge amounts of NPs, and they have excessive cell wall unbreakable capacity. Extracellular enzymes in fungi actively participate in the formation of NPs, playing an essential role in their synthesis. Mishra et al. (2014) demonstrated the utilization of *Trichoderma viride* for the biosynthesis of AuNPs within the size range of 20–30 nm. *Pleurotus ajamore* was utilized for the fabrication of TiO_2, which inhibited the growth of bacteria like *S. aureus* and *P. fluorescence* and showed anticancer activity against the A549 cancer cells (Manimaran et al., 2020). Baskar et al., (2015) reported that *Aspergillus terreus* synthesized ZnONPs size (28–63 nm), and these were spherical in shape. ZnONPs were synthesized from *Aspergillus niger* of size 41–75 nm and confirmation of synthesis of NPs was performed by the FTIR, SEM and UV-Visible spectroscopy, which showed an inhibitory effect against *S. aureus* and *E. coli* (Kalpana et al., 2018).

7.2.3.4 Plant Mediated Synthesis of NPs

Amongst the biological synthesis of NPs, plants are the best agents for the production of NPs. Biosynthesis of NPs by green roots from plants such as flowers, leaves, bark, and fruit extract is very non-toxic and eco-friendly. Plant extract contains biofunctional groups carboxyl and hydroxyl which formed a stabilizing layer and reducing agents (Akhtar et al., 2013). The use of the plant in the fabrication of NPs is very beneficial, non-toxic, environment-friendly, inexpensive, safe, and stable in comparison to fungi and bacteria (Gowramma et al., 2015). The synthesis of ZnONPs from zinc acetate utilizing the leaves of *Ixora coccinia* was conducted, and their purity and stability were confirmed through XRD, DLS, and EDX analysis, revealing a nanoparticle size of 145.1 nm (Yedurkar et al., 2016).

The synthesis of NiNPs utilized *Coriandrum sativum* leaf extract, which served both as a stabilizing and reducing agent in the process (Shanigarapu, 2020). Leaf extract of *Malva sylvestris* was utilized for synthesized CuO, and the synthesized NPs 14 nm in size and crystalline in nature were confirmed by FTIR, SEM, and XRD (Awwad et al., 2015). Leaf extract of papaya was utilized to fabricate CuONPs, resulting in NPs with a crystalline structure. These synthesized NPs demonstrated inhibition of the growth of *Ralstonia solanacearum*, a causal agent of bacterial wilt (Chen et al.,

2019). Citrus lemon fruit extract at room fabricated CuONPs, spherical in shape, crystalline in nature, and with sizes of 30 nm (Amer et a., 2021).

Not only leaf, fruit extract but also seed extract is utilized for the biosynthesis of NPs. Buazar et al. (2019) utilized wheat seed extract for synthesizing crystalline CuONPs, while CuNPs were synthesized using green tea extract (Kiranmai et al., 2017). Further characterization was conducted using FTIR, UV-visible spectroscopy, and XRD techniques. The biochemicals present in the green tea extract played a crucial role in reducing the Cu ions during the synthesis process. The biofabrication of ZnONPs utilized the leaf extract of *Hibiscus subdariffa*. The purity and stability of the compound were confirmed through FTIR, UV-visible spectroscopy, and XRD analysis. According to Vidya et al. (2013), *Calotropis gigantia* synthesized hexagonal ZnONPs which range from 30–35 nm in size, and identification of biosynthesized ZnONPs was made through XRD and SEM. The synthesis of ZnONPs was facilitated by the flower extract of Chamomile. The characterization of the newly synthesized NPs was conducted using XRD,

TABLE 7.1
Exploring natural sources for nanoparticle synthesis: fungi, bacteria, and plant-based approaches

S. No.	Name of Organism	NPs	Size	References
1.	*Bacillus* spp.	AgNPs	10–20 nm	Sunkar and Nhachiyar (2012)
2.	*Klebesiella pneumonia*	SeNPs		Fesharaki et al., (2010)
3.	*Klebesiella pneumonia*	SeNPs		Fesharaki et al., (2010)
4.	*Bacillus licheniformis*	ZnONPs	100 nm	Abinaya et al., (2018)
5.	*Bacillus subtilis*	AgNPs	20–25 nm	Srinath et al., (2018)
6.	*Lactobacillus lantarum* TA4	ZnONPs		Yusof et al., (2020)
7.	*Paenibacillus polymyxa* strain Sx3	MgONPs MnO_2NPs ZnONPs	10–18 nm 19–63 nm 56–110 nm	Ogunyemi et al., (2020)
8.	*Proteus mirabilis* strain 10B	ZnONPs		Eltarahony et al., (2018)
9.	*Trichoderma viride*	AuNPs	20–30 nm	Mishra et al., (2014)
10.	*Pleurotus ajamore*	TiO_2NPs		Manimaran et al., (2020)
11.	*Fusarium solani* ATLOY-8	AuNPs	40–45 nm	Clarance et al., (2020)
12.	*Cladosporium*	AuNPs	28–63 nm	Munawer et al., (2020)
13.	*Aspergillus terreus*	ZnONPs	5–10 nm	Baskar et al., (2015)
14.	*Stereum hirsutum*	CuONPs		Cuevas et al., (2015)
15.	*Aspergillus niger*	ZnONPs	41–75 nm	Kalpana et al., (2018)
16.	*Ixora coccinia*	ZnONPs	145.1 nm	Yedurkar et al., (2016)
17.	*Coriandrum sativum*	NiNPs		Shanigarapu, (2020)
18.	*Sagetaria*	Co_3O_4NPs		Khalil et al., (2020)
19.	*Malva sylvestris*	CuONPs	14 nm	Awwad et al., (2015)
20.	*Pyrifolia*	CuONPs	17–30 nm	Sundaramurthy et al., (2015)
21.	Papaya	CuONPs		Chen et al., (2019)
22.	Citrus lemon fruit	CuONPs	30 nm	Amer et a., (2021)
23.	Seed extract of wheat	CuONPs	20 nm	Buazar et al., (2019)
24.	Green tea extract	CuNPs		Kiranmai et al., (2017)
25.	*Hibiscus subdariffa*	ZnONPs		Bala et al., (2015)
26.	*Calotropis gigantia*	ZnONPs	30–35 nm	Vidya (2013)
27.	Flower extract of Chamomile	ZnONPs		Ogunyemi et al., (2019)
28.	*Salvia spinosa*	AgNPs		Pirtarighat et al., (2019)
29.	Piper betel	AgNPs	6–14 nm	Khan et al., (2020)
30.	*Ocimum sanctum*	NiNPs		Kuchekar et al., (2018)

UV-visible spectroscopy, SEM, and FTIR. These ZnONPs, synthesized from flower extract, exhibited inhibition against rice leaf blight disease (Ogunyemi et al., 2019). Pirtarighat et al. (2019) reported the utilization of *Salvia spinosa* extract in biosynthesizing AgNPs, with XRD and FTIR revealing the formation of AgNPs. Additionally, they employed *Piper betle* leaf extract to fabricate AgNPs, resulting in spherical NPs with sizes ranging from 6–14 nm. These AgNPs effectively reduced the growth of plant pathogenic fungi such as *Fusarium solani* and *Alternaria brassicae* (Khan et al., 2020).

7.3 CHARACTERIZATION OF NPS

Scientists employ various techniques to identify, determine, and characterize NPs. The morphological structure of NPs are determined by scanning electron microscope (Prince et al., 2019). Transmission electron microscope identifies the size and crystallographic nature of NPs (Kuppusamy et al., 2017). Fourier transform infrared spectroscopy determines the functional group and optical characterization of NPs (Wisam et al., 2018), and energy-dispersive X-ray spectroscopy is utilized to determine the elemental analysis and characterize the chemical composition and purity of compounds (Maruthupandy et al., 2017).

7.4 REDUCTION OF INOCULUM LOAD USING NPS IN AGROECOSYSTEMS

A myriad of researchers are working on the role of NPs that enhance the nutrient uptake system, electron transport system, and enzymatic action. Carbon NPs were applied on bitter gourds, which increased the water content, biomass, fruit length, and weight. Silica-based NPs were found to enhance the growth and germination in lentils. TiO_2NPs stimulated the breakdown of an organic compounds thus increasing the photosynthesis and enhancing water uptake (Podder et al., 2018).

The non-judicious use of traditional pesticides, and fungicides has left various residual effects on the environment and on crops, but nanotechnology provides a new hope to researchers for a better environment. NPs are used in the form of metallic oxides, metalloid carbon nanomaterials, and nanomaterials in the form of liposomes, quantum dots, and dendrites that act as biosensors which are used in disease diagnostics and employed as delivery vehicles (Elmer and White 2018). Due to their accumulation in the cytoplasm and interaction with the bacterial cell membranes, nanoparticles exhibit potent effects, such as antifungal, antibacterial, and anti-pathogenic effects. (Kumari et al. 2020) reported that biosynthesized AgNPs, by changing the metabolome and proteome, enhanced plant defence mechanisms through the suppression of pathogens. Thus, NPs function as plant immunity boosters leading to significant improvement in yield characters.

NPs reduce the inoculum load of plant pathogens. The following mechanisms take place and thereby pathogen populations are reduced:

1. NPs discharge toxic ions which changes the permeability and function of membrane protein.
2. NPs help damage the transport and nutrient uptake system of the cell membrane.
3. The generation of reactive oxygen species leads to cell metabolism destruction and induces oxidative stress.

NPs exert a positive impact in the germination of seeds by enhancing the nitrate reductase enzyme, decreasing superoxide radicals and antioxidant stress through H_2O_2 (Bajpai et al., 2020). Khan et al. (2019) reported that NPs showed a toxic effect in plant pathogens.

7.5 ROLE OF NPS IN PLANT DISEASE MANAGEMENT

Although NPs have many roles in agriculture, but their antimicrobial property makes them unique because they enhance the defence mechanism in the plant. A significant number of crops are affected by the non-judicious use of chemical driven products, which are non-ecofriendly to the environment. However, due to NPs application, traditional agriculture transforms into advanced agriculture by developing nano pesticides, nano herbicides, and nano fertilizers (see Table 7.2).

7.5.1 NPs Application on Bacterial Populations

Plant pathogenic bacteria such as *Agrobacterium* spp., *Pseudomonas* spp., *Ralstonia* sp., and *Acidovorax* spp. are remarkably causing great yield loss in agriculture. AgNPs produced from *Phyllanthus emblica* leaf extract inhibited growth of *Acidovorax oryzae* by effluxing of nucleic material from the cell (Masum et al., 2019). The combined effect of Zn and TiO_2NPs reduced the growth of *Xanthomonas* sp. causing leaf spot on roses (Paret et al., 2013). The nanocomposite of chitosan-magnesium and MgO has shown a strong inhibitory effect against the *Acedovorax oryzae* (Ahmed et al., 2021).

7.5.2 NPs Application on Fungi

According to Elmer et al. (2018), the CuNPs limit the growth of leaf blight, *Fusarium* wilt of tomatoes, and *Verticillium* wilt of the eggplant. *Streptomyces greseus* biosynthesized CuNPs, which reduced root rot disease in tea caused by *Poria hypolateritia* (Ponmurugan et al., 2016). ZnONPs reduced the growth of fungal plant pathogens such as *Aternaria alternta*, *Rhizopus stolonifera*, *Fusarium oxysporum*, *Mucor*, and *Rhizoctonia solanii* (Wani et al., 2012). At 256 µg/mL concentration CeONPs decreased the population by 46% and 48% of *Fusarium oxysporum* f. sp. *lycopersici* on leaves and roots of the tomato, respectively (Adisa et al., 2017). Ajaz et al., (2021) recently reported that AgNPs synthesized from the *Bacillus* sp. strain reduced the mycelia growth of *Colletotrichum falcatum* which caused the red rot of sugar cane at 20 ppm concentration. *Trichoderma* sp. biosynthesized SeNPs which inhibited the growth of *Fusarium* sp. and *Alternaria* sp. by destroying the sheath blight disease of rice (Hu et al., 2019). At 50 ppm concentration, AgNPs inhibited 85% mycelial growth of *Rhizoctonia solani* (Soltani et al., 2017). The application of SeNPs reduced the infection of *Phytophthora infestans,* causing late blight, impressively, and registered a remarkable increase of yield (Joshi et al., 2021).

7.5.3 NPs Application on the Plant Pathogenic Nematode Population

Meloidogyne incognita and *M. javanica* have significantly reduced the yield of agriculturally important crops which undermines productivity and profitability. According to Kalaiselvi et al., (2019) AgNPs synthesized from the extract of *Euphorbia tirucalli* exhibited strong nematocidal activity against *M. incognita.* AgNPs synthesized from the bacteria also showed strong activity against the *M. javanica* (Hamed et al. 2019). The CuNPs inhibited the activity of *M. incognita* at their second juvenile stage (Mohamed et al., 2019). The AgNPs exhibited nematocidal activity against the nematode cells which seems to be due to destruction of membrane permeability, ATP formation, and the generation of excessive amounts of oxidative stress (Lim et al., 2012).

7.5.4 Nanopesticides

Nanopesticides formed at micro to nano scales are called nano pesticides. Various researchers have reported Ag and ZnO nano pesticides which are developed by living cells or microorganisms

TABLE 7.2
Role of NPs in the management of plant pathogens

S. No.	NPs	Pathogen	References
1.	AgNPs	*Acidovorax oryzae*	Masum et al. (2019)
2.	ZnNPs and TiO_2NPs	*Xanthomonas* sp.	Paret et al. (2013)
3.	Chitosan-magnesium and MgONPs	*Acedovorax oryzae*	Ahmed et al. (2021)
4.	AgNPs	*Xanthomonas oryzae*	Patil et al. (2020)
5.	CuNPs	*Fusarium* spp. and *Verticillium*	Elmer et al. (2018)
6.	CuNPs	*Poria hypolateritia*	Ponmurugan et al. (2016)
7.	ZnONPs	*Aternaria alternata, Rhizopus stolonifer, Fusarium oxysporum, Mucor, Rhizoctonia solani*	Wani et al. (2012)
8.	CeONPs	*Fusarium oxysporum* f. sp. *lycopersici*	Adisa et al. (2017)
9.	AgNPs	*Colletotrichum falcatum*	Ajaz et al. (2021)
10.	SeNPs	*Fusarium sp., Alternaria sp.*	Hu et al. 2019
11.	AgNPs	*Rhizoctonia solani*	Soltani et al. (2017)
12.	SeNPs	*Phytophthora infestans*	Joshi et al. (2021)
13.	AgNPs	*M. incognita*	Kalaiselvi et al. (2019)
14.	AgNPs	*M. javanica*	Hamed et al. (2019)
15.	CuNPs	*M. incognita*	Lim et al. (2012)

employed against harmful living organisms. Nanopesticides developed through nanoengineering are biodegradable and decrease the environmental threat (Bajpai et al., 2020).

7.5.5 Nanoinsecticides

Nanoparticle-based insecticides were developed in early 2000 during the delivery of insecticides lipids, chitosan, and silica which function as mediators. NPs function as active molecules in comparison to other conventional insecticides. Jenne et al., (2018) observed that ZnO and chitosan NPs combined effect showed a strong inhibitory effect against the *Caryedon serratus.*

7.5.6 Advance Techniques of Nanotechnology in Plant Disease Diagnosis

7.5.6.1 Nanodiagnostic Kit

This kit detects plant pathogens. This technique acts as a sensor and has determined several toxins. It is used by farmers in cash crops like corn, barley, and wheat (Lattanzio et al., 2012; Nezhad et al., 2014).

7.5.6.2 Quantum Dots

This uses three-dimensional NPs. It is much stronger than conventional fluorescence because quantum dots contain fluorescent dyes (Zhao and Zeng, 2015). The quantum dots technique is a very fruitful technique in the agro system because it easily detects the plant virus and is much simpler than PCR and ELISA.

7.5.6.3 Nanobiosensors

This is a device which is employed in the recognition of biological elements. Currently, researchers are developing NPs which can function as nanosensors and help in sensing diseased materials. Nanosensors have a great role in the agrosystem: they can generate a signal for the detection of plant health issues, soil condition, nutrients, reduction of water and nutrients, and also identified environmental conditions and plant physiology (Kanjana, 2017). Antonacci et al., (2018) reported that AuNPs are good nano-bio sensors used in the detection of plant pathogens.

7.6 MECHANISM OF ACTION OF NPS AGAINST PLANT PATHOGENS

NPs are better agents in reducing the phytopathogens due to their adherence nature to microbial cell membranes. CuO, Ag, and ZnONPs are utilized in the management of soil-born phytopathogens (Gupta and Rai, 2017). Dakal et al., (2016) observed that NPs induce various changes in host plants upon contact with pathogens, including cell wall disruption, ATPase inhibition, RNA and protein synthesis inhibition, and intracellular reactive oxygen species production. When CuONPs are applied as seed treatment and foliar spray, they can increase the sakurantin to 96.4%. This is a phytoalexin which protects from *Gibberella fujikuroi*.

7.7 NPS IN CROP HEALTH IMPROVEMENT

Nanotechnology is a fast growing industry specifically in the scientific field which has also had great importance in the minimizing of the load of pathogens such as bacteria, fungi plant-parasitic nematodes, viruses, and insects. NPs based on- TiO_2, CuO, ZnO, and Ag exhibited strong activity against plant pathogens (Kim et al., 2018). Moreover, various researchers have observed that NPs enhanced the nutrient uptake system, electron transport system, and enzymatic action. Carbon driven NPs when applied on bitter gourds, increased water content, biomass, fruit length and weight. Silica-based NPs enhanced growth and germination in lentils, and TiO_2 NPs stimulated the breakdown of an organic compound, increasing photosynthesis and enhancing water uptake (Podder et al., 2018). Rajput et al. (2018) observed that CuONPs enhanced the chloroplasts but decreased the number of chloroplasts in barley. TiO_2NPs shield wheat plants from abiotic stresses such as salinity and drought, while also promoting favourable physiological conditions (Faraji and Sepehri, 2018). Overall, NPs showed a positive impact on plant growth and yield characteristics of different plants.

7.8 ENTRY AND UPTAKE OF NPS IN PLANTS

Researchers reported that NPs can cross the membrane either extracellularly or intracellularly in the xylem. The entry of NPs into the plant depends upon the different properties such as plant species, cell wall, and their size (Singh et al., 2015). Tripathi et al. (2017) reported that water molecules penetrate the cell wall through pores with diameters ranging from 5 to 20 nm. The NPs in plant cells enter through the active transport system by plasma membrane recycling and regulations, signal transfer, and cellular reactions.

7.9 PHYTOTOXICITY AND RESIDUE OF NPS

NPs are delivered into plant cells which interact with plant cell organelles. The concentration of NPs significantly influences their phytotoxicity. As the NPs concentration rises, the likelihood of adverse effects on plant health also increases. In other words, higher concentrations of NPs can lead to greater exposure of plant tissues to potentially toxic substances, resulting in physiological stress, cellular damage, and impaired growth and development (Surendranath and Mohanan, 2020). NPs are transferred to plant cells through a clathrin-independent pathway and a clathrin-dependent pathway. In a clathrin-dependent pathway, NPs are attached to the clathrin protein cell surface and invaginated into the cell. Different studies have revealed that clathrin-independent pathways contain the following mechanisms: (1) lipid raft and caveolae-mediated uptake (2) fluid phase endocytosis (3) phagocytosis (Murphy et al., 2005). The NPs create a toxic effect in plants that may be termed as nano-toxicity or phytotoxicity (Song and Lee, 2016). NPs induce phytotoxicity in plants by reducing root elongation, inhibiting seed germination, and generating reactive oxygen species in both leaf and root cells. The production of reactive oxygen species leads to oxidative damage in plant cell mitochondria, nuclear DNA, and chloroplasts (Imlay and Linn, 1988). Silica NPs induce cytotoxicity by generating reactive oxygen species, which in turn promotes lipid peroxidation of cellular membranes (Fu et al., 2014). Stampoulis et al. (2009) observed that AgNPs remarkably reduced total biomass and germination rate of *Cucurbita pepo*, ultimately inhibiting overall plant growth and yield. CrNPs application enhanced the phytotoxicity in *Pisum sativum* at the seedling stage which significantly generated various oxidant molecules in the plants (Tripathi et al., 2015). AgNPs exhibited dose-dependent toxicity and showed significantly reduced nutrient concentration, plants remained shortened and poor biomass was observed (Yang et al., 2018). Nhan et al. (2015) reported that CeO_2 (Ceric oxide) NPs induced phytotoxicity by the destruction of chloroplasts and decreased the absorption ability of the plants to Mg, Zn, and P to their xylem sap. Germination of ryegrass was inhibited by the toxic activity of ZnONPs and nano zinc (Lin and Xing 2007). Phytotoxicity of AgNPs causes extreme destruction to the vacuole, cell wall, and metabolism (Mazumdar and Ahmed, 2011). AgNPs, Al_2O_3, and ZnONPs have induced phytotoxicity and created alterations in the photosystem and protein degradation (Hossain et al., 2016). The oxidative stress is considered to be one of the causes of phytotoxicity which reduces the amount of plant hormone. Hao et al. (2016) reported that rice plants treated with carbon nanotubes reduced the quality of plant hormones. There are many processes which are used for measuring the phytotoxicity of the plants. The phytotoxicity is determined by the experimental design, interaction ability of NPs with plant cells, concentration, or doses used at the developmental phase, and the physiological and biochemical characters of the plants. In other methods, plant tissue is directly injected by the NPs and spraying of NPs on the whole plant or specific organ: leaves, stem, flower, bud, and so forth (Tarafdar et al., 2012). NPs have specific properties, which can regulate the biochemical and the physiological, such as nitrogen metabolism, solute transport, photosynthesis, and respiration. NPs act as entry and exit elements and influence the cellular, biochemical attributes (Manjunatha et al., 2016). When NPs accumulates at high concentrations in the soil, it reduces the quality and yield of the crop and becomes a major threat to our surroundings (Singh et al., 2018).

7.9.1 Biodegradable NPs

Different biodegradable NPs are utilized for the reduction of phytotoxicity and residue of NPs such as carbon quantum dots, Nano Barium Titanate (NBT), fullerenes 60, superparamagnetic iron oxide, TiO_2, Au, and chitosan NPs. Chitosan NPs are exhibited to be blood viable, nontoxic, antioxidant, antitumour, biodegradable, and antimicrobial (Thandapani et al., 2017). Chitosan NPs are also used in cosmetic goods, metal processing, and freshwater treatment (Mohammed et al., 2017). TiO_2 and CuONPs significantly decreased the biomass in flooded rice fields, and soil microbial activity (Xu et al., 2015). According to You et al., (2018) Fe_3O_4NPs and ZnONPs affected the soil enzyme activity, altered the soil microorganism, and reduced the biological nitrogen fixation.

7.9.2 Monitoring of Soil and Water for NPs Residue Analysis

The residue of NPs depends on soil characteristics such as clay minerals, texture, pH, and soil organic matter, as well as the microbial community. These factors significantly influence the aggregation and transport of NPs in the soil (Khalkhal et al., 2020). Josko et al. (2020) reported that CuO, and ZnONPs are inversely related to soil pH which demonstrate that soil pH has a great impact on the availability of NPs. Soil texture also affects the mobility of NPs. For instance, sandy soil exhibited faster movement and adsorption of AgNPs as compared to fine-textured soil (Mahdi, et al., 2018).

7.10 FUTURE DIRECTIONS AND CHALLENGE

Nanotechnology is increasing and becoming widely usable in sustainable agriculture. NPs have created a revolution in the agricultural development of biosensors, fabrications of xylem vessels, and nanocarriers. NPs significantly impact plant growth and development including root and shoot growth, photosynthesis and seedling vigour. NPs improve plant health and increase the defence mechanism. Biosynthesized NPs have great importance in agriculture, increasing plant growth, phytoremediation capability, fruit yield, nano pesticides, nano fertilizers, and biomass, and also have a role in the preservation and development of seeds and crops (El-Saadony et al., 2022).

NPs have an important role in the agrosystem not only in improving crop quality but also protecting from plant pathogens. The application of NPs has reduced the reliance on chemical-based fertilizers, pesticides, fungicides, insecticides, and nematocides due to their adverse effects on plants and the environment when used indiscriminately. The development of quantum dots, nano biosensors, and nano diagnostic kits is very beneficial in the detection of the well-being of plant health and plant pathogens (Younas et al., 2020).

7.11 CONCLUSIONS AND FUTURE PROSPECTS

Phytopathogens, including bacteria, fungi, and plant parasitic nematodes, have inflicted significant economic losses across a wide range of plant species. Synthesizing NPs holds promise in plant disease management in the form of nematocides, fungicides, bactericides, and even nano-fertilizers for enhancing plant health. The potential future application of NPs in gene delivery to crop plants could confer resistance to pathogens and enhance crop quality. The use of NPs in managing various plant diseases offers hope to researchers, particularly those biologically synthesized, which are perceived as environmentally friendly. Plants treated with NPs have exhibited poor biomass when not used judiciously. Therefore, it is crucial to explore the molecular mechanisms involved in mitigating the adverse effects of NPs and ensuring their proper application.

REFERENCES

Abbas, Q., Yousaf, B., Ali, M. U., Munir, M. A. M., El-Naggar, A., Rinklebe, J. & Naushad, M. (2020). Transformation pathways and fate of engineered nanoparticles (ENPs) in distinct interactive environmental compartments: A review. *Environ Int, 138*, 105646.

Abinaya, M., Vaseeharan, B., Divya, M., Sharmili, A., Govindarajan, M. & Alharbi, N. S. (2018). Bacterial exopolysaccharide (EPS)-coated ZnONPs showed high antibiofilm activity and larvicidal toxicity against malaria and Zika virus vectors. *J Trace Elem Med Biol, 45*, 93–103.

Adisa, I. O., Hernandez-Viezcas, J. A., Elmer, W. H., White, J. C., Peralta-Videa, J. R. & Gardea-Torresdey, J. L. (2017). Evaluating the role of CeO_2 NPs in the suppression of Fusarium wilt disease in tomato plant. Paper presented at the 6th Sustain *Nanotechnol Conference Los Angeles 24*, 5959–5970.

Ahmadi, S. (2020) Mathematical modelling of cytotoxicity of metal oxide nanoparticles using the index of ideality correlation criteria. *J Chemosphere, 242*, 125192.

Ahmed, T., Noman, M., Shahid, M., Shahid, M. S. & Li, B. (2021) Antibacterial potential of green magnesium oxide nanoparticles against rice pathogen *Acidovorax oryzae*. *Mater Lett, 282*, 128839.

Ajaz, S., Ahmed, T., Shahid, M., Noman, M., Shah, A. A., Mehmood, M. A., Abbas, A., Cheema, A. I., Iqbal, M. Z. & Li, B. (2021) Bioinspired green synthesis of silver nanoparticles by using a native *Bacillus* sp. strain AW1–2: Characterization and antifungal activity against *Colletotrichum falcatum* Went. Enzyme *Microb Technol, 144*, 109745.

Akbari, B., Tavandashti, M. P. & Zandrahimi, M. (2011) Particle size characterization of nanoparticles—A practical approach Iran. *J Mater Sci Eng, 8*, 48–56.

Akhtar, M. S., Panwar J. & Yun Y.-S. (2013) Biogenic synthesis of metallic nanoparticles by plant extracts. ACS *Sustain Chem Eng, 1*(6), 591–602.

Alavi, M. & Karimi, N. (2018) Characterization, antibacterial, total antioxidant, scavenging, reducing power and ion chelating activities of green synthesized silver, copper and titanium dioxide nanoparticles using *Artemisia haussknechtii* leaf extract. *Artif Cells Nanomed Biotechnol, 46* (8), 2066–2081.

Ali S., Mehmood A. & Khan N. (2021). Uptake, translocation, and consequences of nanomaterials on plant growth and stress adaptation. *J Nanomat*, Article ID 6677616.

Alzahrani, E., Sharfalddin, A. & Alamodi, M. (2015). Microwave hydrothermal synthesis of ferric oxide doped with cobalt. *Adv Nanopart, 4*, 53–60.

Amer, M. W. & Awwad, A. M. (2021). Green synthesis of copper nanoparticles by Citrus lemon fruits extract, characterization and antibacterial activity. *Chem Int, 7*(1), 1–8.

Antonacci, A., Arduini, F., Moscone, D., Palleschi, G. & Scognamiglio, V. (2018). Nanostructured (Bio) sensors for smart agriculture. *TrAC Trends Anal Chem, 98*, 95–103.

Asharani, P. V., Wu, Y. L., Gong, Z. & Valiyaveettil, S. (2008). Toxicity of silver nanoparticles in zebrafish models. *Nanotechnol, 19*(25), 255102.

Atha, D. H., Wang, H., Petersen, E. J., Cleveland, D., Holbrook, D. & Jaruga, P. (2012). Copperoxide nanoparticles mediated DNA damage in terrestrial plant models. *Environ Sci Technol, 46*(3), 1819–1827.

Avila-Quezada, G. D., Golinska, P. & Rai, M. (2022). Engineered nanomaterials in plant diseases: Can we combat phytopathogens. *Appl Microbiol Biotechnol, 106*(1), 117–129.

Awwad, A. M., Albiss, B. A. & Salem, N. M. (2015). Antibacterial activity of synthesized copper oxide nanoparticles using *Malva sylvestris* leaf extract. *SMU Med J, 2*(1), 91–101.

Bajpai, A., Jadhav, K., Muthu Kumar, M., Kumar, S. & Srivatava, G. (2020). Use of Nanotechnology in Quality Improvement of Economically Important Agricultural Crops. In: *Biogenic Nano-Particles and their Use in Agro-ecosystems* (pp. 39–57).

Bakhtiari-Sardari, A., Mashreghi, M., Eshghi, H., Behnam-Rasouli, F., Lashani, E. & Shahnavaz, B. (2020) Comparative evaluation of silver nanoparticles biosynthesis by two cold-tolerant *Streptomyces strains* and their biological activities. *Biotechnol Lett, 42*, 1985–1999.

Bala, N., Saha, S., Chakraborty, M., Maiti, M., Das, S., Basu, R. & Nandy, P. (2015). Green synthesis of zinc oxide nanoparticles using *Hibiscus subdariffa* leaf extract: Effect of temperature on synthesis, anti-bacterial activity and anti-diabetic activity. RSC *Adv, 5*(7), 4993–5003.

Baskar, G., Chandhuru, J., Sheraz Fahad, K., Praveen, A. S., Chamundeeswari, M. & Muthukumar, T. (2015). Anticancer activity of fungal l-asparaginase conjugated with zinc oxide nanoparticles. *J Mater Sci Mater Med, 26*(1), 5380.

Batley, G. E., Kirby, J. K., McLaughlin, M. J. (2013). Fate and risks of nanomaterials in aquatic and terrestrial environments. Acc *Chem Res 46*(3), 854–862.

Buazar, F., Sweidi, S., Badri, M. & Kroushawi, F. (2019). Biofabrication of highly pure copper oxide nanoparticles using wheat seed extract and their catalytic activity: A mechanistic approach. *Green Processing Synth, 8*(1), 691–702.

Chen, J., Mao, S., Xu, Z. & Ding, W. (2019). Various antibacterial mechanisms of biosynthesized copper oxide nanoparticles against soilborne *Ralstonia solanacearum*. *RSC Adv, 9*(7), 3788–3799.

Cho, E. J., Holback, H., Liu, K. C., Abouelmagd, S. A., Park, J. & Yeo, Y. (2013). Nanoparticle characterization: State of the art, challenges, and emerging technologies. *Mol Pharm, 10*(6), 2093–2110.

Clarance, P., Luvankar, B., Sales, J., Khusro, A., Agastian, P., Tack, J. C., Al Khulaifi, M. M., Al-Shwaiman, H. A., Elgorban, A. M. & Syed, A. (2020) Green synthesis and characterization of gold nanoparticles using endophytic fungi *Fusarium solani* and its in-vitro anticancer and biomedical applications. *Saudi J Biol Sci, 27*(2), 706–712.

Cuevas, R., Durán, N., Diez, M. C., Tortella, G. R. & Rubilar, O. (2015). Extracellular biosynthesis of copper and copper oxide nanoparticles by *Stereum hirsutum*, a native white-rot fungus from Chilean forests. *J Nanomater*, *57*, Article ID 789089.

Dakal T. C., Kumar A., Majumdar R. S. & Yadav V. (2016). Mechanistic basis of antimicrobial actions of silver nanoparticles. *Front Microbiol Robiol*, *7*, 1831.

Elmer, W., Ma, C., & White, J. (2018). Nanoparticles for plant disease management. *Current Opinion in Environmental Science & Health, 6*, 66-70.Wani, A. H. & Shah, M. A. (2012). A unique and profound effect of MgO and ZnO nanoparticles on some plant pathogenic fungi. *J Appl Pharmaceut Sci, 2*(3), 40–44. ISSN: 2231–3354.

Elmer, W. & White, J. C. (2018). The future of nanotechnology in plant pathology. *Annu Rev Phytopathol, 56*, 111–133.

Elmer, W. H., De La Torre-Roche, R., Pagano, L., Majumdar, S. & Zuverza- Mena, N. (2018). Effect of metalloid and metallic oxide nanoparticles on *Fusarium* wilt of watermelon. *Plant Disease, 102*(7), 1394–1401.

El-Saadony, M. T., Saad, A. M., Soliman, S. M., Salem, H. M., Desoky, E. S. M., Babalghith, A. O., ... & AbuQamar, S. F. (2022). Role of nanoparticles in enhancing crop tolerance to abiotic stress: A comprehensive review. Frontiers in plant science, 13, 946717.

Eltarahony, M., Zaki, S., ElKady, M. & Abd-El-Haleem, D. (2018). Biosynthesis, characterization of some combined nanoparticles, and its biocide potency against a broad spectrum of pathogens. *J Nanomater*, Article ID 5263814.

Faraji, J. & Sepehri, A. (2018). Titanium dioxide nanoparticles and sodium nitroprusside alleviate the adverse effects of cadmium stress on germination and seedling growth of wheat (*Triticum aestivum* L.). *Univ Sci, 23*(1), 61–87.

Fesharaki, P. J., Nazari, P., Shakibaie, M., Rezaie, S., Banoee, M., Abdollahi, M. & Shahverdi, A. R. (2010). Biosynthesis of selenium nanoparticles using *Klebsiella pneumoniae* and their recovery by a simple sterilization process. *Braz J Microbiol, 41*(2), 461–466.

Fu, P. P., Xia, Q., Hwang, H. M., Ray, P. C. & Yu, H. (2014). Mechanisms of nanotoxicity: Generation of reactive oxygen species. *J Food Drug Anal, 22*(1), 6475.

Gowramma, B., Keerthi, U., Rafi, M. & Rao, D. M. (2015). Biogenic silver nanoparticles production and characterization from native stain of *Corynebacterium species* and its antimicrobial activity. *Biotech, 5*(2), 195–201.

Griffitt, R. J., Luo, J., Gao, J., Bonzongo, J. C. & Barber, D. S. (2008). Effects of particle composition and species on toxicity of metallic nanomaterials in aquatic organisms. *Environ Toxicol Chem, 27*, 1972–1978.

Gupta, R. & Rai, B. (2017). Effect of size and surface charge of gold nanoparticles on their skin permeability: A molecular dynamics study. *Sci Rep, 7*, 45292.

Hamed, S. M., Hagag, E. S. & El-Raouf, N. A. (2019) Green production of silver nanoparticles, evaluation of their nematocidal activity against *Meloidogyne javanica* and their impact on growth of Faba Bean. Beni-Suef. *Univ. J Basic Appl Sci, 8*, 9.

Hao, Y., Yu, F., Lv, R., Ma, C., Zhang, Z., Rui, Y., Liu, L., Cao, W., Xing, B. & Choi, J. (2016). Carbon nanotubes filled with different ferromagnetic alloys affect the growth and development of rice seedlings by changing the C:N ratio and plant hormones concentrations. *PloS One, 11*, 0157264.

Hasan, S. (2015). A review on nanoparticles: Their synthesis and types biosynthesis. *Mech Res J Recent Sci, 4*, 9–11. ISSN 2277-2502.

Hayles, J., Johnson, L., Worthley, C. & Losic, D. (2017) Nanopesticides: A review of current research and perspectives. In: Grumezescu, A.M. (editor), *New Pesticides and Soil Sensors*, Elsevier, Academic Press, pp. 193–225.

Hazarika, A., Yadav, M., Yadav, D. K. & Yadav, H. S. (2022). An overview of the role of nanoparticles in sustainable agriculture. *Biocatal Agric Biotechnol, 43*(10), 102399.

He, M., Protesescu, L. & Caputo, R. (2015). A general synthesis strategy for monodisperse metallic and metalloid nanoparticles (In, Ga, Bi, Sb, Zn, Cu, Sn and their alloys) via in situ formed metal long-chain amides. *Chem Mat, 27*, 635–647.

Hong, J., Peralta-Videa, J. R., Rico, C., Sahi, S., Viveros, M. N., Bartonjo, J. & Gardea-Torresdey, J. L. (2014). Evidence of translocation and physiological impacts of foliar applied CeO_2 nanoparticles on cucumber (*Cucumis sativus*) plants. *Environ Sci Technol, 48*(8), 4376–4385.

Hossain, Z., Mustafa, G., Sakata, K. & Komatsu, S. (2016). Insights into the proteomic response of soybean towards Al2O3, ZnO, and Ag nanoparticles stress. *J Hazard Mater, 304*, 291–305.

Hu, D., Yu, S., Yu, D., Liu, N., Tang, Y., Fan, Y., Wang, C. & Wu, A. (2019). Biogenic *Trichoderma harzianum*-derived Selenium nanoparticles with control functionalities originating from diverse recognition metabolites against phytopathogens and mycotoxins. *Food Control, 106*(8), 106748.

Hu, P., An, J., Faulkner, M. M., Wu, H., Li, Z., Tian, X. & Giraldo, J. P. (2020) Nanoparticle charge and size control foliar delivery efficiency to plant cells and organelles. *ACS Nano, 14*(7), 7970–7986.

Imlay, J. A. & Linn S. (1988). DNA damage and oxygen radical toxicity. *Sci (Washington,) 240*, 1302–1309.

Jang, G. G., Jacobs, C. B., Gresback, R. G., Ivanov, I. N., Meyer III, H. M., Kidder, M. & Moon, J. W. (2015). Size tunable elemental copper nanoparticles: Extracellular synthesis by thermoanaerobic bacteria and capping molecules. *J Mater Chem, 3*(3), 644–650.

Jenne, M., Kambham, M., Tollamadugu, N. P., Karanam, H. P., Tirupati, M. K., Balam, R. R., Shameer, S. & Yagireddy, M. (2018) The use of slow releasing nanoparticle encapsulated Azadirachtin formulations for the management of *Caryedon serratus O.*(groundnut bruchid). *IET Nanobiotechnol, 12*(7), 963–967.

Joerger, R., Klaus, T. & Granqvist, C. G. (2000). Biologically produced silver–carbon composite materials for optically functional thin-film coatings. *Advanced Materials, 12*(6), 407–409.

Joshi, S. M., De Britto, S. & Jogaiah, S. (2021) Myco-engineered Selenium nanoparticles elicit resistance against tomato late Blight disease by regulating differential expression of cellular, biochemical and defence responsive genes. *J Biotechnol, 20*(21), 196–206.

Josko, I., Dobrzynska, J., Dobrowolski, R. & Kusiak, M. (2020) Terpiłowski, K. The effect of pH and ageing on the fate of CuO and ZnO nanoparticles in soils. *Sci Total Environ, 721*, 137771.

Kalabegishvili, T. L., Kirkesali, E. I., Rcheulishvili, A. N., Ginturi, E. N., Murusidze, I. G., Pataraya, D. T., Gurielidze, M. A., Tsertsvadze, G. I., Gabunia, V. N. & Lomidze, L. G. (2012). Synthesis of gold nanoparticles by some strains of *Arthrobacter* genera. *J Mater Sci Eng A Struct Mater Prop Microstruct Process, 4*(1), 43–49.

Kalaiselvi, D., Mohankumar, A., Shanmugam, G., Nivitha, S. & Sundararaj, P. (2019). Green synthesis of silver nanoparticles using latex extract of *Euphorbia tirucalli*: A novel approach for the management of root knot nematode, *Meloidogyne incognita. Crop Prot, 117*, 108–114.

Kalpana, V. N., Kataru, B. A. S., Sravani, N., Vigneshwari, T., Panneerselvam, A. & Rajeswari, V. D. (2018). Biosynthesis of zinc oxide nanoparticles using culture filtrates of Aspergillus niger: Antimicrobial textiles and dye degradation studies. *Open Nano, 3*, 48–55.

Kanjana, D. (2017) Advancement of nanotechnology applications on plant nutrients management and soil improvement. In: Prasad, R., Kumar, V., & Kumar, M. (eds.*) Nanotech Food and Environ Paradigm* (pp. 209–234). Springer, Singapore.

Khalil, A. T., Ovais, M., Ullah, I., Ali, M., Shinwari, Z. K. & Maaza, M. (2020). Physical properties, biological applications and biocompatibility studies on biosynthesized single phase cobalt oxide (Co_3O_4) nanoparticles via Sageretia thea (Osbeck.). *Arab J Chem, 13*(1), 606–619.

Khalkhal, K., Lajayer, B. A. & Ghorbanpour, M. (2020) An overview on the effect of soil physicochemical properties on the immobilization of biogenic nanoparticles. In: *Biogenic NPs and their Use in Agro-Ecosys* (pp. 133–160). Springer Science and Business Media LLC, Singapore.

Khan, M. R., Ahmad, F. & Rizvi, T. F. (2019) Effect of nanoparticles on plant pathogens. In: *Advances in phytonanotechnology* (pp. 215–240). Academic Press, London.

Khan, S., Singh, S., Gaikwad, S., Nawani, N., Junnarkar, M. & Pawar, S. V. (2020). Optimization of process parameters for the synthesis of silver nanoparticles from Piper betle leaf aqueous extract, and evaluation of their antiphytofungal activity. *Environ Sci Pollut Res, 27*(22), 27221–27233.

Kim, D. Y., Kadam, A., Shinde, S., Saratale, R. G., Patra, J. & Ghodake, G. (2018) Recent developments in nanotechnology transforming the agricultural sector: A transition replete with opportunities. *J Sci Food Agric, 98(3)*, 849–864.

Kiranmai, M., Kadimcharla, K., Keesara, N. R., Fatima, S. N., Bommena, P. & Batchu, U. R. (2017). Green synthesis of stable copper nanoparticles and synergistic activity with antibiotics. *Indian J Pharmaceut Sci, 79*(5), 695–700.

Kuchekar, S. R., Dhage, P. M., Gaikwad, V. B., Aher, H. R. & Han, S. H. (2018). Biosynthesis and characterization of nickel nanoparticle using *Ocimum sanctum* (Tulsi) leaf extract. *Chem Sci Trans, 7*(4), 696–702.

Kumar, Y. R., Suresh, J., Nathanael, A. J. Sundrarajan, M. & Hong, S. I. (2014). Novel green synthetic strategy to prepare ZnO nanocrystals using rambutan (*Nephelium lappaceum* L.) peel extract and its antibacterial applications. *Mater Sci Eng, C, 41*, 17–27.

Kumari, M., Pandey, S., Mishra, S. K., Giri, V. P., Agarwal, L., Dwivedi, S. & Mishra, A. (2020). Omics-based mechanistic insight into the role of bioengineered nanoparticles for biotic stress amelioration by modulating plant metabolic pathways. *Front Bioeng Biotechnol, 8*, 242.

Kuppusamy, P., Ilavenil, S., Srigopalram, S., Maniam, G. P., Yusoff, M. M., Govindan, N. & Choi, K. C. (2017). Treating of palm oil mill effluent using *Commelina nudiflora* mediated copper nanoparticles as a novel bio-control agent. *J Cleaner Prod, 141*, 1023–1029. ISSN: 0959-6526.

Lattanzio, V. M. T., Nivarlet, N., Lippolis, V. & Gatta, S. D. (2012). Multiplex dipstick immunoassay for semi-quantitative determination of *Fusarium* mycotoxins in cereals. *Anal Chim Acta, 718*, 99–108.

Lemire, J. A., Harrison, J. J. & Turner, R. J. (2013 Jun). Antimicrobial activity of metals: Mechanisms, molecular targets and applications. *Nat Rev Microbiol, 11*(6), 371.

Li, H., Zhang, J., Wang, T., Luo, W., Zhou, Q. & Jiang, G. (2008). Elemental selenium particles at nano-size (Nano-Se) are more toxic to Medaka (*Oryzias latipes*) as a consequence of hyper-accumulation of selenium: A comparison with sodium selenite. *Aquatic Toxicol, 89*, 251–256.

Lim, D., Roh, J. Y., Eom, H. J., Choi, J. Y., Hyun, J. W. & Choi, J. (2012). Oxidative stress-related PMK-1 P38 MAPK activation as a mechanism for toxicity of silver nanoparticles to reproduction in the nematode *Caenorhabditis elegans*. *Environ Toxicol Chem, 3*, 585–592.

Lin, D. & Xing, B. (2007). Phytotoxicity of nanoparticles: Inhibition of seed germination and root growth. *Environ Pollut, 150*(2), 243–250.

Luo, K., Jung, S., Park, K. H. & Kim, Y. R. (2018). Microbial biosynthesis of silver nanoparticles in different culture media. *J Agric Food Chem, 66*(4), 957–962.

Ma C., Li Q., Jia W., Shang H., Zhao J. & Hao Y. (2021). Role of nanoscale hydroxyapatite in disease suppression of *Fusarium*-infected tomato. *Environ Sci Technol, 55*(20), 13465–13476.

Mahdi, K. N. M., Peters, R., Van Der Ploeg, M., Ritsema, C. & Geissen, V. (2018). Tracking the transport of silver nanoparticles in soil: A saturated column experiment. *Water Air Soil Pollut, 229*, 1–13.

Manimaran, K., Murugesan, S., Ragavendran, C., Balasubramani, G., Natarajan, D., Ganesan, A. & Seedevi, P. (2020). Biosynthesis of TiO_2 nanoparticles using edible mushroom extract: Mosquito larvicidal, histopathological, *pleurotus djamor* antibacterial and anticancer effect. *J Clust Sci, 32*(5), 1–12.

Manjunatha, S. B., Biradar, D. P. & Aladakatti, Y. R. (2016). Nanotechnology and its applications in agriculture: A review. *J Farm Sci, 29*(1), 1–13.

Maruthupandy, M., Zuo, Y., Chen, J. S., Song, J. M., Niu, H. L., Mao, C. J., Zhang, S. Y. & Shen, Y. H. (2017). Synthesis of metal oxide nanoparticles (CuO and ZnONPs) via biological template and their optical sensor applications. *Appl Surf Sci, 397*, 167–174.

Masum, M. M. I., Siddiqa, M. M., Ali, K. A., Zhang, Y., Abdallah, Y., Ibrahim, E., Qiu, W., Yan, C. & Li, B. (2019). Biogenic synthesis of silver nanoparticles using *Phyllanthus emblica* fruit extract and its inhibitory action against the pathogen *Acidovorax oryzae* strain RS-2 of rice bacterial brown stripe. *Front Microbiol, 10*, 820.

Mazumdar, H. & Ahmed, G. U. (2011). Phytotoxicity effect of Silver nanoparticles on *Oryza sativa*. *Int J Chem Tech Res, 3*(3), 1494–1500.

Mishra, A., Kumari, M., Pandey, S., Chaudhry, V., Gupta, K. & Nautiyal, C. (2014). Biocatalytic and antimicrobial activities of gold nanoparticles synthesized by *Trichoderma* sp. *Bioresour Technol, 166*, 235–242.

Mohamed, E. A., Elsharabasy, S. F. & Abdulsamad, D. (2019). Evaluation of in vitro nematocidal efficiency of copper nanoparticles against root-knot nematode Meloidogyne incognita. *South Asian J Parasitol, 2*(1), 1–6.

Mohammed, M. A., Syeda, J., Wasan, K. M. & Wasan, E. K. (2017). An overview of chitosan nanoparticles and its application in non-parenteral drug delivery. *Pharmaceutics, 9*(4), 53.

Moraru, C. I., Panchapakesan, C. P. & Huang, Q. (2003). Nanotechnology: A new frontier in food science. *Food Technol, 57*(12), 24–29.

Munawer, U., Raghavendra, V. B., Ningaraju, S., Krishna, K. L., Ghosh, A. R., Melappa, G. & Pugazhendhi, A. (2020). Biofabrication of gold nanoparticles mediated by the endophytic *Cladosporium* species: Photodegradation, in vitro anticancer activity and in vivo antitumour studies. *Int J Pharm, 588*, 119729.

Murphy, A. S., Bandyopadhyay, A., Holstein, S. E. & Peer, W. A. (2005). Endocytotic cycling of pm proteins. *Annu Rev Plant Biology, 56*, 221–251.

Nasrollahzadeh., M., Mohaddeseh, S., Jaber, D. & Hossein, G. (2020). Pd-based nanoparticles: Plant-assisted biosynthesis, characterization, mechanism, stability, catalytic and antimicrobial activities. *Adv Colloid Interface Sci, 276*, 102103.

Nezhad, A. S. (2014). Future of portable devices for plant pathogen diagnosis. *J Royal Soci Chem, 14*, 2887–2904.

Nhan, L. V., Ma, C., Rui, Y., Liu, S., Li, X., Xing, B. & Liu, L. (2015). Phytotoxic mechanism of nanoparticles: Destruction of chloroplasts and vascular bundles and alteration of nutrient absorption. *Sci Rep, 5*(1), 1–13.

Ogunyemi, S. O., Abdallah, Y., Zhang, M., Fouad, H., Hong, X., Ibrahim, E. & Li, B. (2019). Green synthesis of zinc oxide nanoparticles using different plant extracts and their antibacterial activity against *Xanthomonas oryzae pv. oryzae*. *Artificial Cells, Nanomed Biotechnol, 47*(1), 341–352.

Ogunyemi, S. O., Zhang, M., Abdallah, Y., Ahmed, T., Qiu, W., Ali, M., Yan, C., Yang, Y., Chen, J. & Li, B. (2020). The bio-synthesis of three metal oxide nanoparticles (ZnO, MnO_2, and MgO) and their antibacterial activity against the bacterial leaf blight pathogen. *Front Microbiol 11*, 3099.

Paret, M. L., Gary E. V., Devron R. A., Jeffrey B. J. & Stephen, M. O. (2013). Photocatalysis: Effect of light-activated nanoscale formulations of TiO_2 on *Xanthomonas perforans* and control of bacterial spot of tomato. *Phytopathol, 103*(11), 228–236.

Patil, G., Patel, D., Patel, P. & Patel, A. (2020). Comparative studies of sunlight mediated green synthesis of silver nanoparaticles from *Azadirachta indica* leaf extract and its antibacterial effect on *Xanthomonas oryzae pv. oryzae*. *Arab J Chem, 3*(1), 2865–2872.

Pirtarighat, S., Ghannadnia, M. & Baghshahi, S. (2019). Green synthesis of silver nanoparticles using the plant extract of *Salvia spinosa* grown in vitro and their antibacterial activity assessment. *J Nanostructure Chem, 9*(1), 1–9.

Podder, K., Vijayan, J., Ray, S. & Adak, T. (2018). Chapter 10: Nanotechnology for sustainable agriculture. In: Singh, R. I. & Mondal, S., (eds). *Biotechnol for Sustain Agri Emerging Approaches and Strategies* (pp. 281–303). Woodhead Publishing, United Kingdom.

Ponmurugan, P., Manjukarunambika, K., Elango, V. & Gnanamangai, B. M. (2016). Antifungal activity of biosynthesized copper nanoparticles evaluated against red root-rot disease in tea plants. *J Exp Nanosci,* 11(13), 1–13.

Prince, Joseph J., John, Abel., M., kumar, Pramoth, A., Archana, V., Senthil kumar, N. & Jothiven katachalam, K. (2019). Facile synthesis of solar light active spinel nickel manganite ($NiMn_2O_4$) by co-precipitation route for photocatalytic application. *Res Chem Intermediates, 46*(7), 3509–3525.

Rai V, Acharya S, & Dey N (2012) Implications of nanobiosensors in agriculture. *J Biomater Nanobiotchnol, 03*(02), 315–324.

Rajput, V. D., Minkina, T., Sushkova, S., Tsitsuashvili, V., Mandzhieva, S., Gorovtsov, A., Nevidomskyaya, D. & Gromakova, N. (2018). Effect of nanoparticles on crops and soil microbial communities. *J Soils Sediments, 18*(18), 2179–2187.

Rajput, V., Minkina, T., Fedorenko, A., Sushkova, S., Mandzhieva, S., Lysenko, V., Duplii, N., Fedorenko, G., Dvadnenko, K. & Ghazaryan, K. (2018). Toxicity of copper oxide nanoparticles on spring barley (*Hordeum sativum* distichum). *Sci Total Environ, 645*, 1103–1113.

Rayapa, K. & Reddy. (2017). Green synthesis, morphological and optical studies of CuONPs *J Mol Struct, 1150*, 553–557.

Rico, C. M., Majumdar, S., Duarte-Gardea, M., Peralta-Videa, J. R., Gardea, Torresdey, J. L. (2011). Interaction of nanoparticles with edible plants and their possible implications in the food chain. *J Agric Food Chem, 59* (8), 3485–3498.

Roco, M. M. (2011).The long view of nanotechnology development: The National nanotechnol initiative at 10 years. *J Nanopart Res, 13*, 427.

Sadiq, R., Khan, Q. M., Mobeen, A., & Hashmat, A. J. (2015). In vitro toxicological assessment of iron oxide, aluminium oxide and copper nanoparticles in prokaryotic and eukaryotic cell types. *Drug and Chemical Toxicology, 38*(2), 152–161.

Shang, H., Ma, C., Li, C., White, J. C., Polubesova, T. & Chefetz, B. (2020). Copper sulfide nanoparticles suppress *Gibberella fujikuroi* infection in rice (*Oryza sativa* l.) by multiple mechanisms: Contact-mortality, nutritional modulation and phytohormone regulation. *Environ Sci Nano, 7* (9), 2632–2643.

Shanigarapu, A. K. S. (2020). Green synthesis of nickel nanoparticles from extract of *Coriandrum Sativum* leaves. *Int J Sci Res Devt, 6*(07), ISSN (online): 2321–0613.

Sharma, D., Kanchi, S. & Bisetty, K. (2019). Biogenic synthesis of nanoparticles: A review. *Arab J Chem, 12*(8), 3576–3600.

Singh, A., Singh, N. B., Afzal, S., Singh, T. & Hussain, I. (2018). Zinc oxide nanoparticles: A review of their biological synthesis, antimicrobial activity, uptake, translocation and biotransformation in plants. *J Mater Sci, 53*, 185–201.

Singh, A., Singh, N. B., Hussain, I., Singh, H. & Singh, S. C. (2015). Plant-nanoparticle interaction: An approach to improve agricultural practices and plant productivity. *Int J Pharm Sci Invent, 4*(8), 25–40. ISSN 2319 – 670X.

Singh, M. D., Chirag, G., Patidar, O. P., Meena, H. M. & Prakasha G, Vishwajith. (2017). Nano-fertilizers is a new way to increase nutrients use efficiency in crop production. *IJAS, 9*(7), 3831–3833. ISSN: 0975-3710&E-ISSN: 0975-9107.

Singh, P., Kim, Y. J., Zhang, D. & Yang, D. C. (2016). Biological synthesis of nanoparticles from plants and microorganisms. *Trends Biotechnol, 34*(7), 588–599.

Smith, C. J., Shaw, B. J. & Handy, R. D. (2007). Toxicity of single walled carbon nanotubes to rainbow trout, (Oncorhynchus mykiss): Respiratory toxicity, organ pathologies, and other physiological effects. *Aquatic Toxicol, 82*, 94–109.

Soltani Nejad, M., Bonjar, G. H. S., Khatami, M., Amini, A. & Aghighi, S. (2017). In vitro and in vivo antifungal properties of silver nanoparticles against *Rhizoctonia solani*, a common agent of rice sheath Blight disease. *IET Nanobiotechnol, 11*(3), 236–240.

Song, U. & Lee, S. (2016). Phytotoxicity and accumulation of zinc oxide nanoparticles on the aquatic plants *Hydrilla verticillata* and *Phragmites australis*: Leaf-type dependent responses. *Environ Sci Pollut Res, 23*(9), 8539–8545.

Srinath, B. S., Namratha, K. & Byrappa, K. (2018). Eco-friendly synthesis of gold nanoparticles by *Bacillus subtilis* and their Environmental Applications. *Adv Sci Lett, 24*(8), 5942–5946.

Stampoulis, D., Sinha, S. K. & White, J. C. (2009). Assay-dependent phytotoxicity of nanoparticles to plants. *Environ Sci Technol, 43*, 9473–9479.

Sundaramurthy, N. & Parthiban, C. (2015). Biosynthesis of copper oxide nanoparticles using Pyrus pyrifolia leaf extract and evolve the catalytic activity. *Int Res J Eng Technol, 2*(6), 332–338. e-ISSN: 2395-0056.

Sunkar, S. & Nachiyar, C. V. (2012). Microbial synthesis and characterization of silver nanoparticles using the endophytic bacterium Bacillus cereus a novel source in the benign synthesis. *Glob J Med Res, 12*, 43–49. Online ISSN: 2249–4618 & Print ISSN: 0975-5888.

Surendranath, A. & Mohanan, P. V. (2020). Impact of nanoparticles in balancing the ecosystem. *Biointer face Res Appl Chem, 11* (3), 10461–10481.

Swamy, M. K. (2015). The green synthesis, characterization, and evaluation of the biological activities of silver nanoparticles synthesized from Leptadenia reticulate leaf extract. *Appl Nanosci, 5*(1), 73–81.

Tarafdar, J., Xiong, Y., Wang, W. N., Quinl, D. & Biswas, P. (2012). Standardization of size, shape and concentration of nanoparticle for plant application. *Appl Biol Res, 14*(2), pp. 138–144. ISSN: 0972-0979.

Thandapani, S. & Sukumaran, A. (2017). Size optimization and in vitros biocompsatibility studies of chitosan nanoparticles. *Int J Biol Macromol, 104*, 1794–1806 Article ID 6090846.

Tolaymat, T., El Badawy, A., Sequeira, R. & Genaidy A. (2015). A system-of-systems approach as a broad and integrated paradigm for sustainable engineered nanomaterials. *Sci Total Environ, 1*, 511:595–607.

Tripathi, D. K., Singh, S., Singh, S., Pandey, R., Singh, V. P., Sharma, N. C., Prasad, S. M., Dubey, N. K. & Chauhan, D. K. (2017). An overview on manufactured nano-particles in plants: Uptake, translocation, accumulation and phytotoxicity. *Plant Physiol Biochem, 110*, 2–12.

Tripathi, D. K., Singh, V. P. & Prasad, S. M. (2015). Silicon nanoparticles (SiNp) alleviate chromium (VI) phytotoxicity in Pisum sativum (L.) seedlings. *Plant Physiol Biochem, 96*, 189–198.

Verma, D. K., Patel, S. & Kushwah, K. S. (2021). Effects of nanoparticles on seed germination, growth, phytotoxicity and crop improvement. *Agri Rev, 42*(1), 1–11.

Vidya, C., Hiremath, S., Chandraprabha, M. N., Antonyraj, M. L., Gopal, I. V., Jain, A. & Bansal, K. (2013). Green synthesis of ZnO nanoparticles by *Calotropis gigantea*. *Int J Curr Eng Technol, 1*(1), 118–120. ISSN 2277–4106.

Vinayagam, R., Selvaraj, R., Arivalagan, P. & Varadavenkatesan, T. (2020). Synthesis, characterization and photocatalytic dye degradation capability of *Calliandra haematocephala*-mediated zinc oxide nanoflowers. *J Photochem Photobio B Biol, 203*, 111760.

Wisam, J. A. & Haneen, A. J. (2018). A new paradigm shift to prepare copper nanoparticles using biological synthesis and evaluation of antimicrobial activity, *Plant Arch, 18* (2), e-ISSN:2581-6063 (online), ISSN:0972-5210.

Worrall, E. A., Hamid, A., Mody, K. T., Mitter, N. & Pappu, H. R. (2018). Nanotechnology for plant disease management. *Agronomy, 8*(12), 285.

Xu, C., Peng, C., Sun, L., Zhang, S., Huang, H., Chen, Y. X. & Shi, J. (2015). Distinctive effects of TiO_2 and CuO nanoparticles on soil microbes and their community structures in flooded paddy soil. *Soil Biol Bioche, 86*, 24–33.

Yang, J., Cao, W. & Rui, Y. (2017). Interactions between nanoparticles and plants: Phytotoxicity and defence mechanisms. *J Plant Interactions, 12*(1), 158–169.

Yang, J., Jiang, F., Ma, C., Rui, Y., Rui, M., Adeel, M., Cao, W. & Xing, B. (2018). Alteration of crop yield and quality of wheat upon exposure to silver nanoparticles in a life cycle study. *J Agric Food Chem, 66*(11), 2589–2597.

Yedurkar, S., Maurya, C. & Mahanwar, P. (2016). Biosynthesis of zinc oxide nanoparticles using *Ixora coccinea* leaf extract—A green approach. Open *J Synthesis Theory Appl, 5*(1), 1–14.

You, T., Liu, D., Chen, J., Yang, Z., Dou, R., Gao, X. & Wang, L. (2018). Effects of metal oxide nanoparticles on soil enzyme activities and bacterial communities in two different soil types. *J. Soils Sediments, 18,* 211–221. ISSN: 1439-0108.

Younas, A., Yousaf, Z., Rashid, M., Riaz, N., Fiaz, S., Aftab, A., & Haung, S. (2020). Nanotechnology and plant disease diagnosis and management. Nanoagronomy, 101–123.

Yusof, H. M., Mohamad, R., Zaidan, U. H. & Samsudin, A. A. (2020). Biosynthesis of zinc oxide nanoparticles by cell-biomass and supernatant of *Lactobacillus plantarum* TA4 and its antibacterial and biocompatibility properties. *Sci Rep, 10*(1), 1–13.

Zhao, M. X. & Zeng, E. Z. (2015). Application of functional quantum dot nanoparticles as fluorescence probes in cell labelling and tumour diagnostic imaging. *Nanoscale Res Lett, 10*, 171.

Zhu, H., Han, J., Xiao, J. Q. & Jin, Y. (2008). Uptake, translocation, and accumulation of manufactured iron oxide nanoparticles by pumpkin plants. *J Environ Monitor, 10*(6), 713–717.

Ziental, D., Czarczynska-Goslinska, B., Mlynarczyk, D. T., Glowacka-Sobotta, A., Stanisz, B., Goslinski, T., Sobotta, L. (2020). Titanium dioxide nanoparticles: Prospects and applications in medicine. *Nanomaterial, 10*(2), 387–400.

8 Exploring the Potential of Microbe-Mediated Gold Nanoparticles in Sustainable Plant Pathogen Management

Abdulmujib Gboyega Yusuf, Tesleem Taye Bello, and Oluwatoyin Adenike Fabiyi

8.1 INTRODUCTION

It is vital to manage plant pathogens to ensure minimal loss of agricultural products to plant diseases before and after harvest. Various managemental strategies are available to support this effort. However, some of the available options have negative effects on the environment and are not sustainable. Therefore, it is important to identify other cost-effective and sustainable plant disease managemental strategies (Fabiyi 2022 a; b; c; Fabiyi 2023a; Bello and Fabiyi 2024). The quantity and quality of crop production, agricultural revenue, and environmental consequences are to be considered in relation to sustainable farming systems (Spadaro & Gullino, 2019). Selecting a strategy for managing plant diseases is an important aspect of sustainable farming systems. Most of the existing plant disease managemental strategies do not perfectly fit the required space for sustainability, thus research is on-going to fine tune the most suitable to enhance productivity and sustainability. For example, the applications of agrochemicals have enjoyed tremendous popularity and success over the years for their instant action and ease of availability. Yet, certain harmful side effects on non-target organisations in the environment, population resurgence, and the development of race resistance are some of the reasons for branding this as not sustainable (Fabiyi 2021b; Fabiyi et al., 2023). Hence, there is a continuous need to develop an efficient alternative method that is reliable, with a less harmful effect on the environment. In this respect, nanotechnology has been an alternative that is currently under the spotlight for sustainable plant disease management (Anees et al., 2024; Bello and Fabiyi 2024). Nanotechnology is an unfolding scientific branch that engages matter's atomic, molecular, and super-molecular structure in the nanometric range for the emergence of characteristics necessary for various applications (Fabiyi, Alabi, et al., 2020; Fabiyi, Olatunji, et al., 2020; Koul et al., 2021). Various industrial applications have exploited the use of innovative techniques for sustainable NanoParticle (NP) production, under the canopy of nanoscience in recent times (Khan et al., 2019; Salem & Fouda, 2021). Therefore, there is an explosion in the interest of researchers from the relevant fields to engage in simpler and more efficient approaches to synthesizing and producing major metal nanoparticles. The diversity of nanoscience applications warrants the synthesis of NPs of befitting features such as size, structure, morphology, and monodispersity (Shah et al., 2015; Fabiyi et al., 2018).

The mode of synthesizing nanoparticle material is crucial in nanoscience and novel techniques are being developed. The major methods of fabricating NPs have been classified into physical, chemical, and biological. Due to the bottleneck associated with the traditional methods of NPs fabrications, namely, the physical and chemical, biological method using natural sources has been identified as being more efficient, reliable, and ecofriendly (Shah et al., 2015; Arshad et al., 2017; Fabiyi

DOI: 10.1201/9781003256762-8

et al., 2021a. The synthesis of inorganic material is possible by both unicellular and multicellular microorganisms following a bottom-up manner leading to the synthesis via the redox of metallic ions by the microbe's secretions (Prabhu & Poulose, 2012; Shah et al., 2015). However, microbe-mediated nanoparticle synthesis is complex because various microorganisms have the ability to engage with specific metallic ions in a particular manner (Prabhu & Poulose, 2012). Popular among the metallic ions are those of gold from which gold nanoparticles are synthesized. These particular NPs have gained much attention due to their suitable physicochemical, and Surface plasmon Plasmon Resonance (SPR) properties (Chowdhury et al., 2022). There has been recent attention to the microbe-mediated synthesis of gold nanoparticles due to their non-toxic and biocompatibility attributes. Similarly, the therapeutic activity and antimicrobial potentials of gold nanoparticles have increased their importance for several applications including plant disease management (Jafari et al., 2018). Despite the increase of plant disease control using nanoparticles, it is still important to know the field-scale assay of the effect on plants and the environment as available reports show that most studies conducted so far were done under in-vitro conditions. This chapter focuses on microbe-mediated AuNPs and their appropriate use in the sustainable management of plant pathogens causing considerable yield losses. Similarly, it attempts to highlight the progress recorded in AuNPs microbial-based synthesis and discuss the potential of AuNPs as a potent sustainable control of plant pathogens (Figure 8.1).

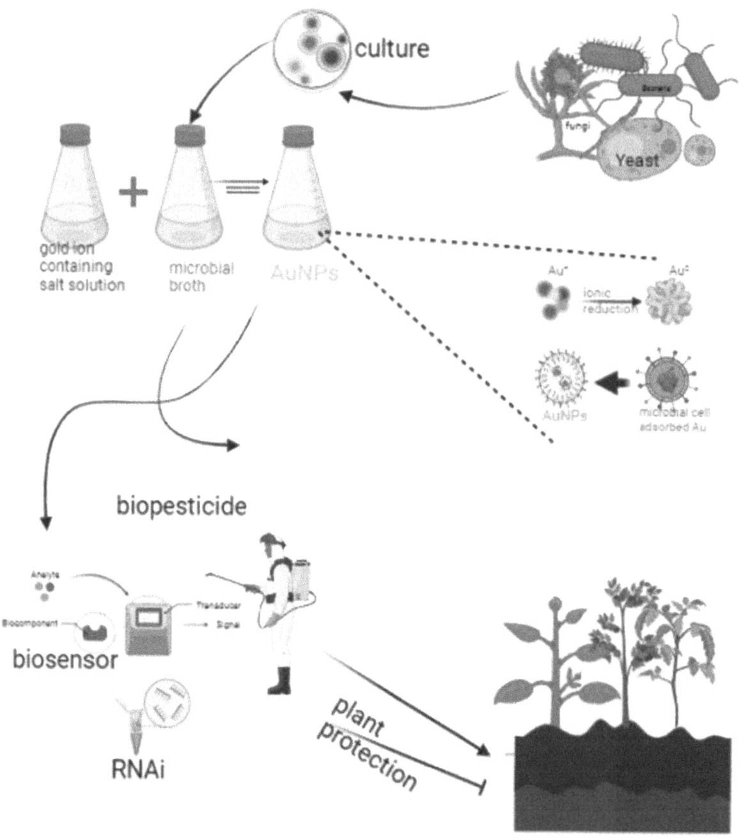

FIGURE 8.1 Schematic representation of microbial-mediated synthesis of AuNPs and their subsequent application in the effective plant pathogen management. The figure outlines the steps of AuNP synthesis facilitated by microbes and highlights the potential utility of these nanoparticles in mitigating issues related to plant pathogen infestations. Such potentials include use in biopesticide formulation, biosensors in detecting presence of pathogens, and as carriers of RNAi for in-planta delivery. AuNPs (Gold Nanoparticles); RNAi (Ribonucleic acid (RNA) interference).

8.2 PLANT DISEASE MANAGEMENT WITH NANOPARTICLES

Although chemical pesticides have proven to be very potent in controlling pathogenic microorganisms, over time, their usage has been accompanied by undesirable outcomes for the environment (Shahid et al., 2018). These pesticide-related concerns warrant the urgent development of alternative means for plant disease management that will be efficient yet with minimal disruption to the ecosystem Fabiyi 2021a; b; c;d Fabiyi 2022d). Various properties displayed by NPs and their successful application in various important fields such as medicine and pharmacology where they play roles in antimicrobial activity and drug delivery (Kaur et al., 2022; Rani et al., 2022) have attracted our interest, as an alternative to chemical pesticides. We believe an item considered relatively safe in human medicine should provide similar confidence when used in food and agriculture. Interestingly, the nanotechnology sector has advanced in the biofabricating of nanoparticles by finding an alternative method for the biological mode of synthesis instead of the physical and chemical approaches which tend to be costly and environmentally unfriendly (Tariq et al., 2022). Biologically synthesized NPs have so far been demonstrated to be useful in plant pathogen management in various aspects such as components of biopesticides, substances for direct crop protection, and in the enhancement of plant resistance to stress (Worrall et al., 2018; Avestan et al., 2019).

The use of NPs in plant disease management can be as a direct protectant or as a carrier for agrochemicals (Worrall et al., 2018). Nanoparticles such as silver, zinc oxide, copper, titanium oxide, and gold have potential for nematocidal, antibacterial and antifungal use that can be applied directly as protectants on plants, seeds, and soil against pests and pathogens (Kah & Hofmann, 2014; Mishra & Singh, 2015; Fabiyi, 2021a). When nanoparticles serve as carriers, they can be formulated as liquids (colloids) that are referred to as nanoemulsions or can be formulated as solid particles that are referred to as nanodispersion (Green & Beestman, 2007; Hayles et al., 2017; Fabiyi et al., 2021).

8.3 NANOPARTICLES AS CARRIERS OF AGROCHEMICALS

Several aspects of agriculture have had the chance to benefit from the application of nanotechnology, like the aspect of seed germination, water management, controlled use of agrochemicals, and agricultural biotechnology (Hayles et al., 2017). Pesticide use has particularly been a focus of improvement with nanotechnology by utilizing the desired features of nanoparticles such as shape and pore size, amongst other surface properties to enhance precise and focused delivery of the pesticide (Khandelwal et al., 2016). The concerns of indiscriminate pesticide use, residual effect, and loss to run-off, which all combine to create pest and disease resistance, reduced element fixation, and biodiversity have received considerable focus that aimed to reduce financial losses, increase efficiency and productivity with minimal environmental consequences. Concerning these, there is the yearning that nanotechnology will yield "nanopesticide" that will enhance pesticide efficacy, minimize dose application, residue run-off, and overall relieve the environment from anticipated pesticide-use-related concerns.

These perceived benefits of nanoparticles can be achieved when nanoparticles act as carriers for agrochemicals. Major features that set aside nanoparticles as enhanced pesticide carriers are their size, shape, chemical composition, and density (Hayles et al., 2017). Formulating pesticides as nanoparticles has advantages in providing formulation stability, increased solubility of the active ingredient, reduced toxicity, enhanced translocation, and upscaled vigor of pesticide potency (Sasson et al., 2007). Nanoparticles similarly can confer evenness in spray coverage, increased plant absorption, and soil adsorption (Cameron & Mitchell, 2007). Generally, agrochemicals are composed of insoluble compounds that require a large dosage of organic solvents to facilitate smooth application. The characteristics of poor solubility of pesticides due to their lipophilic nature reduces the uniformity of the pesticide during application, thereby more quantities are applied in the bid to ensure sufficient dose. Consequently, the toxic residue is abundant in the environment, and of course, increased unnecessary costs. By reducing the particle size to the nanoscale, the water solubility and dissolution rate will be improved (Kalepu & Nekkanti, 2015). The effectivity of agricultural formulations is enhanced when the active ingredient is encapsulated by nanoparticles. Available as common carriers for various pesticides, herbicides, and fungicides are silica nanoparticles, chitosan nanoparticles, solid lipid nanoparticles, and the layer double hydroxide (Park et al., 2010; Brunel et al., 2013).

8.4 NANOPARTICLES AS A DIRECT PREVENTATIVE MEASURE

Nanoparticles have been involved directly in combating the menace of plant pathogens on plants (Dutta et al., 2022). The very common nanoparticles engaged are copper, titanium oxide, silver, and gold nanoparticles. The boom of green synthesis has led to increased enthusiasm for the use of inorganic nanoparticles in various fields particularly in plant disease management due to their eco-friendly anti-microbial properties (Savithramma et al., 2011; Fabiyi, 2023c). There is a growing interest in silver and gold nanoparticles due to the superior characteristic surface zone, biochemical activities, and antibiotic properties that they present (Worrall et al., 2018). Silver nanoparticles have been tested in various studies with positive outcomes while there is still a paucity of evidence for the potential of gold nanoparticles in plant pathogen management. There are various methods of synthesizing nanoparticles but the microbial aspect of green synthesis of NPs has gained wider attention due to its eco-friendly nature and high stability (Debroy et al., 2023).

8.5 MICROBIAL SYNTHESIS OF NANOPARTICLES

In the bid to avoid toxic chemicals, scientists have found a way around the green synthesis of metallic nanoparticles in developing cost-effective and eco-friendly NPs. Green synthesis comprises the use of natural reducing and stabilizing agents such as plants and microbes (Ogunyemi et al., 2019). Microbe-mediated synthesis has come to outshine the traditional methods of nanoparticle synthesis because of its lesser hostility to the environment; the environment being of great concern. The technique employs the service of biological entities such as bacteria, actinomycetes, viruses, algae, and fungi in fabricating the particle to nanosize (Gahlawat and Choudhury, 2019). Similarly, there is an impact of the technique on the shape and forms of the yielded nanoscale structure. The various morphological forms include nanotubes, nanorods, nanoconjugates etc., that present features which make them outstanding as biomedical and antimicrobial resources. Interestingly, some bacteria can fabricate special nanomaterials such as nanocellulose and nanowires (He et al., 2008), and some produce exopolysaccharides with antibiotic features (Li et al., 2017). Nevertheless, most microbes possess the prospect of reducing inorganic metals, and they have high adaptability even under high concentrations of the metal (Salem & Fouda, 2021). It is of note that bacteria are better positioned than other microbes for the synthesizing of NPs because of their dynamic culturing conditions and exponential growth rate (Koul et al., 2021). However, Iranmanesh et al., (2020) claimed that fungi are preferable to other microorganisms due to some factors like safety, dynamism, and ability to yield more NPs volume at a relatively lower cost. The enzymatic reduction occurs, either via an intracellular or extracellular route, when the microbes absorb metal ions within the media which is then followed by a conversion of the metallic ion to elemental form. Intracellularly, the ions are absorbed into the microbial cells where nanoparticles are enzymatically formed, and the extracellular type involves the adhering of the metallic ion onto the cell surface where the reduction occurs in the presence of an enzyme (Zhang et al., 2011). The two strategies for fabricating NPs biologically are simple, direct, and require less energy. The cellular machinery of the organism cells is exploited in the intracellular synthesis of NPs. This technique begins with proper culturing of the organism in a suitable medium and a biomass pellet is obtained from the culture by appropriate centrifugation (Shah et al., 2015). The enzyme activities can be boosted when the culturing factors such as the pH of the medium, its temperature, concentration, light parameter, and centrifugation rate are well optimized (Mukherjee et al., 2001). The aqueous solution of the metal is introduced into the collected biomass and the mixture is incubated until the desired changes are apparent. Likewise, for the extracellular mode, the microbial enzymes are collected from the supernatant following the centrifugation of the microbial culture, which is then introduced to the metal ions. The NPs are then formed through the bio reduction of the ions by the cell-free supernatant (Marooufpour et al., 2019). In general, there is a tendency for extracellular nanoparticle formation in preference to the intracellular mode of formation. This suggestion is made from the process and application point of view whereby there is a further need to recover the nanoparticle from inside the cell. For example, in the intracellular fabrication of AuNps using Verticillium cells, the process requires an extra step of using glass beads to break the cell content free (Corti & Holliday, 2004). Further recovery processes for the intracellularly synthesized NPs include sonication and lysis (Varia et al., 2016), whereas, the filtration step suffices to recover extracellular NPs (Castro et al., 2014).

For a proper grasp on the monodispersity and configuration which aim to achieve NPs of desired properties, bacterial organisms are often employed in this regard. Under the mediation of various microbes, NPs of various kinds of metals and their oxides have been formed (Gahlawat & Roy Choudhury, 2019; Grasso et al., 2020). *Escherichia coli*, *Bacilus cereus*, *Acinebacter*, *Pseudomonas* and *Corynbacterium* sp. have generally been exploited in the synthesis AgNPs. A considerable amount of work has been published on fabricating AgNPs using bacteria. These studies demonstrated that the Ag atoms accumulate in the cellular periplasmic space of the bacteria, and up to 200 nm size of the metal can be accommodated (Prasad et al., 2011). This nanoscale metal has been shown to possess antibacterial and antifungal properties (Kora et al., 2010; Prabhu et al., 2010).

8.6 MICROBIAL SYNTHESIS OF GOLD NANOPARTICLES

Gold nanoparticles (AuNPs) in recent years have received more attention due to their enticing Surface Plasmon Resonance (SPR), optical, electrical, and physicochemical properties. Their quantum size and periodic location confers special useful properties on them which make them unique as antibiotic, drug delivery agents, biocatalysts, and in the bioremediation of environmental toxicity (Chowdhury et al., 2022). AuNPs have been utilized in several applications because of their special features such as high stability under normal environmental conditions, their ability to resist rapid oxidation, and their biocompatibility (Corti & Holliday, 2004). It is worthy of note that the shape and size of NPs are to be considered in their various applications. In an extensive study by Gericke & Pinches (2006), microbial culturing parameters are varied and their influence on the outcome of the nanoparticle is documented. They observed that the quantity of gold particles per cell is inversely proportional to the growth phase of the culture. More particles are yielded at the exponential growth phase of the culture perhaps because the phase contains a higher amount of biologically active enzymes that are responsible for the gold ion reduction. This corroborates the previous suggestion that the enzymes present in the cell wall or cytoplasm of microbes are involved in the reduction of the gold ions (Brown et al., 2000; Mukherjee et al., 2001). The study further indicates the impact of pH on particle morphology where the outcome shows varying particle size and shape at different pH levels.

8.7 BACTERIAL-MEDIATED SYNTHESIS OF GOLD NANOPARTICLES

Since the first demonstration of the capacity of bacterial cells to fabricate AuNPs intracellularly (Beveridge & Murray, 1980), various protocols for microbe-mediated AuNPs synthesis have been documented. AuNPs ranging from 5–25 nm were intracellularly formed when the bacterium *Bacillus subtilis* acted on the gold on. Deposition of gold ion was minimal in cells with a modified carboxyl group suggesting that the carboxyl is influential in the selection of the site of metal deposition (Beveridge & Murray, 1980). A later work using the same bacteria *B. subtilis* to bio-fabricate NPs showed that the bacteria is capable of enhancing proper dispersal of the NPs of size ranges 20–25 nm (Srinath et al., 2018). *Klebsiella pneumoniae*, in a simple biological demonstration, reduced the gold ions exposed to its cellular content to nanosize particles within a colloidal solution (Malarkodi et al., 2013). Through a fourier transform infrared spectroscopy, the study further observed the role of a carboxyl group in the binding of the functional group of the AuNPs to the microbial cells. To confirm the influence of the culturing medium on the size and the site of AuNP synthesis, *Shewanella algae*, grown in a solution of neutral pH, yielded AuNPs of about 20 nm size around the periplasmic space (Konishi et al., 2006). Meanwhile, larger AuNPs were extracellularly produced under a very acidic medium. The mechanism of the microbe-mediated AuNP synthesis requires the transfer of electrons during microbial respiration (Dominguez-Benetton et al., 2018). The pH of the medium is responsible for the necessary protonation required suggesting the reason for the large size of AuNPs yielded at acidic pH which is due to the high electrostatic adsorption of the acidic pH (J. Varia et al., 2014). The gold ions are usually diffused to the microbial cell surface where they bind to the protein recognition motif leading to subsequent processes such as ion exchange, micro-precipitation, chelation, and biosorption (Schiewer & Volesky, 1995; Sağ & Aktay, 2000). *Rhodopseudomonas capsulate* is another bacteria that can synthesize AuNPs from chloroauric acid ($HAuCl_4$). The form and size of the NPs can be varied by manipulating the pH of the biomass and acid medium. In their work, He et al. (2007) observed that gold nanowires obtained from the bacterial biosynthesis process are usually spherical in shape. They also noticed that the bacteria is particularly helpful in stabilizing the product over a reasonable period perhaps as a result of the proteinous secretion of the bacteria that serves as the capping agent (He et al., 2008). The gram-negative bacterium, *Escherichia coli* and its biomolecule have been used in the fabrication of AuNPs using the cell-based and cell-free route respectively. For the cell-based route, the biomass of the *E. coli* culture was directly used to reduce gold ions from chloroauric acid (Du et al., 2007). The average size of the AuNps fabricated by the cells of the *E. coli* was about 20 nm and spherical in form. Enzymes purified from *Escherichia coli* provide a cell-free medium for AuNP production (Gholami-Shabani et al., 2015). The AuNPs fabricated were spherical and were produced up to 10 nm in size. The use of microbial enzymes is a relative aspect of microbe-mediated nanoparticle production which presents several opportunities and more stability over the microbe under storage conditions. The cell-free method also presents a situation whereby the necessary steps of cell lysis, separation, and purification are not required. In addition, the use of biomolecules such as the enzyme will overcome the issue of some microorganisms lacking essential secretory proteins vital for NP synthesis. The enzyme can also be reused, unlike the microbial biomass, creating cost efficiency, and reduction in the time required for culturing the microorganism (Riddin et al., 2009). This approach has also been demonstrated to improve detection sensitivity and selectivity (Xie et al., 2012).

The algae cell extract from the bacterium *Shewanella* has also been employed for its shape-controlling and metal-reducing capacity to synthesize spherical gold NPs and gold nanoplates at room temperature and pH 2.8 (Ogi et al., 2010). Interestingly, the temporal factor is a huge determinant in the formation of AuNPs. AuNPs of size 9.6 nm, formed after an hour, and particle sizes of about 200 nm were detected after 24 hrs. The team also found more AuNPs to be yielded with *S. algae* extract than in the resting phase of the bacterium. To increase the biosynthesis rate of gold NPs, a cell-free extract of *Staphylococcus epidermis* was tried, to synthesize AuNPs from hydrogen tetrachloroaurate ($HAuCl_4$) (Srinath & Ravishankar Rai, 2015), and the outcome was commendable. The duration of synthesis was significantly reduced from 15 h to 10 min at a high incubation temperature of 115°C. The outcome shows an advancement over the unjustifiable longer synthesis duration at room temperature. Recent works also extended the use of gold nanoparticles to the diagnosing of pathogens. Gold nanoparticles aid in the preparation of biosensors because of their capacity to transfer electrons smoothly (Singh et al., 2013). To develop an environmentally friendly route for synthesizing the admired AuNPs, the use of a gram-positive bacterium, *Bacillus stearothermophilis* was given an attempt (Luo et al., 2014). The attempt yielded AuNPs of size 5–30 nm of varying shapes according to transmission electron microscopy. *Streptomyces* sp. is another gram-positive bacteria that have been used successfully in fabricating AuNPs from gold sources (Karthik et al., 2013). The bacteria species yielded up to 5–50 nm AuNPs size from chloroauric acids. AuNPs of up to size 50 nm have been formed using *Pseudomonas fluorescens* (Syed et al., 2016). Colonies from the culture were subjected to large-scale fermentation to produce a cell-free medium and subjected to gold chloroaurate of about 10% of the total volume for gold synthesis. The synthesized particles were triangular and spherical in morphological analysis. This morphology is in accordance with previous works (Verma et al., 2013).

8.8 FUNGI-MEDIATED SYNTHESIS OF GOLD NANOPARTICLES

Several study-based pieces of evidence have demonstrated the ability of bacterial organisms and their associated biomolecules in achieving both cell-based and cell-free synthesis of AuNPs. In the same context, fungal organisms are also a fine repertoire for the microbe-mediated synthesis of AuNPs. In a simpler view, fungi can be exploited to synthesize NPs in two major categories, the in-vivo and the in-vitro method (Molnár et al., 2018). In the in-vivo category otherwise known as the intracellular category, the fungi fabricate the nanoparticles within their mycelia which is an innate survival skill of the fungi because it allows them to process the toxic metal ions to the less-harmful form (Ramanathan et al., 2013). The cell-free extract or filtrate is generated and utilized in metal bio reduction to form nanoparticles. The extracellular route produced particles that are free of unwanted cellular components because of the clean state of their site of formation (Kulkarni et al., 2015). From a broader perspective, the synthesis of nanoparticles from fungi action on gold ions can have four orientations which are dependent on how the fungal mycelia are handled (Molnár et al., 2018). One of the four perspectives aligns with the intracellular or in-vivo categories where the mycelia are diversely utilized. However, there are further reasons that complicate this method such as the need to separate the AuNPs from the cell. Regarding the other three orientations, which are all classified as extracellular synthesis categories, the first orientation is utilizing the fungi secretion such as enzymes and bio-compounds for the synthesis of AuNPs (Castro et al., 2014). A report demonstrated the active site of a secreted protein in biosynthesis to be the carboxylic side of its amino acid, in the presence of other peptide linkages and phosphate group. (Das et al., 2012). Although, the method presents some weaknesses by retaining some unwanted components from the fermentation media (Molnár et al., 2018). Another orientation comes from the approach where cell lysis is carried out to expose its intracellular components which are utilized in synthesizing the AuNPs biologically to produce the BioAuNPs (Das et al., 2012). Lastly, An interesting approach utilizes in synthesizing AuNPs aside from the extra and intracellular approach is the autolysate. The mycelia can be processed to yield an aqueous extract capable of reducing gold ions to AuNPs as well (Mukherjee et al., 2002). This is possibly due to cell autolysis that occurs during mycelia processing at the incubation period leading to membrane protein and surface carbohydrate disintegration. Nevertheless, it is quite unfortunate that the method is prone to bacteria contamination during mycelia processing. A good news is that the simple addition of antibiotic can prevent this problem (Hur et al., 2014). However, large scale use of antibiotics is not encouraged. A contrary role of antibiotics in converting metal ions to NPs has been demonstrated and the antibiotics are reported to have a synergetic effect in nanoparticle formation by acting as a reducing agent in the conversion (Brown et al., 2000; Roshmi et al., 2015).

The success of fungi-mediated AuNPs has been recorded via various bio-assay endeavors. Processing the mycelium of *Fusarium oxysporium* into an aqueous extract will reduce gold chloride into AuNPs (Mukherjee et al., 2002). This is possibly the first time to report a cell-free fungal-mediated approach in producing AuNPs extracellularly. The endeavor was ignited as a result of an earlier work by the same group where *Verticillium* sp. was shown to possess the capacity of producing AuNPs intracellularly by its reduction action on gold ions. (Mukherjee et al., 2001). The fungus

Aspergillus terreus IFO prompted the formation of AuNPs within an eco-friendly protocol that is capable of yielding AuNPs of desirable sizes (Priyadarshini et al., 2014). Nanosizes in the range of 10–19 nm are achievable at pH 10 of the fungal extract. Large biomolecules possessing certain functional groups are an integral part of the extract that executed the bio reduction and stabilization that occurred during the bio-fabrication. Likewise, the cell-free extract of *Aspergillus niger* is capable of reducing gold ions to form AuNPs (Soni & Prakash, 2012). The crystalline AuNPs are similarly formed by the reduction of gold ions due to the enzymatic activities of the extract. A similar report from Verma and colleagues demonstrated the role of another *Aspergillus sp.*; *A. Clavatus*, in producing triangular-shaped AuNPs through the reduction of chloroaurate aqueous solution (Verma et al., 2010). The outcome of the study presented some application opportunities that lie in the shape and size of the nanoparticle. Nanotriangular particles possess special plasmonic features that put them forward as anisotropic which is required in vital applications (Verma et al., 2010; Ortiz-Castillo et al., 2020). The particle size, shape, and composition are all important factors in the success of AuNPs as anisotropic. More recently, *Alternaria chlamydospore*, a marine fungus, was exploited for the first time as a biofactory for synthesizing AuNPs of various nanosizes and shapes (Ameen et al., 2023). The AuNPs were observed to possess properties relevant to biocontrol after it significantly inhibited the growth of various bacteria pathogens at varied concentrations.

Handling fungi is comparatively easier than most pathogens, aside from its capacity to secrete high proteins and enzymes, it also offers an advantage in the large-scale synthesis of nanoparticles. The first suggestion of the possible role of enzyme secretion of a fungus in the reduction of gold ions to nanoparticles arises when a *Verticillium* sp. was exposed to aqueous chloroaurate (Mukherjee et al., 2001). AuNPs of around 20 nm in diameter were assumed to be formed by the action of the enzymes found in the cell wall and the cytoplasmic membrane possibly because the gold nanoparticles were found to be formed on both the internal and external surface of the fungi cell (Mukherjee et al., 2001). The external surface site for AuNP formation implies that the mycelia would be gold-plated following the reaction. However, there are differences in the sizes of the nanoparticle from each site. Fewer but larger AuNPs are found within the cell while smaller particles are found on the wall of the cell. Similar to the role of pH in cutting the size of the nanoparticle, the rate of the cytosolic extract of the fungal organism in the reaction mixture is also observed to be a determining factor in the morphology of gold nanoparticles. A higher concentration of *Candida albican* cytosolic extract will yield a large amount of isomorphic spherical AuNPs of about 40 nm whilst, at lower concentrations of the same extract, more of non-spherical gold nanoparticle are yielded (Chauhan et al., 2011). The non-spherical shapes include the hexagonal and triangular categories that are often larger than the spherical. There are instances whereby the two modes of microbial biosynthesis–the extracellular and the intracellular–are simultaneously presented by some fungi. An instance is when the fungus *Penicillium sp.* achieved this double feat by bio-fabricating AuNPs from gold ions (Du et al., 2011). The intracellularly formed AuNPs of about 50 nm size, after incubating the fungal biomass with gold chloride for a period, were found to be spherical and properly dispersed within the cell. On the other hand, the extracellularly formed AuNPs were achievable by using biomolecule-containing cell filtrate to reduce the gold ion to AuNPs. The group deduced from the work that nanostructures, other than the conventional, such as nanowires were possible when there is a lower amount of cell filtrate compared to the concentration of the gold ion (Du et al., 2011).

Furthermore, the synthesis of AuNPs by the extract of *Sclerotium rolfsii* seems possible through the activity of the NADPH-dependent enzyme present in the extract (Narayanan & Sakthivel, 2011). The enzyme acted on gold ions from chloroauric acid to produce AuNPs of varying shapes with anisotropic features such as triangles, hexagons, decahedral, and rods. The ratio of the metal salt and the cell-free filtrate are determining factors in the morphology observed. Positive observation, using FTIR spectroscopy, of aromatic amino acids such as tryptophan, tyrosine and so on, points to the presence of protein in the cell-free filtrate. In the bid to establish the exact role of the coenzyme in the bio-fabrication, the cell-free extract was subjected to heat denaturation, and the extract with the denatured enzyme was still capable of yielding AuNPs. This is an indication that there are other components, aside from the enzyme, present in the extract that equally participates in fabricating the nanoparticles (Molnár et al., 2018). Ultimately, adequate care is needed when using the extracellular extract of a fermented fungal broth because the supernatant will unavoidably accommodate residues from the culturing media that are capable of taking a role in the gold ion reduction (Molnár et al., 2018). Most growth media contain reducing sugar, and synthetic media similarly contain nonreducing sugar, and there is less certainty that the fungi will consume all the various carbon sources in the different media. The synthetic media can reduce this effect which nevertheless still contains glucose that can eventually participate in AuNP formation at an increased temperature (Qi et al., 2004). Hence, the eventual implication is that it is difficult to separate the exact role of the fungi biomolecule generated extracellularly from the effect of the media residue on AuNP formation. For those interested in clarifying this equivocal status of the extracellular route, a subsequent re-washing and processing of the fermented mycelia should suffice, which must be confirmed to lack the capacity to produce nanoparticles (Molnár et al., 2018).

8.9 OTHER MICROORGANISMS CAPABLE OF BIOSYNTHESIS OF AuNPs

Other organisms such as yeast, algae, and virus have been used for the biosynthesis of nanoparticles. Algae are also bioagents that are efficient in the production of various metabolic NPs including AuNPs from gold ions (Agarwal et al., 2019). Viruses have also been exploited in the synthesis of NPs due to their unique structural integrity and reduced influence on the environment (Steinmetz & Manchester, 2011; Chung et al., 2020).

8.10 GOLD NANOPARTICLES AND SUSTAINABLE PLANT PATHOGEN MANAGEMENT

A wide variety of microorganisms are pathogenic to plants which include several species of bacteria, fungi, viruses, and nematodes. The activities of some of these plant pathogens are peculiar to specific hosts while a whole lot of others are general in their target. However, whether the pathogen is a specific or general parasite, there should be means of managing their destructive lifestyle, and which must consider the effect on the environment, under the canopy of sustainable pest management. Pesticides are a common intervention of pest control in agriculture which is a vital part of Integrated Pest Management (IPM). IPM was introduced to control the usage of pesticides and is accepted as a means to curb the danger of pesticide residual effects (Sinha et al., 2017). IPM involves detailed monitoring, and consciousness of the dynamics of pests and pathogens activities (Pimentel, 2009). IPM is based on knowledge of the ecosystem with a focus on long-term pest control by combining various tactics that are true to the task, and taking necessary precautions in the use of pesticides (Peshin et al., 2009). The use of pesticides is accompanied by several environmental issues which include but not limited to spray drifting and runoff, escaping into the air during the application, off-target delivery, and residual effect on the environment (Castro et al., 2014; Chowdappa & Gowda, 2021). Through nanotechnology, pesticides and other agro allied chemicals can be safely used through efficient administration of nanomaterials. Overall, the amount of toxic effluent discharged to the environment are considerably reduced while increased pesticides efficacy is achieved (Mousavi & Rezaei, 2011).

Development of NPs with antimicrobial properties is possible through nanotechnology and its application in protecting crops against pathogens including proper control of pesticide release, development of nanosensors useful in the early detection of plant pathogen presence, and direct protection of plants against pathogens (Scott & Chen, 2012). The nanoparticle possesses antimicrobial features which is the capacity to disrupt the growth and activities of plant pathogens yet with no destructive impact on the surrounding plant tissues (Panja et al., 2021). Gold nanoparticles biosynthesized by microbial factories have been reported to possess antimicrobial activities against many species of pathogenic microorganisms. Applying AuNPs can take different forms in plant pathogen management such as seed treatment or foliar spray or as a carriage for other chemical-based active ingredients (Khan et al., 2014).

8.11 MECHANISM OF NP ACTION ON PLANT PATHOGEN

Having said much about the production of AuNPs through important microbe-mediated means, let us now have a look into how they can disrupt pathogens foraging activities. Experimental evidence in this regard has pointed to some processes. Various antimicrobial nanoparticles are capable of reducing bacteria load due to their high surface area to volume ratio which leads to the emergence of improved properties that the bacteria cannot resist (Whitesides, 2005; Xia, 2008). Through electrostatic interaction, NPs can attach to a bacterium membrane and thus the membrane gets disrupted (Antoine et al., 2006). Other factors allowing toxicity include the diffusion rate, bacteria growth rate and biofilm formation (Hajipour et al., 2012). Gold nanoparticles have been shown to be toxic and photo-mutagenic against bacteria species by the generation of free radical that is toxic to bacteria species via the action of the radicals on the bacterial genetic composition (Santo et al., 2008; Hajipour et al., 2012). In addition, cell death of bacteria colony occurs when NPs and other antibiotic proves to be synergetic on the bacteria via a process whereby the antibiotic facilitate cell wall damage and facilitate the entrance of the nanoparticle into the bacteria where the genetic material are muddled up (Devi & Joshi, 2012). Reports have also shown that a phenomenon termed ionic or molecular mimicry is possible during the interaction between biomolecules and nanometal whereby biomolecules bind to the metal ion whose structure is similar to a normal co-factor of the biomolecule, this situation will cause cellular dysfunction (Clarkson, 1993; Lemire et al., 2013). The nutrient assimilation ability of microbe can also be disrupted when sulfate uptake is interfered with by metal toxicity (Pereira et al., 2008). When Sulphur uptake is inhibited, as mediated by the metal ion, it will cause the cells to starve and the growth impeded (Lemire et al., 2013). Various metal base NPs have been shown to be effective against the tomato bacterial wilt caused by Ralstonia solanacearum. Significant reduction in the incidence of the tomato bacterial wilt was recorded upon treatment with CuONPs (Jiang et al., 2022). Consequently, there was significant reduction in symptoms level and overall yield improvement, even, the rhizopheric community were enriched. In a similar reaction, it was noticed that low concentration of metals nanocomposite were effective even against soil persisting gram-positive bacteria (Dawwam et al., 2022). Olive knot disease are caused by Pseudomonas savastanoi pv. Savastanoi. This bacterial pathogen has been observed to be inhibited under in-vitro condition by the action of synthesized nanocrystal (Schiavi et al., 2022). Further demonstration of the management of bacterial pathogens of agricultural importance is obvious when bacterial wilt of solanaceaous crops caused by *Ralstonia solanacearum* succumbed to the inhibitory prowess of NPs. The bacterial cells were observed to lose innate shape and flagella, and ruptured ultimately during the interaction with NPs (Khairy et al., 2022). Interestingly, empirical efforts have also revealed the effect of AuNPs of plant pathogenic bacteria. For example, the rate of biofilm formation of *Pseudomonas aeruginosa* was inhibited at AuNPs concentration of 10 ppm (Yu et al., 2016). Biofilms are required for colonization and the loss of which will tamper with the pathogenicity (Mina et al., 2019). Generally, the different pathways leading to bacteria death all facilitate the corruption of membrane permeability and respiratory activities, and damage to the cell component (Alkhataf, 2021). Analogously, such effects have also been displayed by the plant pathogenic fungi.

The effect of the nanoparticle on the fungal organism is also reflected in the cells where the earliest observation was usually the disappearance of the cell wall and the cell membrane, followed by degeneration of cell organelles and ultimate cell death. The basic systems within the organism are disrupted by the nanoparticles, particularly the transport system where the intracellular ions are released and the metal ion are accumulated (Du et al., 2012). Inhibition of the H^+-ATPase-mediated proton pumping is also associated with the susceptibility of fungi isolates to AuNPs (Ahmad et al., 2013). This is a result of the corruption of the integrity of the membrane associated with the transport system when H^+ -dependent symporters, antiporter, and other channel-mediated transport have lost their transmembrane electrochemical potential activities (Beyenbach & Wieczorek, 2006). Consequently, cell growth slows down and the cell gradually dies out. There are pieces of evidence that the NPs size has a role in the antifungal activity of the AuNPs. The observation that the potency of the nanoparticle increases with decreasing particle size indicates the importance of the size in the inhibition. The large surface area of the tinier nanoparticles enhanced their adherence with the binding site of the membrane protein (Ali et al., 2020). It could also be deduced that the diffusion rate of the tinier AuNPs through the cell membrane is faster than the bigger nanoparticles. This deduction is evidenced in the work of Ahmad et al., (2013), where AuNPs of 7 nm was observed to possesses excellent antifungal activities than the 15 nm sized NPs

8.12 CONTROL OF PLANT PATHOGENS BY AuNPs

Nanoparticles have been successfully exploited in controlling plant pathogenic organisms (Choudhary et al., 2021), inspired by the advancement of microbe-mediated synthesis of nanoparticles due to their reduced effect on the environment (Oves et al., 2018). The biosynthesized metallic NPs have been demonstrated to exhibit their potency against gram-positive and gram-negative bacteria alike (Alkhataf, 2021). Similarly, various activities of pathogenic fungi have been inhibited by NPs. Much of the attention and demonstration of NPs effect on pathogenic microbes have been given to silver NPs probably due to the ease of the biosynthesis from various green materials (Ali et al., 2020). Although some organic gold ions are antibacterial, most AuNPs are antifungal, but the outcome of their antibacterial properties lies in the effect of the chemicals that are unable to be completely separated from AuNPs such as gold ion, and surface coating agents that co-existed during the biosynthesis (Brust et al., 1995; Zhang et al., 2015). Nonetheless, AuNPs exhibit synergistic behavior along with antibiotics to stimulate bactericidal activities or express bactericidal activities under special conditions such as radiation (Zharov et al., 2006; Grace & Pandian, 2007). The viable interaction between AuNPs and some antibiotics allows for easy adsorption of the antibiotic to the nanoparticles perimeters. This feature confers on AuNPs an efficient drug delivery enhancement. *Pseudomonas aeruginosa*, or *Escherichia coli*, and other gram-positive and negative bacteria have been successfully managed using colloidal AuNPs capped with antibiotics, and the combination displayed more efficient antibacterial activity than the sole drug (Grace & Pandian, 2007). This suggests that AuNPs are effective carriers in drug delivery. Gold NPs have been utilized in a physical arrangement where the nanoparticle is energized through irradiation to increase their potency on the bacterial cell. This method is effective in killing the gram-positive bacterium *Staphylococcus aureus* and can ever permit selective killing when the NPs are infused with antibodies (Zharov et al., 2006). This technique displays interesting potential for controlling plant pathogenic bacteria as it appears to be environmentally friendly due to its selectivity whereby untargeted organisms will be spared. However, its cost-effectiveness and large-scale production should be looked into. Similarly, the antibacterial effect of AuNPs can be optimized by incorporating certain enhancement conditions such as dye and molecules. For example, methylene blue and toluidine blue have been demonstrated to enhance the antimicrobial activity of AuNPs after incorporation into 2 nm AuNPs (Naik et al., 2011).

Thus far, we have shown the potency of AuNPs as a sole antimicrobial agent or as an improved agent when combined with other agents, it is now necessary to provide study-based evidence of the use of AuNPs in controlling plant pathogens. Root-knot nematodes (RKN) were controlled in pot experiments and in-vitro. The nematocidal effect was efficient when applied directly to the nematode in water whereas the reproduction factor of the nematode was low which is indicative of the efficiency of the AuNPs in controlling RKN (Thakur et al., 2018). Interestingly, the study demonstrated that the biomass of the tomato increased when AuNPs were used to control nematode with no harmful impact on the plant physiology. This work presents a great view into the potential of AuNPs in the sustainable management of plant pathogens and requires an efflux of research queries of AuNPs in this endeavor.

8.13 NANOPARTICLES ENHANCE SOIL MICROBIAL ACTIVITIES

There is a possible impact of AuNPs on soil health improvement through their effect on soil microbial activities. A very interesting aspect of nanotechnology is the effect of nanoparticles on microbes. It has been demonstrated that microbiological reaction rates can be influenced by nanoparticles (Choe et al., 2000). The nitrate reduction rate is increased by a nanoscaled zero-valent ion Fe(O). Nitrate reduction in the soil occurs either through a microbial process of denitrification or dissimilatory nitrate reduction to ammonium, leading to the loss of nitrate (NO_3) while the greenhouse gas, nitrous oxide (N_2O) is given off (Giles et al., 2012). This greenhouse effect warrants the remediation of nitrate. Some techniques of remediation include irrigation, replacement with clean soil, and improvement of cultural practices, but these practices involve high costs and are sometimes less effective (Ravindran et al., 2007). Consequently, it is important to find an alternative technique that is feasible, effective, and with a reduced negative impact on the environment. The success of AuNPs in speeding up the rate of microbiological reaction will present the potential in enhancing the denitrification processes that will ultimately lead to more fixed nitrogen and better soil health. The success of this will go a long way in the history of sustainable agriculture. The demonstration by Shin and Cha (2008) that the efficiency of nitrate-respiring bacteria can be triggered by receiving hydrogen ions from Fe(o) nanoparticles is a gateway to the potential of the nanoparticles for such purpose. Hydrogen gas is a major thermodynamically favorable electron donor for nitrate reduction (Till et al., 1998). Although, since several studies have delved into the application of nanoparticles in enhancing the microbiological reaction rate, there are still many aspects of improving soil health that need scientific dedication. The outcome will be biologically and cost-effective tools, with relative safety to the environment, and also stimulate plant resistance to invading pathogens (Akanmu et al., 2021; Niu & Zhang 2023).

8.14 NANOPARTICLES AND BIOTECHNOLOGY IN PLANT PATHOLOGY

As it is known, the hallmark of plant pathogen management is the use of resistance variety. This method, especially when used within an integrated pest management program, has been considered to be the most endearing and acceptable for achieving sustainable plant disease control. Resistant plants have been developed over the years through interdisciplinary research to reduce crop losses to plant diseases. One of the prominent tools in this advancement is RNA silencing which allows for the interference of the expression of genetic messages vital in the synthesis of virulent genes of plant pathogens (Kuo & Falk, 2020). An important step in RNA interference is delivering siRNA or miRNA into the plant, and viral vectors have been exploited in their delivery of siRNA into the plant (Guo et al., 2019). However, there are some setbacks associated with the use of virus vectors in delivering siRNA which include limited host range and expression of viral disease symptoms by the transgenic plant (Chellappan et al., 2004). Therefore, it is essential to seek an alternative technique for achieving post-transcriptional gene silencing. Incidentally, siRNA delivery mechanism involves a part where the excellent nanoparticles features will be supportive of an improved delivery (Hamid & Saleem, 2022). The ability of nucleic acid to form a condensed complex nanoformulation with polymer nanoparticles has been utilized in facilitating anionic nucleic acid delivery into the cells (Behlke, 2006), for enhanced siRNA protection and delivery. Similarly, nanoscience-dependent biotechnology has been applied in the identification of resistance genes and the development of non-viral gene delivery systems (Bharali et al., 2005; Yetisgin et al., 2020). Nanoscience has advanced to provide AuNPs efficient in the delivery of nucleic acid into cells because of their capacity to be specific in their intracellular activities, desirable size, and biocompatibility (Xu et al., 2006). In addition, AuNPs adsorbed antisense RNA has a higher affinity to bind to its complementary counterpart than other formations (Zhang et al., 2018). So far, evidence-based studies have pointed to the immense role of AuNPs in facilitating transgenic plants with significant benefits for plant pathogen management.

8.15 THE INDIRECT EFFECT OF AuNPs ON SUSTAINABLE PLANT PATHOGENS MANAGEMENT

Gold NPs present several opportunities for application in several plant-pathogen diagnostic platforms such as biosensors (Holzinger et al., 2014). The presence of pathogens in the field is the precursor to disease establishment once the optimizing conditions are favorable. Early detection of such pathogen will enable apt control whereby less of the control option will be utilized or even a simple cultural practice will suffice due to the vulnerable phase of the pathogen. Through biosensors, pathogens of economic importance can be timely detected and this could enable avoidance or prevention of the introduction of exotic pathogenic species (Kulabhusan et al., 2022). Antibodies that are relevant to specifying pathogens are labeled with AuNPs and exploited in developing diagnostic devices (Boonham et al., 2008; Chartuprayoon et al., 2010). The pathogen *Tilletia indica* causing kernel burn of wheat can be rapidly detected using a nano-gold-based immunosensor (Singh et al., 2010). This AuNPs-based diagnostic tool relies on the surface plasmon resonance feature of gold ions. Plants often respond to biotic stress through defensive patterns that are dependent on phytohormones. The level of these phytohormones such as the Jasmonic Acid (JA) and Salicylic acid (SA) can be monitored by nanoparticle-based electrochemical sensors in the bid to perceive pathogen activities (Khan et al., 2019). Recently, this aspect of nanotechnology has been given focus by researchers to develop sensors for detecting pathogens via biochemical changes occurring in the plant and more are still in progress.

8.16 SIGNIFICANCE OF NANOPARTICLES ON THE ENVIRONMENT

Like any other intervention to control plant pathogens or improved soil health, there are the possibilities of environmental consequences associated with the use of NPs, despite their huge success in various applications. Ameliorative inputs usually leave off residue either in the soil, the plant, or the atmosphere through various means such as surface splashing by rain, leaching or by dispersal in air transit. It is a difficult task to calculate the extent of the negative contribution of NPs to the ecosystem (Morales-Díaz et al., 2017). To understand the total emission and their interaction with organisms and the environment, interdisciplinary efforts and strategies will be involved (Nicolodi & Gianello, 2014). Various studies have propounded that the accumulation of NPs in plants part is possible because the particle is taken up by living issues which are then translocated to other plant parts (De la Torre Roche et al., 2015).

To enable sustainable plant pathogen management, judicious use of NPs should be encouraged by establishing an evidence-based usage enough to control the pathogen and yet safe for the environment. It is important to know the proper handling, application, and enabling cultural practice that will facilitate the optimization of NP in controlling pathogens. The same NPs features, such as their ionic form, that facilitate interaction with biomolecules, colloids, and cell constituents, also position them as substances capable of environmental toxicity. Therefore, beyond their primary application, understanding the extent of their unwanted environmental impact should be factored in when deciding their potential usage. Their capacity to modify living cells genetically is enough to believe them to have the ability to transform harmless substances into harmful ones (Stone et al., 2007; Du et al., 2015). Consequently, depending on the temporal and spatial factors, there is the possibility of a negative interaction emerging from the alliance between NPs and cells (Hu et al., 2014; Martínez-Fernández et al., 2016). Some effects of NPs on plants have been reported to be phytotoxic (Ali & Ali, 2019). Some studies have demonstrated the presence of metal-base toxicity in plants (Lin & Xing, 2007; Faizan et al., 2018), therefore, the mechanism of translocation in-plantae should be elucidated. NPs have been suggested to be persistent in crops via a study that seeks to determine how NPs can affect seed germination (Lin & Xing, 2007). The study demonstrated that inhibition of seed germination occurred at the later incubation stage and not the former soaking stage, implying that the effect did not manifest at the contact phase but at a later phase. It is notable from this observation that NPs containing materials should not be disposed carelessly into the environment. In addition, NPs can also affect plant growth due to its impact on plant growth regulatory hormones and metabolites via its interference with the radicals detoxifying process of the plant (Pagano et al., 2016; Sanzari et al., 2019). This implies that elaborate studies focusing on deep analysis of the metabolic and genetic processes of plants challenged with NPs are required to illuminate the many facets of NPs interactions with plants. consequently, it is justifiable to state that only certified safe NMs should be considered for future application in plant and animal systems (Sanzari et al., 2019). Effect of NPs on untargeted soil biota especially those considered beneficial to the ecosystem should be assessed (Arshad Ali et al., 2020). This could be achieved by monitoring the respiration and enzymatic activities of the soil bulk, via a microbial ecotoxicological approach (Simonin & Richaume, 2015). Thus, the reaction of soil biodiversities to the presence of NPs can be appraised.

8.17 CONCLUSIONS AND FUTURE OUTLOOKS

Agricultural production is threatened by several factors including plant pathogen which require considerable management that will guarantee minimal loss due to the menace. There are a repertoire of methods by which the control of plant pathogen can be achieved. Unfortunately, many of these methods come with certain negative consequences on man and the environment. Consequently, a method or group of methods that will help curb the menace of plant pathogens, reduced crop loss and have very low negative imprint on the environment will be plausible. Nanotechnology has uncovered various distinguished methods valuable in crop protection under the spotlight of its capacity to serve as nanocarriers for active ingredient relevant to pesticides or nanopesticides. Various nanomaterial have been developed to possess great antimicrobial properties. Nanomaterials can be metal based, metal-based oxide or non-metal, all with a vast range of application in crop protection. AuNPs are proven to possess remarkable properties including high surface area-to-volume ratio, surface functionalization and biocompatibility. Despite the opportunity granted by nanotechnology to assist overcome plant pathogen management's bottlenecks such as toxicity experienced by non-target organism, and environmental safety, control efficacy and ultimate sustainability of the crop production system, some recent studies have demonstrated the presence of metal-base toxicity in plants and NPs have been suggested to be persistent in crops. Therefore, it has become imperative that the mechanism of translocation in-plantae should be elucidated. Future research should therefore focus more on the phytotoxic aspects of nanoparticles so as to guide growers on the appropriate applications of these NPs in order to achieve the desired results with little or no detrimental effects on human and environmental health.

REFERENCES

Agarwal, P., Gupta, R., & Agarwal, N. (2019). Advances in synthesis and applications of microalgal nanoparticles for wastewater treatment. *Journal of Nanotechnology*, *2019*, e7392713.

Ahmad, T., Wani, I. A., Lone, I. H., Ganguly, A., Manzoor, N., Ahmad, A., Ahmed, J., & Al-Shihri, A. S. (2013). Antifungal activity of gold nanoparticles prepared by solvothermal method. *Materials Research Bulletin*, *48*(1), 12–20.

Akanmu, A. O., Babalola, O. O., Venturi, V., Ayilara, M. S., Adeleke, B. S., Amoo, A. E., Sobowale, A. A., Fadiji, A. E., & Glick, B. R. (2021). Plant disease management: Leveraging on the plant-microbe-soil interface in the biorational use of organic amendments. *Frontiers in Plant Science, 12*. www.frontiersin.org/articles/10.3389/fpls.2021.700507

Ali, A., Ovais, M., Cui, X., Rui, Y., & Chen, C. (2020). Safety assessment of nanomaterials for antimicrobial applications. Chemical Research in Toxicology, 33(5), 1082–1109.

Ali, S., & Ali, S. (2019). Nanotechnology is the potential cause of phytotoxicity. *Journal of Biomaterials, 3*, 1.

Alkhataf, F. (2021). Gold and silver nanoparticles: Green synthesis, microbes, mechanism, factors, plant disease management and environmental risks. *Saudi Journal of Biological Sciences, 28*.

Ameen, F., Al-Maary, K. S., Almansob, A., & AlNadhari, S. (2023). Antioxidant, antibacterial and anticancer efficacy of Alternaria chlamydospora-mediated gold nanoparticles. *Applied Nanoscience, 13*(3), 2233–2240.

Anees, A., Gupta, R., Phanindra, P. V., Fabiyi, O. A., Thera, U. K., Bello, T. T., & Ahmad, F. (2024). Green Synthesis of Nanoparticles and Applications. In *Advanced Nanotechnology in Plants* (pp. 89–122) CRC Press.

Antoine Thill, Ophélie Zeyons, Olivier Spalla, Franck Chauvat, Jerôme Rose, Mélanie Auffan, & Flank, A. M. (2006, August 31). *Cytotoxicity of CeO2 Nanoparticles for Escherichia coli. Physico-Chemical Insight of the Cytotoxicity Mechanism* (world) [Research-article]. ACS Publications; American Chemical Society.

Arshad, A., Iqbal, J., Mansoor, Q., & Ahmed, I. (2017). Graphene/SiO2 nanocomposites: The enhancement of photocatalytic and biomedical activity of SiO2 nanoparticles by graphene. *Journal of Applied Physics, 121*(24), 244901.

Arshad Ali, M., Ahmed, T., Wu, W., Wang, Y., An, Q., Sun, G., Li, B., Hossain, A., Masum, Md. M. I., & R., H. (2020). Advancements in plant and microbe-based synthesis of metallic nanoparticles and their antimicrobial activity against plant pathogens. *Nanomaterials, 10*, 1146.

Avestan, S., Ghasemnezhad, M., Esfahani, M., & Byrt, C. S. (2019). Application of nano-silicon dioxide improves salt stress tolerance in strawberry plants. *Agronomy, 9*(5), Article 5.

Behlke, M. A. (2006). Progress towards in vivo use of siRNAs. *Molecular Therapy: The Journal of the American Society of Gene Therapy, 13*(4), 644–670.

Bello, T. T., & Fabiyi, O. A. (2024). Facile Biosynthesis of Gold Nanoparticles Using Microbes and Their Applications in the Management of Economically Important Plant Pathogens. In Rai, M., & Avila-Quezada, G. (Eds.) Nanotechnology in Plant Health (1st ed.) (pp. 164–179) CRC Press.

Beveridge, T. J., & Murray, R. G. (1980). Sites of metal deposition in the cell wall of Bacillus subtilis. *Journal of Bacteriology, 141*(2), 876–887.

Beyenbach, K. W., & Wieczorek, H. (2006). The V-type H+ ATPase: Molecular structure and function, physiological roles and regulation. *Journal of Experimental Biology, 209*(4), 577–589.

Bharali, D. J., Klejbor, I., Stachowiak, E. K., Dutta, P., Roy, I., Kaur, N., Bergey, E. J., Prasad, P. N., & Stachowiak, M. K. (2005). Organically modified silica nanoparticles: A nonviral vector for in vivo gene delivery and expression in the brain. *Proceedings of the National Academy of Sciences, 102*(32), 11539–11544.

Boonham, N., Glover, R., Tomlinson, J., & Mumford, R. (2008). Exploiting Generic Platform Technologies for the Detection and Identification of Plant Pathogens. In D. B. Collinge, L. Munk, & B. M. Cooke (Eds.), *Sustainable Disease Management in a European Context* (pp. 355–363). Springer Netherlands.

Brown, S., Sarikaya, M., & Johnson, E. (2000). A genetic analysis of crystal growth. *Journal of Molecular Biology, 299*(3), 725–735.

Brunel, F., El Gueddari, N. E., & Moerschbacher, B. M. (2013). Complexation of copper(II) with chitosan nanogels: Toward control of microbial growth. *Carbohydrate Polymers, 92*(2), 1348–1356.

Brust, M., Fink, J., Bethell, D., J. Schiffrin, D., & Kiely, C. (1995). Synthesis and reactions of functionalised gold nanoparticles. *Journal of the Chemical Society, Chemical Communications, 0*(16), 1655–1656.

Cameron, N., & Mitchell, M. E. (2007). *Nanoscale: Issues and Perspectives for the Nano Century*. John Wiley & Sons.

Castro, L., Blázquez, M. L., Muñoz, J. ángel, González, F. G., & Ballester, A. (2014). Mechanism and applications of metal nanoparticles prepared by bio-mediated process. *Reviews in Advanced Sciences and Engineering, 3*(3), 199–216.

Castro, M., Ojeda, C., & Cirelli, A. (2014). Advances in surfactants for agrochemicals. *Environmental Chemistry Letters, 12*.

Chartuprayoon, N., Rheem, Y., Chen, W., & Myung, N. (2010). Detection of plant pathogen using LPNE grown single conducting polymer nanoribbon. *ECS Meeting Abstracts*, *MA2010-02*(40), 2278.

Chauhan, A., Zubair, S., Tufail, S., Sherwani, A., Sajid, M., Raman, S. C., Azam, A., & Owais, M. (2011). Fungus-mediated biological synthesis of gold nanoparticles: Potential in detection of liver cancer. *International Journal of Nanomedicine*, *6*, 2305–2319.

Chellappan, P., Vanitharani, R., & Fauquet, C. M. (2004). Short interfering RNA accumulation correlates with host recovery in DNA virus-infected hosts, and gene silencing targets specific viral sequences. *Journal of Virology*, *78*(14), 7465–7477.

Choe, S., Chang, Y.-Y., Hwang, K.-Y., & Khim, J. (2000). Kinetics of reductive denitrification by nanoscale zero-valent iron. *Chemosphere*, *41*(8), 1307–1311.

Choudhary, M., Paret, M., & Jones, J. (2021). Natural or Green Synthesis Nanomaterials and Impact on Plant Pathogens. In *Nanotechnology-Based Sustainable Alternatives for the Management of Plant Diseases* (pp. 5–29).

Chowdappa, P., & Gowda, S. (2013). Nanotechnology in crop protection: status and scope. Pest Management in Horticultural Ecosystems, 19(2), 131–151.

Chowdhury, N. K., Choudhury, R., Gogoi, B., Chang, C.-M., & Pandey, R. P. (2022). Microbial synthesis of gold nanoparticles and their application. *Current Drug Targets*, *23*(7), 752–760.

Chung, Y. H., Cai, H., & Steinmetz, N. F. (2020). Viral nanoparticles for drug delivery, imaging, immunotherapy, and theranostic applications. *Advanced Drug Delivery Reviews*, *156*, 214–235.

Clarkson, T. W. (1993). Molecular and ionic mimicry of toxic metals. *Annual Review of Pharmacology and Toxicology*, *33*(1), 545–571.

Corti, C. W., & Holliday, R. J. (2004). Commercial aspects of gold applications: From materials science to chemical science. *Gold Bulletin*, *37*(1), 20–26.

Das, S. K., Dickinson, C., Lafir, F., Brougham, D. F., & Marsili, E. (2012). Synthesis, characterization and catalytic activity of gold nanoparticles biosynthesized with Rhizopus oryzae protein extract. *Green Chemistry*, *14*(5), 1322–1334.

Dawwam, G. E., Al-Shemy, M. T., & El-Demerdash, A. S. (2022). Green synthesis of cellulose nanocrystal/ZnO bio-nanocomposites exerting antibacterial activity and downregulating virulence toxigenic genes of food-poisoning bacteria. *Scientific Reports*, *12*(1), Article 1.

De la Torre Roche, R., Servin, A., Hawthorne, J., Xing, B., Newman, L. A., Ma, X., Chen, G., & White, J. C. (2015). Terrestrial trophic transfer of bulk and nanoparticle La2O3 does not depend on particle size. *Environmental Science & Technology*, *49*(19), 11866–11874.

Debroy, A., Joshi, S., Yadav, M., & George, N. (2023). Chapter 15 – Green Synthesis of Nanoparticles from Bio-Waste for Potential Applications: Current Trends, Challenges, and Prospects. In C. M. Hussain, A. Kushwaha, R. N. Bharagava, & L. Goswami (Eds.), *Bio-Based Materials and Waste for Energy Generation and Resource Management* (Vol. 5, pp. 431–466). Elsevier.

Devi, L. S., & Joshi, S. R. (2012). Antimicrobial and synergistic effects of silver nanoparticles synthesized using soil fungi of high altitudes of Eastern Himalaya. *Mycobiology*, *40*(1), 27–34.

Dobias, J., Suvorova, E. I., & Bernier-Latmani, R. (2011). Role of proteins in controlling selenium nanoparticle size. *Nanotechnology*, *22*(19), 195605.

Dominguez-Benetton, X., Varia, J. C., Pozo, G., Modin, O., Ter Heijne, A., Fransaer, J., & Rabaey, K. (2018). Metal recovery by microbial electro-metallurgy. *Progress in Materials Science*, *94*, 435–461.

Du, H., Lo, T.-M., Sitompul, J., & Chang, M. W. (2012). Systems-level analysis of Escherichia coli response to silver nanoparticles: The roles of anaerobic respiration in microbial resistance. *Biochemical and Biophysical Research Communications*, *424*(4), 657–662.

Du, J., Hu, X., & Zhou, Q. (2015). Graphene oxide regulates the bacterial community and exhibits property changes in soil. *RSC Advances*, *5*(34), 27009–27017.

Du, L., Jiang, H., Liu, X., & Wang, E. (2007). Biosynthesis of gold nanoparticles assisted by Escherichia coli DH5α and its application on direct electrochemistry of hemoglobin. *Electrochemistry Communications*, *9*(5), 1165–1170.

Du, L., Xian, L., & Feng, J.-X. (2011). Rapid extra-/intracellular biosynthesis of gold nanoparticles by the fungus Penicillium sp. *Journal of Nanoparticle Research*, *13*(3), 921–930.

Dutta, P., Kumari, A., Mahanta, M., Biswas, K. K., Dudkiewicz, A., Thakuria, D., Abdelrhim, A. S., Singh, S. B., Muthukrishnan, G., Sabarinathan, K. G., Mandal, M. K., & Mazumdar, N. (2022). Advances in nanotechnology as a potential alternative for plant viral disease management. *Frontiers in Microbiology, 13*. www.frontiersin.org/articles/10.3389/fmicb.2022.935193

Fabiyi, O., Olatunji, G., Atolani, O., & Olawuyi, O. (2020). Preparation of bio-nematocidal nanoparticles of Eucalyptus officinalis for the control of cyst nematode (Heterodera sacchari). *Journal of Animals Plant Science, 30*(5), 1172–1177.

Fabiyi, O. A. (2021a). Sustainable management of Meloidogyne incognita infecting carrot (Daucus carota): Green synthesis of silver nanoparticles with Cnidoscolus aconitifolius. *Vegetos, 34*(2), 277–285.

Fabiyi, O. A. (2021b). Evaluation of Nematocidal activity of *Terminalia glaucescens* fractions against *Meloidogyne incognita* on *Capsicum chinense. Journal of Horticulture Research, 29*(1), 67–74.

Fabiyi, O. A. (2021c). Evaluation of plant materials as root-knot nematode (*Meloidogyne incognita*) suppressant in okro (*Abelmuscous esculentus*). *Agriculturae Conspectus Scientificus, 86*(1), 51–56.

Fabiyi, O. A. (2021d). Application of furfural in sugarcane nematode pest management. *Pakistan Journal of Nematology, 39*(2), 151–155.

Fabiyi, O. A., Alabi, R. O., & Ansari, R. A. (2020). Nanoparticles' synthesis and their application in the management of phytonematodes: An overview. In: Ansari, R., Rizvi, R., Mahmood, I. (eds) Management of Phytonematodes: Recent Advances and Future Challenges. Springer, Singapore.125–140.

Fabiyi, O. A., Claudius-Cole, A. O., Olatunji, G. A., Abubakar, D. O., & Adejumo, O. A. (2021). Response of Meloidogyne javanica to silver nanoparticle liquid from agricultural wastes. *AGRIVITA, Journal of Agricultural Science, 43*(3), 507–517.

Fabiyi, O. A., Olatunji, G. A., & Saadu, A. O. (2018). Suppression of Heterodera sacchari in rice with agricultural waste-silver nano particles. *The Journal of Solid Waste Technology and Management, 44*(2), 87–91.

Fabiyi, O. A. (2022a). Fractions from *Mangifera indica* as an Alternative in *Meloidogyne incognita* Management. *Pakistan Journal of Nematology, 40*(1), 65–74.

Fabiyi, O. A. (2022b). Cytotoxicity and nematocidal potential of leaf extracts of *Adansonia digitata* and *Khaya senegalensis* on root knot nematode (*Meloidogyne incognita*) associated with cabbage *(Brassica oleracea). Journal of Agricultural Sciences-Sri Lanka, 17*(3), 425–436.

Fabiyi, O. A. (2022c). Evaluation of weeds against root-knot nematode (*Meloidogyne incognita*) in vegetables. *Sarhad Journal of Agriculture, 38*(4), 1289–1299.

Fabiyi, O. A. (2022d). Application of composited municipal refuse dump site soil, orange, potato and pineapple peels in the control of root knot nematode (*Meloidogyne incognita*) infecting carrots (*Daucus carota* L.). *The Journal of Solid Waste Management, 48*(3), 474–485.

Fabiyi, O. A. (2023a). Influence of biodegraded bark extracts of *Anacardium occidentale* on survival of *Meloidogyne incognita* infecting cabbage plants. *Journal of Tropical Agriculture, 61*(2), 167–174.

Fabiyi, O. A. (2023). Nano Clay Materials and Their Role in Plant Disease Management. In Avila-Quezada, G. D., & Rai, M. (Eds.), *Nanophytopathology* (pp. 145–156).

Fabiyi, O. A. (2023c). The Perspective of Nanotechnology in Plant Protection. In Avila-Quezada, G. D., & Rai, M. (Eds.), *Nanophytopathology* (pp. 23–38).

Fabiyi, O. A., Adebisi, O. O., Falore, S. O., & Claudius-Cole, A. O. (2023). Publisher correction Potential inhibition of entomopathogenic nematodes and plant growth-promoting bacteria with exposure to selected herbicides and insecticides.

Faizan, M., Faraz, A., Yusuf, M., Khan, S. T., & Hayat, S. (2018). Zinc oxide nanoparticle-mediated changes in photosynthetic efficiency and antioxidant system of tomato plants. *Photosynthetica, 56*(2), 678–686.

Gahlawat, G., & Roy Choudhury, A. (2019). A review on the biosynthesis of metal and metal salt nanoparticles by microbes. *RSC Advances, 9*(23), 12944–12967.

Gericke, M., & Pinches, A. (2006). Microbial production of gold nanoparticles. *Gold Bulletin, 39*(1), 22–28.

Gholami-Shabani, M., Shams-Ghahfarokhi, M., Gholami-Shabani, Z., Akbarzadeh, A., Riazi, G., Ajdari, S., Amani, A., & Razzaghi-Abyaneh, M. (2015). Enzymatic synthesis of gold nanoparticles using sulfite reductase purified from Escherichia coli: A green eco-friendly approach. *Process Biochemistry, 50*(7), 1076–1085.

Giles, M., Morley, N., Baggs, E., & Daniell, T. (2012). Soil nitrate reducing processes – Drivers, mechanisms for spatial variation, and significance for nitrous oxide production. *Frontiers in Microbiology, 3*. www.frontiersin.org/articles/10.3389/fmicb.2012.00407

Grace, A. N., & Pandian, K. (2007). Quinolone antibiotic-capped gold nanoparticles and their antibacterial efficacy against gram positive and gram negative organisms. *Journal of Bionanoscience, 1*(2), 96–105.

Grasso, G., Zane, D., & Dragone, R. (2020). Microbial nanotechnology: Challenges and prospects for green biocatalytic synthesis of nanoscale materials for sensoristic and biomedical applications. *Nanomaterials*, *10*(1), Article 1.

Green, J. M., & Beestman, G. B. (2007). Recently patented and commercialized formulation and adjuvant technology. *Crop Protection*, *26*(3), 320–327.

Guo, Z., Li, Y., & Ding, S.-W. (2019). Small RNA-based antimicrobial immunity. *Nature Reviews Immunology*, *19*(1), Article 1.

Hajipour, M. J., Fromm, K. M., Akbar Ashkarran, A., Jimenez de Aberasturi, D., Larramendi, I. R. de, Rojo, T., Serpooshan, V., Parak, W. J., & Mahmoudi, M. (2012). Antibacterial properties of nanoparticles. *Trends in Biotechnology*, *30*(10), 499–511.

Hamid, A., & Saleem, S. (2022). Role of nanoparticles in management of plant pathogens and scope in plant transgenics for imparting disease resistance. *Plant Protection Science*, *58*(3), 173–184.

Hayles, J., Johnson, L., Worthley, C., & Losic, D. (2017). 5 – Nanopesticides: A Review of Current Research and Perspectives. In A. M. Grumezescu (Ed.), *New Pesticides and Soil Sensors* (pp. 193–225). Academic Press.

He, S., Guo, Z., Zhang, Y., Zhang, S., Wang, J., & Gu, N. (2007). Biosynthesis of gold nanoparticles using the bacteria Rhodopseudomonas capsulata. *Materials Letters*, *61*(18), 3984–3987.

He, S., Zhang, Y., Guo, Z., & Gu, N. (2008). Biological synthesis of gold nanowires using extract of Rhodopseudomonas capsulata. *Biotechnology Progress*, *24*(2), 476–480.

Holzinger, M., Le Goff, A., & Cosnier, S. (2014). Nanomaterials for biosensing applications: A review. *Frontiers in Chemistry*, *2*. www.frontiersin.org/articles/10.3389/fchem.2014.00063

Hu, X., Kang, J., Lu, K., Zhou, R., Mu, L., & Zhou, Q. (2014). Graphene oxide amplifies the phytotoxicity of arsenic in wheat. *Scientific Reports*, *4*(1), Article 1.

Hur, Y. E., Kim, S., Kim, J.-H., Cha, S.-H., Choi, M.-J., Cho, S., & Park, Y. (2014). One-step functionalization of gold and silver nanoparticles by ampicillin. *Materials Letters*, *129*, 185–190.

Iranmanesh, S., Shahidi Bonjar, G. H., & Baghizadeh, A. (2020). Study of the biosynthesis of gold nanoparticles by using several saprophytic fungi. *SN Applied Sciences*, *2*(11), 1851.

Jafari, M., Rokhbakhsh-Zamin, F., Shakibaie, M., Moshafi, M. H., Ameri, A., Rahimi, H. R., & Forootanfar, H. (2018). Cytotoxic and antibacterial activities of biologically synthesized gold nanoparticles assisted by Micrococcus yunnanensis strain J2. *Biocatalysis and Agricultural Biotechnology*, *15*, 245–253.

Jiang, H., Lv, L., Ahmed, T., Jin, S., Shahid, M., Noman, M., Osman, H.-E. H., Wang, Y., Sun, G., Li, X., & Li, B. (2022). Effect of the nanoparticle exposures on the tomato bacterial wilt disease control by modulating the Rhizosphere bacterial community. *International Journal of Molecular Sciences*, *23*(1), Article 1.

Kah, M., & Hofmann, T. (2014). Nanopesticide research: Current trends and future priorities. *Environment International*, *63*, 224–235.

Kalepu, S., & Nekkanti, V. (2015). Insoluble drug delivery strategies: Review of recent advances and business prospects. *Acta Pharmaceutica Sinica B*, *5*(5), 442–453.

Karthik, L., Kumar, G., Keswani, T., Bhattacharyya, A., Reddy, B. P., & Rao, K. V. B. (2013). Marine actinobacterial mediated gold nanoparticles synthesis and their antimalarial activity. *Nanomedicine: Nanotechnology, Biology and Medicine*, *9*(7), 951–960.

Kaur, K., Ahmed, B., Singh, J., Rawat, M., Kaur, G., AlKahtani, M., Alhomaidi, E. A. H., & Lee, J. (2022). Bryonia laciniosa Linn mediated green synthesized Au NPs for catalytic and antimicrobial applications. *Journal of King Saud University – Science*, *34*(4), 102022.

Khairy, A. M., Tohamy, M. R. A., Zayed, M. A., Mahmoud, S. F., El-Tahan, A. M. M., El-Saadony, M. T., & Mesiha, P. K. (2022). Eco-friendly application of nano-chitosan for controlling potato and tomato bacterial wilt. *Saudi Journal of Biological Sciences*, *29*(4), 2199–2209.

Khan, I., Saeed, K., & Khan, I. (2019). Nanoparticles: Properties, applications and toxicities. *Arabian Journal of Chemistry*, *12*(7), 908–931.

Khan, M. R., Haque, Z., & Kausar, N. (2014). Management of the root-knot nematode Meloidogyne graminicola infesting rice in the nursery and crop field by integrating seed priming and soil application treatments of pesticides. *Crop Protection*, *63*, 15–25.

Khandelwal, N., Barbole, R. S., Banerjee, S. S., Chate, G. P., Biradar, A. V., Khandare, J. J., & Giri, A. P. (2016). Budding trends in integrated pest management using advanced micro- and nano-materials: Challenges and perspectives. *Journal of Environmental Management*, *184*, 157–169.

Konishi, Y., Tsukiyama, T., Ohno, K., Saitoh, N., Nomura, T., & Nagamine, S. (2006). Intracellular recovery of gold by microbial reduction of AuCl4− ions using the anaerobic bacterium Shewanella algae. *Hydrometallurgy*, *81*(1), 24–29.

Kora, A. J., Sashidhar, R. B., & Arunachalam, J. (2010). Gum kondagogu (Cochlospermum gossypium): A template for the green synthesis and stabilization of silver nanoparticles with antibacterial application. *Carbohydrate Polymers*, *82*(3), 670–679.

Koul, B., Poonia, A. K., Yadav, D., & Jin, J.-O. (2021). Microbe-mediated biosynthesis of nanoparticles: Applications and future prospects. *Biomolecules*, *11*(6), Article 6.

Kulabhusan, P. K., Tripathi, A., & Kant, K. (2022). Gold nanoparticles and plant pathogens: An overview and prospective for biosensing in forestry. *Sensors (Basel, Switzerland)*, *22*(3), 1259.

Kulkarni, R. R., Shaiwale, N. S., Deobagkar, D. N., & Deobagkar, D. D. (2015). Synthesis and extracellular accumulation of silver nanoparticles by employing radiation-resistant Deinococcus radiodurans, their characterization, and determination of bioactivity. *International Journal of Nanomedicine*, *10*, 963–974.

Kuo, Y.-W., & Falk, B. W. (2020). RNA interference approaches for plant disease control. *BioTechniques*, *69*(6), 469–477.

Lemire, J. A., Harrison, J. J., & Turner, R. J. (2013). Antimicrobial activity of metals: Mechanisms, molecular targets and applications. *Nature Reviews Microbiology*, *11*(6), Article 6.

Li, C., Zhou, L., Yang, H., Lv, R., Tian, P., Li, X., Zhang, Y., Chen, Z., & Lin, F. (2017). Self-assembled exopolysaccharide nanoparticles for bioremediation and green synthesis of noble metal nanoparticles. *ACS Applied Materials & Interfaces*, *9*(27), 22808–22818.

Lin, D., & Xing, B. (2007). Phytotoxicity of nanoparticles: Inhibition of seed germination and root growth. *Environmental Pollution (Barking, Essex: 1987)*, *150*(2), 243–250.

Luo, P., Liu, Y., Xia, Y., Xu, H., & Xie, G. (2014). Aptamer biosensor for sensitive detection of toxin A of Clostridium difficile using gold nanoparticles synthesized by Bacillus stearothermophilus. *Biosensors and Bioelectronics*, *54*, 217–221.

Malarkodi, C., Rajeshkumar, S., Vanaja, M., Paulkumar, K., Gnanajobitha, G., & Annadurai, G. (2013). Eco-friendly synthesis and characterization of gold nanoparticles using Klebsiella pneumoniae. *Journal of Nanostructure in Chemistry*, *3*(1), 30.

Marooufpour, N., Alizadeh, M., Hatami, M., & Asgari Lajayer, B. (2019). Biological Synthesis of Nanoparticles by Different Groups of Bacteria. In R. Prasad (Ed.), *Microbial Nanobionics: Volume 1, State-of-the-Art* (pp. 63–85). Springer International Publishing.

Martínez-Fernández, D., Barroso, D., & Komárek, M. (2016). Root water transport of Helianthus annuus L. under iron oxide nanoparticle exposure. *Environmental Science and Pollution Research*, *23*(2), 1732–1741.

Mina, I. R., Jara, N. P., Criollo, J. E., & Castillo, J. A. (2019). The critical role of biofilms in bacterial vascular plant pathogenesis. *Plant Pathology*, *68*(8), 1439–1447.

Mishra, S., & Singh, H. B. (2015). Biosynthesized silver nanoparticles as a nanoweapon against phytopathogens: Exploring their scope and potential in agriculture. *Applied Microbiology and Biotechnology*, *99*(3), 1097–1107.

Molnár, Z., Bódai, V., Szakacs, G., Erdélyi, B., Fogarassy, Z., Sáfrán, G., Varga, T., Kónya, Z., Tóth-Szeles, E., Szűcs, R., & Lagzi, I. (2018). Green synthesis of gold nanoparticles by thermophilic filamentous fungi. *Scientific Reports*, *8*(1), Article 1.

Morales-Díaz, A. B., Ortega-Ortíz, H., Juárez-Maldonado, A., Cadenas-Pliego, G., González-Morales, S., & Benavides-Mendoza, A. (2017). Application of nanoelements in plant nutrition and its impact in ecosystems. *Advances in Natural Sciences: Nanoscience and Nanotechnology*, *8*(1), 013001.

Mousavi, S. R., & Rezaei, M. (2011). Nanotechnology in agriculture and food production. *Journal of Applied Environmental and Biological Sciences*, *1*, 414–419.

Mukherjee, P., Ahmad, A., Mandal, D., Senapati, S., Sainkar, S. R., Khan, M. I., Ramani, R., Parischa, R., Ajayakumar, P. V., Alam, M., Sastry, M., & Kumar, R. (2001). Bio reduction of AuCl4− ions by the fungus, Verticillium sp. and surface trapping of the gold nanoparticles formed. *Angewandte Chemie International Edition*, *40*(19), 3585–3588.

Mukherjee, P., Senapati, S., Mandal, D., Ahmad, A., Khan, M. I., Kumar, R., & Sastry, M. (2002). Extracellular synthesis of gold nanoparticles by the fungus Fusarium oxysporum. *ChemBioChem*, *3*(5), 461–463.

Naik, A. J. T., Ismail, S., Kay, C., Wilson, M., & Parkin, I. P. (2011). Antimicrobial activity of polyurethane embedded with methylene blue, toluidene blue and gold nanoparticles against Staphylococcus aureus; illuminated with white light. *Materials Chemistry and Physics*, *129*(1), 446–450.

Narayanan, K. B., & Sakthivel, N. (2011). Facile green synthesis of gold nanostructures by NADPH-dependent enzyme from the extract of Sclerotium rolfsii. *Colloids and Surfaces A: Physicochemical and Engineering Aspects*, *380*(1), 156–161.

Nicolodi, M., & Gianello, C. (2014). Understanding soil as an open system and fertility as an emergent property of the soil system. *Sustainable Agriculture Research.*

Niu, B., & Zhang, G. (2023). Effects of different nanoparticles on microbes. *Microorganisms*, *11*(3), 542.

NVKV Prasad, T., Subba Rao Kambala, V., & Naidu, R. (2011). A critical review on biogenic silver nanoparticles and their antimicrobial activity. *Current Nanoscience*, *7*(4), 531–544.

Ogi, T., Saitoh, N., Nomura, T., & Konishi, Y. (2010). Room-temperature synthesis of gold nanoparticles and nanoplates using Shewanella algae cell extract. *Journal of Nanoparticle Research*, *12*(7), 2531–2539.

Ogunyemi, S. O., Abdallah, Y., Zhang, M., Fouad, H., Hong, X., Ibrahim, E., Masum, Md. M. I., Hossain, A., Mo, J., & Li, B. (2019). Green synthesis of zinc oxide nanoparticles using different plant extracts and their antibacterial activity against Xanthomonas oryzae pv. Oryzae. *Artificial Cells, Nanomedicine, and Biotechnology*, *47*(1), 341–352.

Ortiz-Castillo, J. E., Gallo-Villanueva, R. C., Madou, M. J., & Perez-Gonzalez, V. H. (2020). Anisotropic gold nanoparticles: A survey of recent synthetic methodologies. *Coordination Chemistry Reviews*, *425*, 213489.

Oves, M., Aslam, M., Rauf, M. A., Qayyum, S., Qari, H. A., Khan, M. S., Alam, M. Z., Tabrez, S., Pugazhendhi, A., & Ismail, I. M. I. (2018). Antimicrobial and anticancer activities of silver nanoparticles synthesized from the root hair extract of Phoenix dactylifera. *Materials Science and Engineering: C*, *89*, 429–443.

Pagano, L., Servin, A. D., De La Torre-Roche, R., Mukherjee, A., Majumdar, S., Hawthorne, J., Marmiroli, M., Maestri, E., Marra, R. E., Isch, S. M., Dhankher, O. P., White, J. C., & Marmiroli, N. (2016). Molecular response of crop plants to engineered nanomaterials. *Environmental Science & Technology*, *50*(13), 7198–7207.

Panja, A., Mishra, A. K, Dash, M, Pandey, N. K., Singh, S. K., & Kumar. (2021). Silver nanoparticles – A review. *Eurasian Journal of Medicine and Oncology.*

Park, M., Lee, C.-I., Seo, Y. J., Woo, S. R., Shin, D., & Choi, J. (2010). Hybridization of the natural antibiotic, cinnamic acid, with layered double hydroxides (LDH) as green pesticide. *Environmental Science and Pollution Research*, *17*(1), 203–209.

Pereira, Y., Lagniel, G., Godat, E., Baudouin-Cornu, P., Junot, C., & Labarre, J. (2008). Chromate causes sulfur starvation in yeast. *Toxicological Sciences*, *106*(2), 400–412.

Peshin, R., Bandral, R. S., Zhang, W., Wilson, L., & Dhawan, A. K. (2009). Integrated Pest Management: A Global Overview of History, Programs and Adoption. In R. Peshin & A. K. Dhawan (Eds.), *Integrated Pest Management: Innovation-Development Process: Volume 1* (pp. 1–49). Springer Netherlands.

Pimentel, D. (2009). Pesticides and Pest Control. In R. Peshin & A. K. Dhawan (Eds.), *Integrated Pest Management: Innovation-Development Process: Volume 1* (pp. 83–87). Springer Netherlands.

Prabhu, S., & Poulose, E. K. (2012). Silver nanoparticles: Mechanism of antimicrobial action, synthesis, medical applications, and toxicity effects. *International Nano Letters*, *2*(1), 32.

Prasad, T. N. V. K. V., Elumalai, E. K., & Khateeja, S. (2011). Evaluation of the antimicrobial efficacy of phytogenic silver nanoparticles. *Asian Pacific Journal of Tropical Biomedicine*, *1*(1), S82–S85.

Priyadarshini, E., Pradhan, N., Sukla, L. B., & Panda, P. K. (2014). Controlled synthesis of gold nanoparticles using *Aspergillus terreus* IF0 and its antibacterial potential against gram negative pathogenic bacteria. *Journal of Nanotechnology*, *2014*, e653198.

Qi, Z., Zhou, H., Matsuda, N., Honma, I., Shimada, K., Takatsu, A., & Kato, K. (2004). Characterization of gold nanoparticles synthesized using sucrose by seeding formation in the solid phase and seeding growth in aqueous solution. *The Journal of Physical Chemistry B*, *108*(22), 7006–7011.

Ramanathan, R., Field, M. R., O'Mullane, A. P., Smooker, P. M., Bhargava, S. K., & Bansal, V. (2013). Aqueous phase synthesis of copper nanoparticles: A link between heavy metal resistance and nanoparticle synthesis ability in bacterial systems. *Nanoscale*, *5*(6), 2300–2306.

Rani, P., Ahmed, B., Singh, J., Kaur, J., Rawat, M., Kaur, N., Matharu, A. S., AlKahtani, M., Alhomaidi, E. A. H., & Lee, J. (2022). Silver nanostructures prepared via novel green approach as an effective platform for biological and environmental applications. *Saudi Journal of Biological Sciences*, *29*(6), 103296.

Ravindran, K. C., Venkatesan, K., Balakrishnan, V., Chellappan, K. P., & Balasubramanian, T. (2007). Restoration of saline land by halophytes for Indian soils. *Soil Biology and Biochemistry, 39*(10), 2661–2664.

Riddin, T. L., Govender, Y., Gericke, M., & Whiteley, C. G. (2009). Two different hydrogenase enzymes from sulphate-reducing bacteria are responsible for the bioreductive mechanism of platinum into nanoparticles. *Enzyme and Microbial Technology, 45*(4), 267–273.

Roshmi, T., Soumya, K. R., Jyothis, M., & Radhakrishnan, E. K. (2015). Effect of biofabricated gold nanoparticle-based antibiotic conjugates on minimum inhibitory concentration of bacterial isolates of clinical origin. *Gold Bulletin, 48*(1), 63–71.

Sağ, Y., & Aktay, Y. (2000). Mass transfer and equilibrium studies for the sorption of chromium ions onto chitin. *Process Biochemistry, 36*(1), 157–173.

Salem, S. S., & Fouda, A. (2021). Green synthesis of metallic nanoparticles and their prospective biotechnological applications: An overview. *Biological Trace Element Research, 199*(1), 344–370.

Santo, C. E., Taudte, N., Nies, D. H., & Grass, G. (2008). Contribution of copper ion resistance to survival of Escherichia coli on metallic copper surfaces. *Applied and Environmental Microbiology, 74*(4), 977–986.

Sanzari, I., Leone, A., & Ambrosone, A. (2019). Nanotechnology in plant science: To make a long story short. *Frontiers in Bioengineering and Biotechnology, 7*. www.frontiersin.org/articles/10.3389/fbioe.2019.00120

Sasson, Y., Levy-Ruso, G., Toledano, O., & Ishaaya, I. (2007). Nanosuspensions: Emerging Novel Agrochemical Formulations. In I. Ishaaya, A. R. Horowitz, & R. Nauen (Eds.), *Insecticides Design Using Advanced Technologies* (pp. 1–39). Springer.

Savithramma, N., Rao, M. L., Rukmini, K., & Devi, P. S. (2011). Antimicrobial activity of silver nanoparticles synthesized by using medicinal plants. *International Journal of ChemTech Research, 3*(3), 1394–1402.

Schiavi, D., Francesconi, S., Taddei, A. R., Fortunati, E., & Balestra, G. M. (2022). Exploring cellulose nanocrystals obtained from olive tree wastes as sustainable crop protection tool against bacterial diseases. *Scientific Reports, 12*(1), Article 1.

Schiewer, Silke., & Volesky, Bohumil. (1995). Modeling of the proton-metal ion exchange in biosorption. *Environmental Science & Technology, 29*(12), 3049–3058.

Scott, N., & Chen, H. (2012). Nanoscale science and engineering for agriculture and food systems. *Industrial Biotechnology, 8*(6), 340–343.

Shah, M., Fawcett, D., Sharma, S., Tripathy, S. K., & Poinern, G. E. J. (2015). Green synthesis of metallic nanoparticles via biological entities. *Materials, 8*(11), Article 11.

Shahid, M., Ahmed, B., & Khan, M. S. (2018). Evaluation of microbiological management strategy of herbicide toxicity to greengram plants. *Biocatalysis and Agricultural Biotechnology, 14*, 96–108.

Shin, K.-H., & Cha, D. K. (2008). Microbial reduction of nitrate in the presence of nanoscale zero-valent iron. *Chemosphere, 72*(2), 257–262.

Simonin, M., & Richaume, A. (2015). Impact of engineered nanoparticles on the activity, abundance, and diversity of soil microbial communities: A review. *Environmental Science and Pollution Research, 22*(18), 13710–13723.

Singh, S., Jain, D. V. S., & Singla, M. L. (2013). Sol–gel based composite of gold nanoparticles as matix for tyrosinase for amperometric catechol biosensor. *Sensors and Actuators B: Chemical, 182*, 161–169.

Singh, S., Singh, M., Agrawal, V. V., & Kumar, A. (2010). An attempt to develop surface plasmon resonance based immunosensor for Karnal bunt (Tilletia indica) diagnosis based on the experience of nano-gold based lateral flow immuno-dipstick test. *Thin Solid Films, 519*(3), 1156–1159.

Sinha, K., Ghosh, J., & Sil, P. C. (2017). New pesticides: A cutting-edge view of contributions from nanotechnology for the development of sustainable agricultural pest control. In A. M. Grumezescu (Ed.), *New Pesticides and Soil Sensors* (pp. 47–79). Academic Press.

Soni, N., & Prakash, S. (2012). Synthesis of gold nanoparticles by the fungus <em>Aspergillus niger</em> and its efficacy against mosquito larvae. *Reports in Parasitology, 2*, 1–7.

Spadaro, D., & Gullino, M. L. (2019). Sustainable Management of Plant Diseases. In M. Farooq & M. Pisante (Eds.), *Innovations in Sustainable Agriculture* (pp. 337–359). Springer International Publishing.

Srinath, B. S., Namratha, K., & Byrappa, K. (2018). Eco-friendly synthesis of gold nanoparticles by Bacillus subtilis and their environmental applications. *Advanced Science Letters, 24*(8), 5942–5946.

Srinath, B. S., & Ravishankar Rai, V. (2015). Rapid biosynthesis of gold nanoparticles by Staphylococcus epidermidis: Its characterisation and catalytic activity. *Materials Letters, 146*, 23–25.

Steinmetz, N. F., & Manchester, M. (2011). *Viral Nanoparticles: Tools for Material Science and Biomedicine*. Pan Stanford Publishing.

Stone, V., Johnston, H., & Clift, M. J. D. (2007). Air pollution, ultrafine and nanoparticle toxicology: Cellular and molecular interactions. *IEEE Transactions on NanoBioscience*, *6*(4), 331–340.

Syed, B., Prasad, N. M. N., & Satish, S. (2016). Endogenic mediated synthesis of gold nanoparticles bearing bactericidal activity. *Journal of Microscopy and Ultrastructure*, *4*(3), 162–166.

Tariq, M., Mohammad, K. N., Ahmed, B., Siddiqui, M. A., & Lee, J. (2022). Biological synthesis of silver nanoparticles and prospects in plant disease management. *Molecules*, *27*(15), Article 15.

Thakur, R., Shirkot, P., & Dhirta, B. (2018). *Studies on Effect of Gold Nanoparticles on Meloidogyne Incognita and Tomato Plants Growth and Development*.

Till, B. A., Weathers, L. J., & Alvarez, P. J. J. (1998). Fe(0)-supported autotrophic denitrification. *Environmental Science & Technology*, *32*(5), 634–639.

Varia, J. C., Zegeye, A., Velasquez-Orta, S., & Bull, S. (2016). Process analysis of AuCl4− sorption leading to gold nanoparticle synthesis by Shewanella putrefaciens. *Chemical Engineering Journal*, *288*, 482–488.

Varia, J., Zegeye, A., Roy, S., Yahaya, S., & Bull, S. (2014). Shewanella putrefaciens for the remediation of Au^{3+}, Co^{2+} and Fe^{3+} metal ions from aqueous systems. *Biochemical Engineering Journal*, *85*, 101–109.

Verma, V. C., Anand, S., Ulrichs, C., & Singh, S. K. (2013). Biogenic gold nanotriangles from Saccharomonospora sp., an endophytic actinomycetes of *Azadirachta indica* A. Juss. *International Nano Letters*, *3*(1), 21.

Verma, V. C., Singh, S. K., Solanki, R., & Prakash, S. (2010). Biofabrication of anisotropic gold nanotriangles using extract of endophytic Aspergillus clavatus as a dual functional reductant and stabilizer. *Nanoscale Research Letters*, *6*(1), 16.

Whitesides, G. M. (2005). Nanoscience, nanotechnology, and chemistry. *Small*, *1*(2), 172–179.

Worrall, E. A., Hamid, A., Mody, K. T., Mitter, N., & Pappu, H. R. (2018). Nanotechnology for plant disease management. *Agronomy*, *8*(12), Article 12.

Xia, Y. (2008). Nanomaterials at work in biomedical research. *Nature Materials*, *7*(10), Article 10.

Xie, X., Xu, W., & Liu, X. (2012). Improving colorimetric assays through protein enzyme-assisted gold nanoparticle amplification. *Accounts of Chemical Research*, *45*(9), 1511–1520.

Xu, Z. P., Zeng, Q. H., Lu, G. Q., & Yu, A. B. (2006). Inorganic nanoparticles as carriers for efficient cellular delivery. *Chemical Engineering Science*, *61*(3), 1027–1040.

Yetisgin, A. A., Cetinel, S., Zuvin, M., Kosar, A., & Kutlu, O. (2020). Therapeutic nanoparticles and their targeted delivery applications. *Molecules*, *25*(9), Article 9.

Yu, Q., Li, J., Zhang, Y., Wang, Y., Liu, L., & Li, M. (2016). Inhibition of gold nanoparticles (AuNPs) on pathogenic biofilm formation and invasion to host cells. *Scientific Reports*, *6*(1), Article 1.

Zhang, X., Yan, S., Tyagi, R. D., & Surampalli, R. Y. (2011). Synthesis of nanoparticles by microorganisms and their application in enhancing microbiological reaction rates. *Chemosphere*, *82*(4), 489–494.

Zhang, Y., Newton, B., Lewis, E., Fu, P. P., Kafoury, R., Ray, P. C., & Yu, H. (2015). Cytotoxicity of organic surface coating agents used for nanoparticles synthesis and stability. *Toxicology in Vitro*, *29*(4), 762–768.

Zhang, Z., Liu, S., Zhou, T., Zhang, H., Wang, F., Zhang, G., Wang, X., & Liu, T. (2018). Salt-induced assembly transformation of DNA–AuNP conjugates based on RCA Origami: From linear arrays to nanorings. *Langmuir*, *34*(30), 8904–8909.

Zharov, V. P., Mercer, K. E., Galitovskaya, E. N., & Smeltzer, M. S. (2006). Photothermal nanotherapeutics and nanodiagnostics for selective killing of bacteria targeted with gold nanoparticles. *Biophysical Journal*, *90*(2), 619–627.

9 Bio-Fabrication of Nanomaterials and Their Application in Plant Disease Management

Scope, Challenges and Strategies

Dina S. S. Ibrahim, Rizwan Ali Ansari, Sozan E. El-Abeid, Manal Mostafa, Fayrouz M. Saeed, and Ahmed Ghannouchi

9.1 INTRODUCTION

Because of unique characteristics and the distinct physicochemical, biological, and morphological differences NanoParticles (NPs) exhibit from those at the macroscale. NPs are classified based on morphology and composition into NPs, nanobars, nanocables, nanotubes, nanobelts, nanofibers, nanospheres, and quantum dots and as either organic or inorganic, respectively (Hou et al., 2013). NPs are fascinatingly diverse, falling into two main categories. The first group comprises carbon-based structures like fullerenes, polymers, and composite materials, offering a rich array of possibilities. The second category includes metallic wonders such as Au, Ag, Cu, Al, Zn, and Cd, alongside semi-metallic marvels like Se and Si. Magnetic materials like Se and Si, as well as ceramics, round off this captivating world of nanoscale innovation (Mo et al., 2017; Rafique et al., 2017). The surface area to volume aspect ratio is considered the most substantial feature of NPs which allows them to easily interact with other particles (Narayanan and Sakthive, 2010; Thakkar et al., 2010). Despite their extensive application in diverse scientific fields, including the management of plant diseases, methods for synthesizing NPs such as sol-gel processes, electrodeposition, combustion routes, thermal methods, and co-precipitation techniques have exhibited certain detrimental effects (Thema et al., 2016). The reason is that the physical methods require considerable energy, and the chemical ones produce environmentally toxic chemicals and non-biodegradable materials (Ismail et al, 2016). Therefore, scientists have developed specific effective approaches which are assumed to be eco-friendly (Abbasi et al., 2023; Sudha et al., 2017; Uddin et al., 2021). Using various biological entities such as trees, algae, and microorganisms to produce metal NPs offers a clean, cost-effective, renewable, and eco-friendly approach, considered a viable option (Hussen and Mohammed, 2021). Thus, this chapter deals with the synthesis of NPs by employing various available approaches. We have addressed also how the application of NPs in managing plant diseases, which are causing significant yield losses to a wide range of agricultural crop commodities, can be pivotal.

DOI: 10.1201/9781003256762-9

9.2 NP PRODUCTION TECHNIQUES

The common methods include top-down and bottom-up methods that study interacting nanomaterials and interfacing nanostructures (Tripathy et al. 2023). Physical, chemical, or biological methods represent potential methods to produce metallic NPs. Different forms of nano-formulation including polymer-metal nanocomposite, biopolymer, and nano-capsulation are synthesized through this technique. There are various types of methods, for example, attrition and pyrolysis are from physical methods, whereas solid-state interaction and wet chemical are examples of chemical methods. Despite being used on a wide scale, the processes of NP synthesis through different methods have certain limitations (Ranjan et al., 2016). The most commonly used methods such as the physical and chemical methods consume huge energy to keep the high pressure and temperature constant, and they also result in harmful biproducts due to the utilization of toxic substances. Therefore, attention has been shifted towards affordable costs, non-toxicity, and eco-friendly methods for producing NPs (Thakkar et al. 2010). Various microorganisms including bacterial species such as *Lactobacillus, Pseudomonas, Lactobacillus, Thermomonospora* and *Escherichia*, and a diverse range of fungi such as *Aspergillus* have been frequently used for producing NPs. Besides, plant species, such as Lemon Grass and *Aloe vera* and various algae are used to biologically produce metallic NPs.

Fungi are considered the most important biological constituents in NP synthesis due to their ability to tolerate the metals and their feature of bioaccumulation (Shankar et al. 2003; Ansari et al., 2024; Ansari and Mahmood 2017b; Ansari et al., 2017; Ansari et al., 2023; Boyno et al., 2024; Fabiyi et al., 2024;). Fungi produce multiple different extracellular enzymes that show high efficiency in the large-scale production of environmentally safe, economically viable and easy to handle NPs (Jain et al. 2011). Because of the environmental concerns about chemical synthesis of NPs, green NP production has become increasingly essential (Malik et al. 2023). Optimal application and use of the precise function of nanoparticles is entirely dependent on the thorough information obtained through various characterization approaches. Green production of NPs requires the use of natural biodegradable substances such as plant extracts, enzymes, vitamins and polysaccharides (Aswathi et al. 2023). To develop green manufacturing processes, there is a need to modulate benign approaches for production that utilize mild reaction conditions and non-toxic precursors. The green chemistry is used, for example, to decrease chemical factors accompanied with environmental toxicity (Ranjan et al 2016). The enzymatic method is considered to be a better one than other biological techniques because enzymes' structures are well recognized, and additionally, they are available in pure form (Ranjan et al., 2016). The hydroxyl groups present in polysaccharides and the hemiacetal reducing end possess the capability to reduce precursor salts, initiating the synthesis of NPs. Soluble starch served as both a reducing and stabilizing agent during the synthesis process of NPs. When the salt was reduced using sodium hydroxide in ethylene glycol, several aggregations were seen (Shervani et al., 2008). Mixtures of anisotropic and spherical shapes were generated by adding 5% polyvinylpyrrolidone material to 1% solution of starch (aqueous) for 1 hour at 70°C (Ranjan et al., 2016).

9.2.1 Type of Salt and Plant Using in Phyto-Synthesis

There are significant number of salts which are used to synthesize NPs. They are easy to obtain and handle during the course of experiment. *Calotropis procera* latex and *Aloe vera* flower broth were used to synthesize CuNPs between 15 and 40 nm via copper acetate as a substrate. While $CuSO_4$ was used with leaf extract from *Tabernaemontana divaricate, Syzygium aromaticum*; brown alga, *Bifurcaria bifurcata, Vitis vinifera*, and *citrus medica* in the phytosynthesis of CuNPs. By mixing $CuSO_4$ with *Cassia alata* flower extract, CuONPs were synthesized between 110 to 280 nm (Harne et al., 2012; Karimi & Mohsenzadeh, 2015;). The CuONPs of (50–100 nm) have been synthesized utilizing copper nitrate, extracts from tea leaf, coffee powder, and different parts of Andean blackberry; fruit and leaf. CuNPs less than 10 nm were produced from cupric nitrate by utilizing *Gloriosa superba* extract (Naika et al,. 2015). Copper NPs with diameter less than 5 nm were synthesized by mixing copper acetate with *Centella asiatica* leaf extract (Devi and Singh 2014; Sutradhar et al., 2014).

9.2.2 Challenges in NP Synthesis

Despite of the use of numerous techniques for developing NPs, including chemical deposition, mechanical milling, spinning, and laser ablation, the process is unsuitable for continuous production due to the hazardous chemicals used, and the high cost (Ying et al. 2022). Development of sustainable approach for the synthesis of NPs at commercial level has been realized and strident research in this direction has gained momentum. One of the green strategies used for sustainable production is microbial synthesis of NPs which connects microbiology and nanotechnology. Also, plant extract use is one of the methods of green synthesis that is easy and remarkably effective. Several microorganisms have been investigated for the generation of NPs in microbial communities. Limitations of the biosynthesis process include a lack of monodispersity and a lengthy synthesis period, both of which can be resolved by improving microbial cultivation and extraction procedures. Saravanan et al. (2021) mentioned that chemical synthesis gives concern about the long-term viability of physicochemically produced nanomaterials, which frequently involve energy-demanding procedures, high temperatures, hazardous solvents, or chemical waste. Researchers have thus worked to replace chemical mixtures with ecologically friendly and sustainable technology. The use of living methods have been grown appreciably for the synthesis of NPs. Ermolin et al (2021) used nano liquids with improved application methods. However, there are significant issues that must be resolved before nano liquid is commercialized. Nano fluid operating performance and stability represent key concerns. The stability of nano fluid is crucial for maintaining their thermophysical characteristics after production for a long time. Enhancing nano liquid stability and comprehending nano liquid behaviour are therefore essential steps in the chain required to commercialise this kind of sophisticated fluid.

9.3 BIOLOGICAL SYNTHESIS

Bioengineered NPs driven from organic materials or plant extracts are very important because of their many useful properties and variety of bioactive reducing metabolites. Plants are more congenial and resistant to heavy metals compared to a wide range of bacteria and algae. This attribute renders them a more congenial option for AgNP synthesis. Isn't it fascinating how changing the amounts of chemicals and the conditions of reactions can affect the basic traits of nanoparticles? For instance, factors like temperature and pH can influence their size and shape. Metal NPs have been created using a green approach. This method allows for the incorporation of various biological components like plant extracts, bacteria, fungi and algae. Utilizing plant extract is considered a quite straightforward and easy technique to produce nanoparticles at high levels in comparison to utilizing biological components. Several factors such as solvent, temperature, and the like, influence the nature of nanoparticles. These NPs have the ability to convert salts into metal NPs (Doble and Kruthiventi 2007. The fundamental properties of such NPs have been studied for use in antimicrobials and biological diagnostics. Because pathogenic bacteria might contaminate NPs employed in the medical field, it has been demonstrated that using plant extract for NPs production has more advantages than using microbial procedures (Singh et al., 2018).

9.3.1 Plants

Plants can produce various secondary metabolites such as alkaloids, phenolics and terpenoids that feature reduction and stabilization, which have been used to synthesize nanoparticles, especially biosynthesis (Huang and Yang, 2004). Biomolecules like enzymes, proteins and secondary metabolites that occur in microorganisms are reducing agents in NP synthesis (Siddiqi et al., 2018). Plant extracts are a more sustainable method of nanosynthesis than other forms of biosynthesis, the focus of this section is on plant-mediated nanoparticle synthesis. To synthesize green NPs, the plant extract is mixed with a precursor solution for a brief duration.

Certainly, to synthesize stable and well-characterized green NPs, several factors need careful consideration. These factors include the type and amount of plant extracts used, the choice of precursors, the pH level of the reaction mixture, and the duration of the reaction. Each of these elements plays a vital role in determining the properties and characteristics of the resulting NPs. Additionally, pH directly affects the durability of nanomaterials since an acidic environment (pH 5) encouraged colloid aggregation. AuNPs demonstrated high stability in both acidic and basic environments throughout a wide pH range 5–11. Plant extracts provide a congenial approach for producing NPs. It offers various benefits, such as low cost, safety, no requirement for cell care or culture, long-term storage at ambient temperature, high synthesis rate, and response time ranging from minutes to several hours (Chung et al. 2016). When using microorganisms for NP synthesis, the rate of production often suffers. This is particularly noticeable when the cultures require a medium that takes two or more days to develop. Consequently, this process becomes time-consuming.

Different parts of plants such as leaves, roots, bark, stems and fruits play crucial roles in synthesizing distinct NPs with various shapes and sizes. Plant extracts contain a plethora of components. These include proteins, flavonoids, amino acids, vitamins, minerals, polysaccharides, polyphenols, tannins, terpenoids and organic acids. These components make plant extracts more preferred for NPs synthesis. Studies have shown that not just reducing ions to nano size, all of the compounds play a major role during the synthesis and capping of nanomaterials (Mahmoud 2020). Stable Au and AgNPs were synthesized based on the extraction of *Emblica officinalis*.

The production of ZnONPs using leaves or flowers began with washing the plant parts in running tap water. This was followed by adding sterile water to the process. The goal was to ensure cleanliness and remove any impurities present on the plant surfaces. Then the dried plants are weighed and pulverized with the use of a mortar and pestle. According to the desired concentration, the plant part was given Milli-Q H_2O, and then heated while being continuously stirred with a magnetic stirrer (Heinlaan et al., 2008; Rajeshkumar et al., 2016). To get a clear solution, the solution is filtered with filter-paper, then it is used as a plant extract. Any source of zinc salt (nitrate or sulfate or acetate) 0.5 nM was hydrated and mixed with some of the extract, which is then boiled at a considerable temperature and for the desired time to obtain efficient mixing (Ochieng et al., 2015; Rajeshkumar et al., 2016). Some optimizations may be performed at this point with the help of different degrees of temperature, pH, extract concentration, and time periods. When the colour turns yellow during the incubation period, this is considered as partial visual confirmation of the synthesized NPs (Agarwal, et al., 2017). Agarwal et al. (2017) observed aggregate formation in NPs. These NPs were synthesized from extracts of *Moringa oleifera, pongamia pinnata, Calatropis gigantean, Nephelium lappaceum, Plectranthus amboinicus,* and *Agathosma betulina*. Other scientists have produced metallic NPs by combining banana peel extract and papaya callus (Saini et al 2008).

9.3.2 Bacteria

Microorganisms have developed some mechanisms to evade metal toxicity that may result because of high levels of metals present in some environments, as an adapting mechanism to survive and to cope with their habitat (Pantidos and Horsfall 2014). One of these mechanisms includes reducing the toxicity of these metals through altering their chemical structure and the production of them in the form of NPs. Many different types of microorganisms have the capacity to create inorganic materials. Some produce these materials within their cells, while others do so outside their cellular structures. Prokaryotic bacteria are of great interest in the biosynthesis of metal NPs. They exhibit unique capabilities in this process. Their intricate biochemical machinery enables them to efficiently produce metal NPs (Alloosh et al., 2021). Tsekhmistrenko et al., (2020) reported that bacteria which are considered as nano-factories cannot only remove metal or metalloid ions, nonetheless, they can produce materials with unique characteristics. This explains why the nanomaterial-based mimetics exhibit more stability than the naturally occurring enzymes under harsh conditions. Being able to modify their catalytic activity and because their manufacture is quite simple and financially viable, makes them greatly useful (Wei and Wang, 2013). Bacteria possess the remarkable ability to produce metal NPs both inside and outside their cells. A wide range of microorganisms, including *Pseudomonas stutzeri, Escherichia coli, Salmonella typhi*, and the like, have been observed to engage in this process. They utilize their natural metabolic processes and enzymatic machinery to reduce metal ions and create NPs (Klaus et al., 1999;).

Research conducted in the past has shed light on the capability of certain bacteria, vital for maintaining environmental balance and ecological health, to produce gold nanoparticles. Various bacterial species, including *Rhodopseudomonas capsulata* (He et al., 2007), *Brevibacillus formosus* (Srinath et al., 2017), *Arthrobacter nitroguajacolicus* (Dehnad et al., 2015), *Klebsiella pneumoniae* (Srinath and Rai, 2015), and *E. coli* (Du et al., 2008), have been identified for their involvement in this process. Despite ongoing research efforts, the exact mechanism behind the synthesis of gold NPs remains undisclosed (Mukherjee et al., 2002). According to Kaur (2018) the process of bio-reducing Au^{3+} to Au^0 by bacteria in order to produce AuNPs has shown the participation of $NADH^-$ and NADH dependent enzymes. *Lactobacillus plantarum* and *Aeromonas hydrophila* respectively in the process of ZnONPs bio-synthesis has been well documented (Selvarajan and Mohanasrinivasan 2013;). According to Agarwal et al. (2017) *B. licheniformis* produced eco-friendly ZnONPs that displayed photocatalytic activity and helped in degradation of methylene blue. *Rhodococcus pyridinivorans* was used along with $ZnSO_4$ as a substrate in the formation of spherical shaped NPs of 100–130 nm (Otari et al., 2012). *Halomonas elongata* was reported to synthesize CuONPs (Rad et al., 2018)., and *Bacillus cereus* to create FeONPs (Fatemi et al., 2018). Different bacteria showed the ability to produce AgNPs including *Bacillus licheniformis* (Kalishwaralal et al. 2010) and lactic acid bacteria (Sintubin et al., 2009) and cyanobacteria which synthesize AgNPs that can fix ammonia (Tomer et al., 2019). $AgNO_3$ was used as a precursor in the process of AgNP biosynthesis using *Bacillus amyloliquefaciens* and *Bacillus subtilis. Bacillus pumilus, Bacillus paralicheniformis* and *Sphingomonas paucimobilis*. The newly synthesized AgNPs were found to be spherical and oval, and <20 nm (Allam et al., 2019). *Bacillus endophyticus* and *Deinococcus radiodurans* can produce AgNPs (Gan et al. 2018; Li et al. 2018). Gahlawat and Choudhury (2019) reported that bio-synthesized NPs are more resistant to oxidation, enabling them to be used in different agroclimatic zones.

9.3.3 Fungi

Mycosynthesis is the process of NPs biosynthesis by using fungi. They are widely used in NP synthesis, because of its various chemical structures and sizes, well-defined dimensions (Golhani et al., 2020). Among the different mechanisms involved in mycosynthesis for metal NPs production, fungal nitrate reductase enzyme is the most common and well known one. Fungi exhibit greater adaptability to growth conditions and metal tolerance compared to bacterial populations (Sangappa and Thiagarajan 2012). Fungi have a number of benefits over other species when it comes to the manufacture of green NPs, in terms of isolation and identification of enzymes which are extracellular in nature (Prasad et al. 2016). For instance, a number of fungi secreted proteins have the ability to quickly and safely change metal ions. Thus, these microorganisms have gained significant interest as they can potentially provide speedy and harmless production processes of metallic NPs (Mahendra et al. 2009). This has the inherent capability to synthesize large quantities of proteins, fungi that are not showing considerable metal concentrations can withstand high metal ion concentrations, helping to reduce metal ions into less dangerous ones (Prasad et al 2016). Several species of fungi like *Verticillium, Fusarium, Aspergillus* and *Penicillium* represent a great source of NPs (Hosseini and Sarvi 2015). Singh et al. (2017) reported that AgNPs formed when endophytic fungus supernatant as *Alternaria* sp. formed spherical size NPs with a size less than 30 nm. (Devi and Joshi, 2015) isolated three endophytes including *Penicillium ochrochloron* PFR8, *Aspergillus niger* PFR6, and *Aspergillus tamarii* PFL2 which have been used to biofabricate NPs. The fascinating filamentous fungus, *Neurospora crassa* took centre stage in the creation of both Ag and AuNPs (Castro-Longoria et al., 2011). Different *Aspergillus* spp. like *Aspergillus terreus* CZR1 and *Aspergillus terreus* TFR2, along with *A. flavus* CZR2 and TFR1, *A. tubengensis* TFR3A and *A. japonicas* AJP01 were also utilized for the synthesis of myriad NPs (Tarafdar, 2015). A screening of the filamentous fungus *Neurospora crassa* when added to aqueous solutions of ($HAuCl_4$), showcased its aptitude for (mono and bi) Au/Ag metallic NP production with a diameter 32 nm approximately (Castro-Longoria et al., 2011). Hence, *N. crassa* represents a potential 'Nano-factory' which may synthesize metallic NPs.

9.3.4 Algae

Algae are assumed to be a valuable source for the biosynthesis of NPs (Azizi et al., 2014). Algae stand out for their readily available and diverse nature, offering a cost-effective means to produce NPs. Their inherent characteristics yield highly stable and biologically beneficial NPs, making them an attractive choice for nanoparticle synthesis. Moreover, algae are faster than other bio-synthesized nanoparticles (Dağlıoğlu and Öztürk 2019). With the help of biochemicals or enzymes already found in them, algae may synthesize extracellular or intracellular NPs from any metal salt. Nevertheless, many studies reported that for metal NP synthesis from metal salts, enzyme and reducing substances are considered the essential components of microorganisms and fungi (Papaspyrides et al., 2009; Riddin et al., 2009). Reports by Siddiqi and Husen (2016b) highlight the usage of different species of algae including *Ecklonia cava, Stoechospermum marginatum, Pithophora oedogonium, Sargassum wightii, Fucus vesiculosus, Chlamydomonas reinhardtii, Bifurcaria bifurcata, Oscillatoria willei, Chlorella vulgaris, Sargassum muticum* and *Spirulina platensis* for the biological synthesis of NPs. Rajeshkumar et al. (2013) conducted research showing that AgNPs produced from *Sargassum longifolium* exhibited promising antimicrobial effects. Within an hour, the colour of the reaction mixture changed from yellow to brown, indicating the formation of AgNPs. This mixture comprised algal aqueous extract and $AgNO_3$, suggesting the synthesis of NPs with potent antimicrobial properties. It took the reaction up to 32 hours to complete, and its severity varied over time. The generation of polydispersed AgNPs was verified by the 440 nm absorption peak. Additionally, it was shown that the acidity/alkalinity concentration (PH) of the reaction mixture was crucial in the creation of AgNPs. While the reaction mixture's colour changed more slowly at the acidic pH 6.2 than it did at alkaline pH 8.4, the pH rise also increased the colour intensity that followed the reduction process (Rajeshkumar et al., 2013). Sinha et al. (2015). documented that the biosynthesis of AgNPs with diameter equal to 25–44 nm, the NPs were fabricated utilizing fresh water green alga, *Pithophora oedogonium*, and exhibited antibacterial activity. The quantitative analysis and IR spectrum of algal extraction revealed the existence of different constituents such as proteins, carbohydrates, steroids and saponins which reduce $AgNO_3$ to AgNPs. Biological mediated AgNPs from some aquatic marine alga, *Caulerpa racemosa* have been also reported (Kathiraven et al., 2015). The produced crystalline nanoparticles had a distinctive geometry being as face-centred cubes with sizes of 5–25 nm. Patel et al. (2015) reported the production of AgNPs where 14 bacteria and microalgae were used. Moreover, different shapes of AgNPs; spherical, elongated, and irregular ones with varying morphologies and diameters were produced from a variety of species.

Suganya et al. (2015) found small AuNPs synthesized with an average dimension of approximately 5 nm using *Spirulina platensis*. The synthesis of AuNPs commenced with incubating the algal proteinaceous extract with $HAuCl_4$ in a 1:1 ratio. This mixture was treated with NaOH leading to a colour change from green to greyish yellow. Eventually, the solution transitioned to a vibrant ruby red, signalling the successful production of AuNPs (Mulvaney, 1996). Using a maximum excitation at wavelength 620 nm, three significant peaks were seen at 685, 524, and 385 nm. The frequencies resulted from a secondary class of amines, OH and COO groups that have been paramount in the stabilizing of the AuNPs (Leelavathi et al., 2011). The biosynthesis of CuONPs with dimension ranges between 5–45 nm from the algal extract of *B. bifurcate* have also been reported (Abboud et al., 2014). The resulting NPs were Cu (I, II) oxides that are mixed and crystallized in nature. Due to their use in energy conversion, electrical and catalysis, and magnetic storage media, transition metal oxide NPs are recognized as a significant class of semiconductors (Hossain et al., 2023). Upon the addition of a 1 mM solution of $CuSO_4$ to *B. bifurcate* extract at room temperature, a discernible change in colour became evident (Borgohain et al. 2002; Yin et al. 2005).

9.4 APPLICATION OF NPS IN CROP HEALTH IMPROVEMENT UNDER BIOTIC STRESSORS

9.4.1 NPs in Designing of Nanobiosensors

Sun et al. (2006) defined biosensors as a mixed system of receiver-transducers, used to identify any chemical and physical characteristics of certain environments, in addition to being able to sense the presence of biological analyte via a biological or an organic identification element. Analytes in extremely low amounts could be recognized by nanobiosensors which are very compact biosensors attached to a sensitized element. This technology enables efficient management of land, water, pesticides and fertilizers, which aid in the fast detection of problems and instant decision making, leading to crop improvements. Equally important, the longevity, ratio of high surface to volume, good sensitivity and stability and the rapid kinetics of electron-transfer, provided nanobiosensors with competitive advantages against any of the generations of previous sensors (Scognamiglio 2013). Nano-sized materials which are embedded inside the nanobiosensors, represent the bioreceptor carried on a transducer that send signals to a recognition element for single or multiplex analyte detection (Naresh and Lee 2021). Nano-sized materials are characterized with distinct properties; immobilization, fictionalization and miniaturization (Singh et al. 2021). To improve the analytical performance of NMs, these properties make it possible to integrate transduction system biocomponents into a more intricate design (Arduini et al., 2016) of these biosensors' work by turning them on and off, which gives them the ability to detect analytes at extremely low concentration (part per trillion) and to contain the matrix they evaluated using nano-formulation (Antonacci et al. 2018). Additionally, there are a number of additional uses for nanotechnology, including a smart automatically analytically minded system frequently employed in the agricultural field for production monitoring and evaluation of insect infestations, contamination of soil and water and plant diseases. Electronic noses are regarded as the next generation of sensors. (Hu et al 2019; Usman et al., 2020).

9.4.2 NPs in Plant Disease Management

Elsharkawy et al. (2022) evaluated the effectiveness of chitosan driven NPs and salicylic acid on wheat leaf rust. Results showed significant reductions in leaf rust symptoms in treated plants compared to untreated ones, with chitosan NPs inducing abnormalities in *Puccinia triticina* urediniospores and enhancing the peroxidase and catalase activities. Both treatments led to the appearance of superoxide (O_2^-) and hydrogen peroxide (H_2O_2) discolorations, with chitosan NP and salicylic acid treated plants exhibiting more pronounced effects than untreated plants. Anatomical examinations revealed enhancements in various leaf characteristics with chitosan NP treatment. Genes like PR1-PR5 and PR10 were upregulated, indicating activation of defence mechanisms. The study suggests that chitosan NPs hold promise as a sustainable approach for enhancing yield and controlling wheat leaf rust disease. Imada et al. (2016) reported the potential of magnesium oxide NPs that improved the resistance level of the crop against *Ralstonia solanacearum*, causing wilt on tomatoes. Treatment with magnesium oxide NPs significantly reduced disease incidence when applied before pathogen inoculation, indicating its efficacy in combating bacterial infection. Analysis revealed speedy formation of reactive oxygen species in treated roots which seems to be due to a reaction between magnesium oxide and polyphenols in the roots. Moreover, magnesium oxide NPs enhanced the expression of defence mechanisms in both roots and hypocotyls, while histochemical analyses showed the accumulation of defence enzymes in specific tissues. The findings suggest that magnesium oxide NPs induce systemic resistance in tomato plants against bacterial wilt, offering a promising approach for disease management in agriculture.

NPs can be additionally exploited as bio-nanonematicides (or green chemicals) to inhibit the presence and development of nematodes in the soil. They are organic compounds extracted from plants that have no residual effect, a high reach within roots that allows them to target nematodes in different stages, and a fast spread in irrigation water that enables them to maintain their presence in the soil for a long time (Haroon, 2021). NPs and their use in phytoparasitic nema suppression has opened up a door which is assumed to be economical and sustainable. The study proves the promising potential of plant mediated AgNPs as a means of *M. incognita* control in carrot production (Fabiyi, 2021). The use of synthetic nematicides remains the most common and most effective measure to tackle this problem. However, their application also leaves several unwanted effects on biota, thus rendering them dangerous in the long run (Jones et al. 2017).

9.5 CONCERNS

Soil microorganisms represent reliable identifiers of soil health, due to their role in multiple vital soil phenomena (Dinesh et al., 2012; Fabiyi et al., 2020; Patil et al., 2024; Solanki et al., 2021; Sumbul et al., 2020). There is a necessity to understand the methodology of NP application on soils which could influence the diversity, activities and even the structure of microorganisms. It has been reported that toxicity of NPs could directly affect soil microorganisms, and indirectly impact them through increasing toxic materials in soil naturally (Haghighatdoost et al., 2023). Limited doses of NM toxicity is correlated with their nature; inorganic nanoparticles such as metal and metal oxides are more toxic than organic ones such as fullerenes and carbon nanotubes (Rajput et al., 2018). Moreover, NPs produced from carbon have been found to have a significant effect on genomes and on the biochemical pathway of microbial population in soil particularly (Wu et al., 2020). Dinesh et al. (2012) found that metal and oxide of this metal exhibited potential to act as antimicrobial agents. As a result of their application in pure culture, they stopped the development of plant growth promoting rhizobacteria and different bacterial species included in nitrogen cycles (Mishra and Kumar, 2009). Furthermore, ZnO and TiO_2, reduced microbial carbon biomass especially for gram negative bacteria (Rashid et al., 2017b; Rajput et al., 2018). When adding different ultra-small doses of TiO_2 and ZnONPs to the soil for over 60 days, a reduction in the microbial carbon biomass and an adverse effect on substrate induced respiration, was shown, thus indicating a reduction in the activity of microbial communities. Whereas, both diversity and structure of bacterial pools and were altered and regressed due to the effect of NMs on enzymatic activities (Ge et al., 2011). Similarly, the microbial carbon biomass, the fungal colony forming units and heterotrophic bacterial able to be cultivated, were significantly reduced by the presence of ZnO and Fe_2O_3NPs (Rashid et al., 2017a, 2017b).

Further studies were undertaken to evaluate the potential impact of nanomaterials on microbial biomass carbon and microbial diversity. Recent investigations have proved that the use of biogenic nanomaterials is less devastating for the microbes present in the soil when compared to their chemically manufactured analogues. However, the use of these materials is becoming more and more used to address nanotoxicity in soils. Only a very few studies were conducted on biogenic NMs which supports the need for further research that explore their application potential and associated impact (Mishra et al., 2020; Ottoni et al., 2020). Tong et al. (2016) reported the lowest effects of different particle sized of C_{60}NMs on microbial biomass carbon and enzymatic activities. Other metal oxide NPs (CeO_2, Fe_3O_4 and SnO_2) failed to show a significant impact on MBC while significantly affecting biodiversity (Antisari et al., 2013). Contrasting results were also observed with FeONPs as they particularly stimulated the growth of actinobacteria by boosting activity of both the urease and the invertase enzyme (He et al., 2011). Other studies in the longer term have showcased the positive impacts of ZnO and CuONPs (10 mg kg^{-1}) on microbial communities and their enzymatic activity, when applied in low concentrations (Jośko et al., 2019), as well as the absence of a significant inhibition of denitrification under low concentrations of CuONPs (Zhao et al., 2020). Sometimes contradicting impacts of NPs on microbial population and their activities are mainly related to the applied dose to the soil, which can be clearly observed with AgNPs having a negative effect on microbial functioning only at high doses (Rahmatpour et al., 2017). The majority of NPs have been found to negatively affect soil microbial communities. Examples include Ag, Al_2O_3, TiO_2, CuO and ZnO. However, there are some NPs that have minimal effects on these communities. These include Si, Fe, Au, Pd NPs (Suresh et al., 2013). The differences of impact and nature of action from one NP to another support the necessity to consider several factors when investigating the reactivity of NPs with the soil environment and the toxicity on microorganisms, particularly the NPs nature, size, shape, dose, and concentration as well as soil characteristics such as type of soil (Chen et al., 2019). Producing reactive oxygen species is another technique that NPs can trigger depending on properties of their structure such as the surface area, size and shape and (Abdal Dayem et al., 2017). When *Allium cepa* root hairs interacted with different sizes of AuNPs, it triggered a response. This reaction was dependent on both the dosage and size of the NPs involved. As a result, there was an increase in the production of reactive oxygen species. This increase led to heightened levels of lipid peroxidation and chromosome aberrations within the plant cells (Rajeshwari et al., 2016). The silver NPs can release Ag^+ ions. There is uncertainty as to whether the toxicity of NPs is the result of these ion releases or of the interactions of NPs themselves. To clear the confusion, the toxicity of NPs and Ag ions were assessed and compared in *Arabidopsis thaliana*. It was found that the NPs decreased root elongation and growth of vegetative parts and induced a stronger reduction of photosynthetic pigment content compared to Ag^+ which had a lesser effect and weaker results overall (Qian et al., 2013). The treatment of *Arabidopsis thaliana* AgNPs also led to changes in plasma membrane and the cessation of root growth and expansion of leaves (Sosan et al., 2016). AgNPs applied on barley improved the length of root and shoot systems and enzyme activities at concentrations of 0.1 mM, while a higher concentration (0.5–1–mM) reduced these processes (Fayez et al., 2017; Paramo et al., 2020).

NPs exhibit the ability to permeate the root plasma membrane through transporter mechanisms, both passively and actively (Lv et al., 2014; Rossi et al., 2018). The apoplastic pathway facilitates the entry of larger NPs (200 nm) into cells, allowing them to pass through the porous cell wall and traverse the intracellular space (Shukla et al., 2016). Once NPs enter the plant, they are transported by some conductive tissue to other sections (Hu et al., 2020; Karny et al., 2018; Lv et al., 2014).

9.6 LIMITATIONS AND CHALLENGES

Biosynthesis of NPs using microbial organisms and their derivatives, has become the centre of attention due to simple and low-cost production. Because of their unique properties, microbial based NPs are widely used in environmental remediation. There are multiple factors to be considered during the selection of microbes for the NPs synthesis. Factors such as replication process, biochemical activity and growth rate also affect the biosynthesis process. The discovery of certain biomolecules which are able to reduce and stabilize metal NPs from their precursor is another problem. The majority of research was conducted at lab level, which allowed for an easy maintenance of microbial growth conditions (Korbekandi et al., 2014). It is more appropriate to improve the features, nature and types of NPs (Singh et al., 2018). However, these challenges pave the way for new opportunities in this recent field of research. To overcome these challenges, a new era of 'green synthesis' approaches are now the focus of attention (Kumar et al. 2021). Utilizing green synthesis in the production process of nanomaterials and by enhancing aspects of regulation, control, clean up, and remediation processes, their environmental friendliness will boost our understanding.

9.7 CONCLUSIONS AND FUTURE LINES OF ACTION

Improved agriculture and healthier environment is expected from the judicious use of NPs. Application of NPs have ensured increased nutrient uptake capacity, efficient and accurate usage of inputs, diseases diagnosis and management, solving environmental stresses and enhanced manufacturing, storage and packaging processes. This could be used as either as an identification tool in the diagnosis process of plant diseases as nanosensors or as delivery system of functional molecules. Among various methods for synthesizing NPs such as physical, chemical, and biological approaches, the physical and chemical processes are often deemed costly and have the potential to be phytotoxic. Therefore, attention is now on starting to search for alternative methods and to initiating the production of NPs using eco-friendly, non-toxic and reliable processes to widen their use in agriculture. Production of NPs from biological means; microorganisms or plant extracts is a viable available option to achieve the targeted number of NPs of desired quality.

REFERENCES

Abdal Dayem, A., Hossain, M. K., Lee, S. B., Kim, K., Saha, S. K., Yang, G. M., ... & Cho, S. G. (2017). The role of reactive oxygen species (ROS) in the biological activities of metallic nanoparticles. *International Journal of Molecular Sciences*, *18*(1), 120.

Abbasi, M., Gholizadeh, R., Kasaee, S. R., Vaez, A., Chelliapan, S., Fadhil Al-Qaim, F., ... & Kamyab, H. (2023). An intriguing approach toward antibacterial activity of green synthesized Rutin-templated mesoporous silica nanoparticles decorated with nanosilver. *Scientific reports*, *13*(1), 5987.

Abboud, Y., Saffaj, T., Chagraoui, A., El Bouari, A., Brouzi, K., Tanane, O., & Ihssane, B. (2014). Biosynthesis, characterization and antimicrobial activity of copper oxide nanoparticles (CONPs) produced using brown alga extract (Bifurcaria bifurcata). *Applied nanoscience*, *4*, 571–576.

Abdullah, H. S. T. S. H., Asseri, S. N. A. R. M., Mohamad, W. N. K. W., Kan, S. Y., Azmi, A. A., Julius, F. S. Y., & Chia, P. W. (2021). Green synthesis, characterization and applications of silver nanoparticle mediated by the aqueous extract of red onion peel. *Environmental Pollution*, *271*, 116295.

Agarwal, H., Kumar, S. V., & Rajeshkumar, S. (2017). A review on green synthesis of zinc oxide nanoparticles–An eco-friendly approach. *Resource-Efficient Technologies*, *3*(4), 406–413.

Allam, N. G., Ismail, G. A., El-Gemizy, W. M., & Salem, M. A. (2019). Biosynthesis of silver nanoparticles by cell-free extracts from some bacteria species for dye removal from wastewater. *Biotechnology Letters*, *41*(3), 379–389.

Alloosh, M., I Khaddam, W., & K Almuhammady, A. (2021). Biosynthesis of metal nanoparticles using microorganisms and its medicinal applications. *Novel Research in Microbiology Journal*, *5*(1), 1077–1090.

Ansari, R. A., & Khan, T. A. (2012a). Parasitic association of root-knot nematode, *Meloidogyne incognita* on guava. *e- Journal of Science Technology*, *5*(12), 65–67.

Ansari, R. A., & Khan, T. A. (2012b). Diversity and community structure of phytonematodes associated with guava in and around Aligarh, Uttar Pradesh, India. *Trends in Bioscience*, *5*(3), 202–204.

Ansari, R. A., & Mahmood, I. (2017a). Determination of disease incidence caused by *Meloidogyne* spp. and or *Fusarium udum* on pigeonpea in Aligarh district: a survey. *Trends in Biosciences*, 10(24), 5239–5243.

Ansari, R. A., & Mahmood, I. (2017b). Optimization of organic and bio-organic fertilizers on soil properties and growth of pigeon pea. *Scientia Horticulturae*, *226*, 1–9.

Ansari, R. A., & Mahmood, I. (2019a). *Plant Health Under Biotic Stress – Volume II, Microbial Interactions*, Springer Nature, Singapore. ISBN: 978-981-13-6040-4; DOI: 10.1007/978-981-13-6040-4

Ansari, R. A., & Mahmood, I. (2019b). *Plant Health Under Biotic Stress – Volume I, Organic Strategies*, Springer Nature, Singapore. ISBN: 978-981-13-6043-5; DOI: 10.1007/978-981-13-6043-5

Ansari, R. A., Mahmood, I., Rizvi, R., Sumbul, A., & Safiuddin. (2017). Siderophores: augmentation of soil health and crop productivity. In: Kumar V., Kumar, M., Prasad, R. and Sharma, S. (eds.) *Probiotics in Agroecosystem*, Springer Nature Singapore, Singapore, pp. 291–312.

Ansari, R. A., Rizvi, R., & Mahmood, I. (2023). *Mycorrhizal Symbiosis and Agroecosystem Restoration*, Springer Nature Singapore, Singapore.

Ansari, R. A., Rizvi, R., Sumbul, A., & Mahmood, I. (2020). Plant-growth-promoting rhizobacteria (PGPR)-based sustainable management of phytoparasitic nematodes: current understandings and future challenges. In Ansari, R.A., Rizvi, R. & Mahmood, I. (eds.) *Management of Phytonematodes: Recent Advances and Future Challenges* Springer Nature Singapore, Singapore, (pp. 51–71).

Ansari, R. A., Sumbul, A., Rizvi, R., & Mahmood, I. (2019). Organic soil amendments: potential tool for soil and plant health management. In: Ansari, R.A. & Mahmood, I. (eds.) *Plant Health Under Biotic Stress: Volume 1: Organic Strategies* Springer Nature Singapore, Singapore (pp. 1–35).

Ansari, R. A., Sumbul, A., Rizvi, R., & Mahmood, I. (2024). Impact of environmental gases on mycorrhizal symbiosis and its influence on ecosystem functioning under the current climate change scenario. In Ansari, R.A., Rizvi, R. & Mahmood, I. (eds.) *Mycorrhizal Symbiosis and Agroecosystem Restoration* (pp. 51–76). Singapore: Springer Nature Singapore.

Ansari, R. A., Rizvi, R., & Mahmood, I. (2020). *Management of Phytonematodes: Management of Phytonematodes: Recent Advances and Future Challenges*, Springer Nature, Singapore. ISBN: 978-981-15-4087-5; DOI: 10.1007/978-981-15-4087-5

Antisari, L. V., Carbone, S., Gatti, A., Vianello, G., & Nannipieri, P. (2013). Toxicity of metal oxide (CeO_2, Fe_3O_4, SnO_2) engineered nanoparticles on soil microbial biomass and their distribution in soil. *Soil Biology and Biochemistry*, *60*, 87–94.

Antonacci, A., Arduini, F., Moscone, D., Palleschi, G., & Scognamiglio, V. (2018). Nanostructured (Bio) sensors for smart agriculture. *TrAC Trends in Analytical Chemistry*, *98*, 95–103.

Arduini, F., Micheli, L., Moscone, D., Palleschi, G., Piermarini, S., Ricci, F., & Volpe, G. (2016). Electrochemical biosensors based on nanomodified screen-printed electrodes: Recent applications in clinical analysis. *TrAC Trends in Analytical Chemistry*, *79*, 114–126.

Aswathi, V. P., Meera, S., Maria, C. A., & Nidhin, M. (2023). Green synthesis of nanoparticles from biodegradable waste extracts and their applications: a critical review. *Nanotechnology for Environmental Engineering*, *8*(2), 377–397.

Azizi, S., Ahmad, M. B., Namvar, F., & Mohamad, R. (2014). Green biosynthesis and characterization of zinc oxide nanoparticles using brown marine macroalga Sargassum muticum aqueous extract. *Materials Letters*, *116*, 275–277.

Ben-Moshe, T., Frenk, S., Dror, I., Minz, D., & Berkowitz, B. (2013). Effects of metal oxide nanoparticles on soil properties. *Chemosphere*, *90*(2), 640–646.

Borgohain, K., Murase, N., & Mahamuni, S. (2002). Synthesis and properties of Cu2O quantum particles. *Journal of applied physics*, *92*(3), 1292–1297.

Boyno, G., Ansari, R. A., Durak, E. D., Güneş, H., Çevik, R., & Demir, S. (2024). Arbuscular mycorrhizal technology in sustainable agriculture: current knowledge and challenges in agroforestry. In: Ansari, R.A., Rizvi, R. & Mahmood, I. (eds.) *Mycorrhizal Symbiosis and Agroecosystem Restoration* Springer Nature Singapore, Singapore (pp. 173–195).

Castro-Longoria, E., Vilchis-Nestor, A. R., & Avalos-Borja, M. (2011). Biosynthesis of silver, gold and bimetallic nanoparticles using the filamentous fungus Neurospora crassa. *Colloids and Surfaces B: Biointerfaces*, *83*(1), 42–48.

Chen, Y. Z., Gu, B., Uchida, T., Liu, J., Liu, X., Ye, B. J., ... & Jiang, H. L. (2019). Location determination of metal nanoparticles relative to a metal-organic framework. *Nature communications*, *10*(1), 3462.

Chung, I. M., Park, I., Seung-Hyun, K., Thiruvengadam, M., & Rajakumar, G. (2016). Plant-mediated synthesis of silver nanoparticles: their characteristic properties and therapeutic applications. *Nanoscale Research Letters*, *11*, 1–14.

Dağlıoğlu, Y., & Öztürk, B. Y. (2019). A novel intracellular synthesis of silver nanoparticles using Desmodesmus sp.(Scenedesmaceae): different methods of pigment change. *Rendiconti Lincei. Scienze Fisiche e Naturali*, *30*(3), 611–621.

de Freitas Bueno, R. C. O., Ansari, R. A., Lima, G. P. P., & Sakate, R. K. (2019). Phytosanitation: a novel approach toward disease management. In Ansari, R.A. & Mahmood, I. (eds.) *Plant Health Under Biotic Stress: Volume 1: Organic Strategies* Springer Nature Singapore, Singapore (pp. 73–90).

Dehnad, A., Hamedi, J., Derakhshan-Khadivi, F., & Abuşov, R. (2015). Green synthesis of gold nanoparticles by a metal resistant Arthrobacter nitroguajacolicus isolated from gold mine. *IEEE Transactions on Nanobioscience*, *14*(4), 393–396.

Devi, H. S., & Singh, T. D. (2014). Synthesis of copper oxide nanoparticles by a novel method and its application in the degradation of methyl orange. *Advance in Electronic and Electric Engineering*, *4*(1), 83–88.

Devi, L. S., & Joshi, S. R. (2015). Ultrastructures of silver nanoparticles biosynthesized using endophytic fungi. *Journal of Microscopy and Ultrastructure*, *3*(1), 29–37.

Devi, T. S., Das, D., Ansari, R. A., Rizvi, R., Sumbul, A., & Mahmood, I. (2020). Role of organic additives in the sustainable management of phytoparasitic nematodes. In: Ansari, R.A., Rizvi, R. & Mahmood, I. (eds.) *Management of Phytonematodes: Recent Advances and Future Challenges* Springer Nature Singapore, Singapore (pp. 279–295).

Díaz, J. A., Garza-García, J. J. O., García-Gaytán, V., León-Morales, J. M., Zamudio-Ojeda, A., & García-Morales, S. (2021). Beneficial elements improve the growth, biomass production, and photosynthetic pigments of *Calendula officinalis* L. seedlings. *Horticultural Science and Technology*, *39*(4), 413–423.

Dinesh, R., Anandaraj, M., Srinivasan, V., & Hamza, S. (2012). Engineered nanoparticles in the soil and their potential implications to microbial activity. *Geoderma*, *173*, 19–27.

Doble M, Kruthiventi AK (2007) Green chemistry and engineering. Academic Press, Cambridge

Doble, M., Rollins, K., & Kumar, A. (2010). *Green Chemistry and Engineering*. Academic Press.

Du, W. L., Xu, Y. L., Xu, Z. R., & Fan, C. L. (2008). Preparation, characterization and antibacterial properties against E. coli K88 of chitosan nanoparticle loaded copper ions. *Nanotechnology*, *19*(8), 085707.

Elsharkawy, M. M., Omara, R. I., Mostafa, Y. S., Alamri, S. A., Hashem, M., Alrumman, S. A., & Ahmad, A. A. (2022). Mechanism of wheat leaf rust control using chitosan nanoparticles and salicylic acid. *Journal of Fungi*, *8*(3), 304.

Ermolin, M. S., Ivaneev, A. I., Fedyunina, N. N., Karandashev, V. K., Burmistrov, A. A., & Fedotov, P. S. (2021). Natural silicate nanoparticles: separation, characterization, and assessment of stability and perspectives of their use as reference nanomaterials. *Analytical and Bioanalytical Chemistry*, *413*(15), 3999–4012.

Fabiyi, O. A. (2021). Sustainable management of *Meloidogyne incognita* infecting carrot (*Daucus carota*): green synthesis of silver nanoparticles with Cnidoscolus aconitifolius. *Vegetos*, *34*(2), 277–285.

Fabiyi, O. A., Alabi, R. O., & Ansari, R. A. (2020). Nanoparticles' synthesis and their application in the management of phytonematodes: an overview. In: Ansari, R.A., Rizvi, R. & Mahmood, I. (eds.) *Management of Phytonematodes: Recent Advances and Future Challenges* Springer Nature Singapore, Singapore (pp. 125–140).

Fabiyi, O. A., Ansari, R. A., Rizvi, R., Sumbul, A., & Mahmood, I. (2024). Influence of arbuscular mycorrhizal fungi on soil health amelioration and plant fitness under hostile environment. In: Ansari, R.A., Rizvi, R. & Mahmood, I. (eds.) *Mycorrhizal Symbiosis and Agroecosystem Restoration* (pp. 227–248). Singapore: Springer Nature Singapore.

Fatemi, M., Mollania, N., Momeni-Moghaddam, M., & Sadeghifar, F. (2018). Extracellular biosynthesis of magnetic iron oxide nanoparticles by *Bacillus cereus* strain HMH1: characterization and in vitro cytotoxicity analysis on MCF-7 and 3T3 cell lines. *Journal of Biotechnology*, *270*, 1–11.

Fayez, K. A., El-Deeb, B. A., & Mostafa, N. Y. (2017). Toxicity of biosynthetic silver nanoparticles on the growth, cell ultrastructure and physiological activities of barley plant. *Acta Physiologiae Plantarum*, *39*, 1–13.

Fu, H., Yang, Y., Zhu, R., Liu, J., Usman, M., Chen, Q., & He, H. (2018). Superior adsorption of phosphate by ferrihydrite-coated and lanthanum-decorated magnetite. *Journal of Colloid and Interface Science*, *530*, 704–713.

Gahlawat, G., & Choudhury, A. R. (2019). A review on the biosynthesis of metal and metal salt nanoparticles by microbes. *RSC Advances*, *9*(23), 12944–12967.

Gan, L., Zhang, S., Zhang, Y., He, S., & Tian, Y. (2018). Biosynthesis, characterization and antimicrobial activity of silver nanoparticles by a halotolerant Bacillus endophyticus SCU-L. *Preparative Biochemistry and Biotechnology*, *48*(7), 582–588.

Ge, Y., Schimel, J. P., & Holden, P. A. (2011). Evidence for negative effects of TiO_2 and ZnO nanoparticles on soil bacterial communities. *Environmental Science & Technology*, *45*(4), 1659–1664.

Golhani, D. K., Khare, A., Burra, G. K., Jain, V. K., & Rao Mokka, J. (2020). Microbes induced biofabrication of nanoparticles: a review. *Inorganic and Nano-Metal Chemistry*, *50*(10), 983–999.

Haghighatdoost, F., Mohammadifard, N., & Sarrafzadegan, N. (2023). Trace Minerals. In *Handbook of Food Bioactive Ingredients: Properties and Applications* (pp. 1–42). Cham: Springer International Publishing.

Harne, S., Sharma, A., Dhaygude, M., Joglekar, S., Kodam, K., & Hudlikar, M. (2012). Novel route for rapid biosynthesis of copper nanoparticles using aqueous extract of Calotropis procera L. latex and their cytotoxicity on tumor cells. *Colloids and Surfaces B: Biointerfaces*, *95*, 284–288.

Haroon, S. A. (2021). Manipulation of bio-nanotechnology in agri-sector for pest control with special emphasis to nematode problems in Egypt. *Annals of Agricultural Science, Moshtohor*, *59*(2), 33–36.

He, S., Feng, Y., Ren, H., Zhang, Y., Gu, N., & Lin, X. (2011). The impact of iron oxide magnetic nanoparticles on the soil bacterial community. *Journal of Soils and Sediments*, *11*, 1408–1417.

He, S., Guo, Z., Zhang, Y., Zhang, S., Wang, J., & Gu, N (2007) Biosynthesis of gold nanoparticles using the bacteria *Rhodopseudomonas capsulata*. *Mater Lett*, *61*, 3984–3987.

Heinlaan, M., Ivask, A., Blinova, I., Dubourguier, H. C., & Kahru, A. (2008). Toxicity of nanosized and bulk ZnO, CuO and TiO_2 to bacteria Vibrio fischeri and crustaceans Daphnia magna and Thamnocephalus platyurus. *Chemosphere*, *71*(7), 1308–1316.

Hossain, N., Mobarak, M. H., Mimona, M. A., Islam, M. A., Hossain, A., Zohura, F. T., & Chowdhury, M. A. (2023). Advances and significances of nanoparticles in semiconductor applications–A review. *Results in Engineering*, *19*, 101347.

Hosseini, M. R., & Sarvi, M. N. (2015). Recent achievements in the microbial synthesis of semiconductor metal sulfide nanoparticles. *Materials Science in Semiconductor Processing*, *40*, 293–301.

Hou, W. C., Westerhoff, P., & Posner, J. D. (2013). Biological accumulation of engineered nanomaterials: a review of current knowledge. *Environmental Science: Processes & Impacts*, *15*(1), 103–122.

Hu, P., An, J., Faulkner, M. M., Wu, H., Li, Z., Tian, X., & Giraldo, J. P. (2020). Nanoparticle charge and size control foliar delivery efficiency to plant cells and organelles. *ACS Nano*, *14*(7), 7970–7986.

Hu, W., Wan, L., Jian, Y., Ren, C., Jin, K., Su, X., ... & Wu, W. (2019). Electronic noses: from advanced materials to sensors aided with data processing. *Advanced Materials Technologies*, *4*(2), 1800488.

Huang, H., & Yang, X. (2004). Synthesis of polysaccharide-stabilized gold and silver nanoparticles: a green method. *Carbohydrate Research*, *339*(15), 2627–2631.

Hussein, B. Y., & Mohammed, A. M. (2021). Biosynthesis and characterization of nickel oxide nanoparticles by using aqueous grape extract and evaluation of their biological applications. *Results in Chemistry*, *3*, 100142.

Ibrahim, D. S. S., Elderiny, M. M., Ansari, R. A., Rizvi, R., Sumbul, A., & Mahmood, I. (2020). Role of *Trichoderma* spp. in the management of plant-parasitic nematodes infesting important crops. In Ansari, R.A., Rizvi, R. & Mahmood, I. (eds.) *Management of Phytonematodes: Recent Advances and Future Challenges* Springer Nature Singapore, Singapore (pp. 259–278).

Ibrahim, D. S. S., Mostafa, M., Abdellatif, A. A., & Ansari, R. A. (2022). Sustainable management of soil-borne plant pathogens. In *Earth Systems Protection and Sustainability: Volume 1* (pp. 35–56). Cham: Springer International Publishing.

Imada, K., Sakai, S., Kajihara, H., Tanaka, S., & Ito, S. (2016). Magnesium oxide nanoparticles induce systemic resistance in tomato against bacterial wilt disease. *Plant Pathology*, *65*(4), 551–560.

Jain, N., Bhargava, A., Majumdar, S., Tarafdar, J. C., & Panwar, J. (2011). Extracellular biosynthesis and characterization of silver nanoparticles using Aspergillus flavus NJP08: a mechanism perspective. *Nanoscale*, *3*(2), 635–641.

Jones, J. G., Kleczewski, N. M., Desaeger, J., Meyer, S. L., & Johnson, G. C. (2017). Evaluation of nematicides for southern root-knot nematode management in lima bean. *Crop protection*, *96*, 151–157.

Jones, R.K. (2017). Nematode Control and Nematicides: Developments Since 1982 and Future Trends. In Fourie, H., Spaull, V., Jones, R., Daneel, M., De Waele, D. (eds) Nematology in South Africa: A View from the 21st Century (pp. 129–150). Springer, Cham. https://doi.org/10.1007/978-3-319-44210-5_6

Kalishwaralal, K., Deepak, V., Pandian, S. R. K., Kottaisamy, M., BarathManiKanth, S., Kartikeyan, B., & Gurunathan, S. (2010). Biosynthesis of silver and gold nanoparticles using Brevibacterium casei. *Colloids and Surfaces B: Biointerfaces*, *77*(2), 257–262.

Jośko, I., Kusiak, M., & Oleszczuk, P. (2021). The chronic effects of CuO and ZnO nanoparticles on Eisenia fetida in relation to the bioavailability in aged soils. *Chemosphere*, *266*, 128982.

Karimi, J., & Mohsenzadeh, S. (2015). Rapid, green, and eco-friendly biosynthesis of copper nanoparticles using flower extract of *Aloe vera*. *Synthesis and Reactivity in Inorganic, Metal-Organic, and Nano-Metal Chemistry*, *45*(6), 895–898.

Karny, A., Zinger, A., Kajal, A., Shainsky-Roitman, J., & Schroeder, A. (2018). Therapeutic nanoparticles penetrate leaves and deliver nutrients to agricultural crops. *Scientific Reports*, *8*(1), 1–10.

Kathiraven, T., Sundaramanickam, A., Shanmugam, N., & Balasubramanian, T. (2015). Green synthesis of silver nanoparticles using marine algae Caulerpa racemosa and their antibacterial activity against some human pathogens. *Applied Nanoscience*, *5*(4), 499–504.

Kaur, P. (2018). Biosynthesis of nanoparticles using eco-friendly factories and their role in plant pathogenicity: a review. *Biotechnology Research and Innovation*, *2*(1), 63–73.

Keshari, N., Kranti, K. V. V. S. K., Gunda, N. K., & Ansari, R. A. (2024). Arbuscular mycorrhizal fungi: a potential agent for phytonematodes management in diverse agro-climatic zones. In *Mycorrhizal Symbiosis and Agroecosystem Restoration* (pp. 147–169). Singapore: Springer Nature Singapore.

Klaus, T., Joerger, R., Olsson, E., & Granqvist, C. G. (1999). Silver-based crystalline nanoparticles, microbially fabricated. *Proceedings of the National Academy of Sciences*, *96*(24), 13611–13614.

Korbekandi, H., Chitsazi, M. R., Asghari, G., Najafi, R. B., Badii, A., & Iravani, S. (2014). Green biosynthesis of silver nanoparticles using Azolla pinnata whole plant hydroalcoholic extract. *Green Processing and Synthesis*, *3*(5), 365–373.

Kumar, J. A., Krithiga, T., Manigandan, S., Sathish, S., Renita, A. A., Prakash, P., ... & Crispin, S. (2021). A focus to green synthesis of metal/metal based oxide nanoparticles: various mechanisms and applications towards ecological approach. *Journal of Cleaner Production*, *324*, 129198.

Kumar, V., Sharma, M., Khare, T., & Wani, S. H. (2018). Impact of nanoparticles on oxidative stress and responsive antioxidative defense in plants. In: Tripathi, D.K., Ahmad, P., Sharma, S., Chauhan, D.K., and Dubey, N.K. (eds.)*Nanomaterials in Plants, Algae, and Microorganisms: Concepts and Controversies: Volume One* (pp. 393–406). Academic Press, Elsevier.

Kumpiene, J., Lagerkvist, A., & Maurice, C. (2008). Stabilization of As, Cr, Cu, Pb and Zn in soil using amendments–a review. *Waste Management*, *28*(1), 215–225.

Lee, W. M., An, Y. J., Yoon, H., & Kweon, H. S. (2008). Toxicity and bioavailability of copper nanoparticles to the terrestrial plants mung bean (Phaseolus radiatus) and wheat (Triticum aestivum): plant agar test for water-insoluble nanoparticles. *Environmental Toxicology and Chemistry: An International Journal*, *27*(9), 1915–1921.

Leelavathi, A., Rao, T. U. B., & Pradeep, T. (2011). Supported quantum clusters of silver as enhanced catalysts for reduction. *Nanoscale Research Letters*, *6*(1), 1–9.

Li, J., Li, Q., Ma, X., Tian, B., Li, T., Yu, J., ... & Hua, Y. (2016). Biosynthesis of gold nanoparticles by the extreme bacterium Deinococcus radiodurans and an evaluation of their antibacterial properties. *International Journal of Nanomedicine*, *11*, 5931.

Li, J., Tian, B., Li, T., Dai, S., Weng, Y., Lu, J., ... & Hua, Y. (2018). Biosynthesis of Au, Ag and Au–Ag bimetallic nanoparticles using protein extracts of Deinococcus radiodurans and evaluation of their cytotoxicity. *International Journal of Nanomedicine*, *13*, 1411.

Lv, J., Zhang, S., Luo, L., Zhang, J., Yang, K., & Christie, P. (2015). Accumulation, speciation and uptake pathway of ZnO nanoparticles in maize. *Environmental Science: Nano*, *2*(1), 68–77.

Lv, S., Tai, F., Guo, J., Jiang, P., Lin, K., Wang, D., ... & Li, Y. (2021). Phosphatidylserine synthase from Salicornia europaea is involved in plant salt tolerance by regulating plasma membrane stability. *Plant and Cell Physiology*, *62*(1), 66–79.

Mahendra, R., Alka, Y., Bridge, P., & Aniket, G. (2009). Myconanotechnology: a new and emerging science. In Mahendra, R., and Bridge, P. (eds.) *Applied Mycology* (pp. 258–267), CAB International, U.K. ISBN 978-1-84593-534-4.

Mahmood, I., Rizvi, R., Sumbul, A., & Ansari, R. A. (2019). Potential role of plant growth promoting rhizobacteria in alleviation of biotic stress. *Plant Health Under Biotic Stress: Volume 2: Microbial Interactions*, 177–188.

Mahmoud, A. E. D. (2020). Nanomaterials: green synthesis for water applications. In: Kharissova, O.V., Kharisov, B.I. & Torres-Martínez, L.M. (eds.) *Handbook of Nanomaterials and Nanocomposites for Energy and Environmental Applications* (pp. 1–21), Springer Cham.

Malik, A. Q., Mir, T. U. G., Kumar, D., Mir, I. A., Rashid, A., Ayoub, M., & Shukla, S. (2023). A review on the green synthesis of nanoparticles, their biological applications, and photocatalytic efficiency against environmental toxins. *Environmental Science and Pollution Research*, *30*(27), 69796–69823.

Mishra, S., Yang, X., & Singh, H. B. (2020). Evidence for positive response of soil bacterial community structure and functions to biosynthesized silver nanoparticles: an approach to conquer nanotoxicity? *Journal of Environmental Management*, *253*, 109584.

Mishra, V. K., & Kumar, A. (2009). Impact of metal nanoparticles on the plant growth promoting rhizobacteria. *Digest Journal of Nanomaterials & Biostructures*, *4*(3), 587–592.

Mo, D., Hu, L., Zeng, G., Chen, G., Wan, J., Yu, Z., ... & Cheng, M. (2017). Cadmium-containing quantum dots: properties, applications, and toxicity. *Applied Microbiology and Biotechnology*, *101*(7), 2713–2733.

Mukherjee, P., Senapati, S., Mandal, D., Ahmad, A., Khan, M. I., Kumar, R., & Sastry, M. (2002). Extracellular synthesis of gold nanoparticles by the fungus *Fusarium oxysporum*. *ChemBioChem*, *3*(5), 461–463.

Mulvaney, P. (1996). Surface plasmon spectroscopy of nanosized metal particles. *Langmuir*, *12*(3), 788–800.

Nadkarni, V. D., Pervin, A., & Linhardt, R. J. (1994). Directional immobilization of heparin onto beaded supports. *Analytical Biochemistry*, *222*(1), 59–67.

Naika, H. R., Lingaraju, K., Manjunath, K., Kumar, D., Nagaraju, G., Suresh, D., & Nagabhushana, H. (2015). Green synthesis of CuO nanoparticles using Gloriosa superba L. extract and their antibacterial activity. *Journal of Taibah University for Science*, *9*(1), 7–12.

Narayanan, K. B., & Sakthivel, N. (2010). Biological synthesis of metal nanoparticles by microbes. *Advances in Colloid and Interface Science*, *156*(1–2), 1–13.

Naresh, V., & Lee, N. (2021). A review on biosensors and recent development of nanostructured materials-enabled biosensors. *Sensors*, *21*(4), 1109.

Ochieng, P. E., Iwuoha, E., Michira, I., Masikini, M., Ondiek, J., Githira, P., & Kamau, G. N. (2015). Green route synthesis and characterization of ZnO nanoparticles using Spathodea campanulata. *International Journal of Biochemistry & Physiology*, *23*, 53–61.

Otari, S. V., Patil, R. M., Nadaf, N. H., Ghosh, S. J., & Pawar, S. H. (2012). Green biosynthesis of silver nanoparticles from an actinobacteria Rhodococcus sp. *Materials Letters*, *72*, 92–94.

Ottoni, C. A., Neto, M. L., Léo, P., Ortolan, B. D., Barbieri, E., & De Souza, A. O. (2020). Environmental impact of biogenic silver nanoparticles in soil and aquatic organisms. *Chemosphere*, *239*, 124698.

Pantidos, N., & Horsfall, L. E. (2014). Biological synthesis of metallic nanoparticles by bacteria, fungi and plants. *Journal of Nanomedicine & Nanotechnology*, *5*(5), 1.

Papaspyrides, C. D., Pavlidou, S. A., & Vouyiouka, S. N. (2009). Development of advanced textile materials: natural fibre composites, anti-microbial, and flame-retardant fabrics. *Proceedings of the Institution of Mechanical Engineers, Part L: Journal of Materials: Design and Applications*, *223*(2), 91–102.

Paramo, L. A., Feregrino-Pérez, A. A., Guevara, R., Mendoza, S., & Esquivel, K. (2020). Nanoparticles in agro-industry: applications, toxicity, challenges, and trends. *Nanomaterials*, *10*(9), 1654.

Patel, V., Berthold, D., Puranik, P., & Gantar, M. (2015). Screening of cyanobacteria and microalgae for their ability to synthesize silver nanoparticles with antibacterial activity. *Biotechnology Reports*, *5*, 112–119.

Patil, S., Ansari, R. A., Gurikar, C., Bhavan, S., & Hariprasad, K. (2024). Propagation of mycorrhizal fungi and their strength in plant health ameliorations. In *Mycorrhizal Symbiosis and Agroecosystem Restoration* (pp. 77–98). Singapore: Springer Nature Singapore.

Peng, Z., Liu, X., Zhang, W., Zeng, Z., Liu, Z., Zhang, C., ... & Yuan, X. (2020). Advances in the application, toxicity and degradation of carbon nanomaterials in environment: a review. *Environment International*, *134*, 105298.

Prasad, B. N., Subramanyam, B., Lakshmipathi, R. N., Ansari, R. A., Rizvi, R., Sumbul, A., ... & Rachmi, C. M. (2020). Utilization of beneficial microorganisms in sustainable control of phytonematodes. In: Ansari, R.A., Rizvi, R., and Mahmood, I. (eds.) *Management of Phytonematodes: Recent Advances and Future Challenges* (pp. 317–337) Springer Nature Singapore, Singapore.

Prasad, R., Pandey, R., & Barman, I. (2016). Engineering tailored nanoparticles with microbes: quo vadis. *WIREs Nanomedicine Nanobiotechnology, 8*, 316–330.

Qasim, S., Zafar, A., Saif, M. S., Ali, Z., Nazar, M., Waqas, M., ... & Hasan, M. (2020). Green synthesis of iron oxide nanorods using Withania coagulans extract improved photocatalytic degradation and antimicrobial activity. *Journal of Photochemistry and Photobiology B: Biology*, *204*, 111784.

Qian, H., Peng, X., Han, X., Ren, J., Sun, L., & Fu, Z. (2013). Comparison of the toxicity of silver nanoparticles and silver ions on the growth of terrestrial plant model Arabidopsis thaliana. *Journal of Environmental Sciences*, *25*(9), 1947–1956.

Rad, M., Taran, M., & Alavi, M. (2018). Effect of incubation time, CuSO4 and glucose concentrations on biosynthesis of copper oxide (CuO) nanoparticles with rectangular shape and antibacterial activity: Taguchi method approach. *Nano Biomed Eng*, *10*(1), 25–33.

Rafique, M., Sadaf, I., Rafique, M. S., & Tahir, M. B. (2017). A review on green synthesis of silver nanoparticles and their applications. *Artificial Cells, Nanomedicine, and Biotechnology*, *45*(7), 1272–1291.

Rahmatpour, S., Shirvani, M., Mosaddeghi, M. R., Nourbakhsh, F., & Bazarganipour, M. (2017). Dose–response effects of silver nanoparticles and silver nitrate on microbial and enzyme activities in calcareous soils. *Geoderma*, *285*, 313–322.

Rajeshkumar, S., Malarkodi, C., Paulkumar, K., Vanaja, M., Gnanajobitha, G., & Annadurai, G. (2013). Intracellular and extracellular biosynthesis of silver nanoparticles by using marine bacteria Vibrio alginolyticus. *Nanoscience Nanotechnology*, *3*(1), 21–25.

Rajeshkumar, S., Malarkodi, C., Vanaja, M., & Annadurai, G. (2016). Anticancer and enhanced antimicrobial activity of biosynthesizd silver nanoparticles against clinical pathogens. *Journal of Molecular Structure*, *1116*, 165–173.

Rajeshwari, A., Suresh, S., Chandrasekaran, N., & Mukherjee, A. (2016). Toxicity evaluation of gold nanoparticles using an Allium cepa bioassay. *RSC Advances*, *6*(29), 24000–24009.

Rajput, V. D., Minkina, T., Sushkova, S., Tsitsuashvili, V., Mandzhieva, S., Gorovtsov, A., ... & Gromakova, N. (2018). Effect of nanoparticles on crops and soil microbial communities. *Journal of Soils and Sediments*, *18*(6), 2179–2187.

Rajput, V., Minkina, T., Sushkova, S., Behal, A., Maksimov, A., Blicharska, E., ... & Barsova, N. (2020). ZnO and CuO nanoparticles: a threat to soil organisms, plants, and human health. *Environmental Geochemistry and Health*, *42*(1), 147–158.

Raliya, R., Franke, C., Chavalmane, S., Nair, R., Reed, N., & Biswas, P. (2016). Quantitative understanding of nanoparticle uptake in watermelon plants. *Frontiers in Plant Science*, *7*, 1288.

Ranjan, S., Dasgupta, N., & Lichtfouse, E. (Eds.). (2016). *Nanoscience in Food and Agriculture* (Vol. 20). Cham: Springer International Publishing.

Rashid, M. I., Shahzad, T., Shahid, M., Imran, M., Dhavamani, J., Ismail, I. M., ... & Almeelbi, T. (2017a). Toxicity of iron oxide nanoparticles to grass litter decomposition in a sandy soil. *Scientific Reports*, *7*(1), 1–11.

Rashid, M. I., Shahzad, T., Shahid, M., Ismail, I. M., Shah, G. M., & Almeelbi, T. (2017b). Zinc oxide nanoparticles affect carbon and nitrogen mineralization of Phoenix dactylifera leaf litter in a sandy soil. *Journal of Hazardous Materials*, *324*, 298–305.

Riddin, T. L., Govender, Y., Gericke, M., & Whiteley, C. G. (2009). Two different hydrogenase enzymes from sulphate-reducing bacteria are responsible for the bioreductive mechanism of platinum into nanoparticles. *Enzyme and Microbial Technology*, *45*(4), 267–273.

Rossi, L., Sharifan, H., Zhang, W., Schwab, A. P., & Ma, X. (2018). Mutual effects and in planta accumulation of co-existing cerium oxide nanoparticles and cadmium in hydroponically grown soybean (Glycine max (L.) Merr.). *Environmental Science: Nano*, *5*(1), 150–157.

Saini, A., Sharma, S., & Chhibber, S. (2008). Protective efficacy of Emblica officinalis against Klebsiella pneumoniae induced pneumonia in mice. *Indian Journal of Medical Research*, *128*(2), 188.

Sangappa, M., & Thiagarajan, P. (2012). Mycobiosynthesis and characterization of silver nanoparticles from Aspergillus niger: a soil fungal isolate. *International Journal of Life Sciences Biotechnology and Pharma Research*, *1*(2), 282–289.

Saravanan, A., Kumar, P. S., Karishma, S., Vo, D. V. N., Jeevanantham, S., Yaashikaa, P. R., & George, C. S. (2021). A review on biosynthesis of metal nanoparticles and its environmental applications. *Chemosphere*, 264, 128580.

Scognamiglio, V. (2013). Nanotechnology in glucose monitoring: Advances and challenges in the last 10 years. *Biosensors and Bioelectronics*, *47*, 12–25.

Selvarajan, E., & Mohanasrinivasan, V. (2013). Biosynthesis and characterization of ZnO nanoparticles using Lactobacillus plantarum VITES07. *Materials Letters*, *112*, 180–182.

Shankar, S. S., Ahmad, A., Pasricha, R., & Sastry, M. (2003). Bioreduction of chloroaurate ions by geranium leaves and its endophytic fungus yields gold nanoparticles of different shapes. *Journal of Materials Chemistry*, *13*(7), 1822–1826.

Shervani, Z., Ikushima, Y., Sato, M., Kawanami, H., Hakuta, Y., Yokoyama, T., ... & Aramaki, K. (2008). Morphology and size-controlled synthesis of silver nanoparticles in aqueous surfactant polymer solutions. *Colloid and Polymer Science*, *286*(4), 403–410.

Shukla, P.K., Misra, P., Kole, C. (2016). Uptake, translocation, accumulation, transformation, and generational transmission of nanoparticles in plants. In: Kole, C., Kumar, D., Khodakovskaya, M. (eds) Plant Nanotechnology. Springer, Cham. pp. 183–218.

Siddiqi, K. S., & Husen, A. (2016a). Green synthesis, characterization and uses of palladium/platinum nanoparticles. *Nanoscale Research Letters*, *11*(1), 1–13.

Siddiqi, K. S., & Husen, A. (2016b). Fabrication of metal and metal oxide nanoparticles by algae and their toxic effects. *Nanoscale Research Letters*, *11*(1), 1–11.

Siddiqi, K. S., Husen, A., & Rao, R. A. (2018). A review on biosynthesis of silver nanoparticles and their biocidal properties. *Journal of Nanobiotechnology*, *16*(1), 1–28.

Simonin, M., & Richaume, A. (2015). Impact of engineered nanoparticles on the activity, abundance, and diversity of soil microbial communities: a review. *Environmental Science and Pollution Research*, *22*(18), 13710–13723.

Singaravelu, G., Arockiamary, J. S., Kumar, V. G., & Govindaraju, K. (2007). A novel extracellular synthesis of monodisperse gold nanoparticles using marine alga, Sargassum wightii Greville. *Colloids and Surfaces B: Biointerfaces*, *57*(1), 97–101.

Singh, A., Dar, M. Y., Joshi, B., et al. (2018). Phytofabrication of silver nanoparticles: novel drug to overcome hepatocellular ailments. *Toxicol Rep.*, *5*, 333–342.

Singh, A. K., Das, A., & Kumar, P. (2021). Nanobiosensors and their applications. In *Nanotechnology* (pp. 249–288). Jenny Stanford Publishing.

Singh, J., Dutta, T., Kim, K. H., Rawat, M., Samddar, P., & Kumar, P. (2018). 'Green'synthesis of metals and their oxide nanoparticles: applications for environmental remediation. *Journal of Nanobiotechnology*, *16*(1), 1–24.

Singh, T., Jyoti, K., Patnaik, A., Singh, A., Chauhan, R., & Chandel, S. S. (2017). Biosynthesis, characterization and antibacterial activity of silver nanoparticles using an endophytic fungal supernatant of Raphanus sativus. *Journal of Genetic Engineering and Biotechnology*, *15*(1), 31–39.

Sinha, S. N., Paul, D., Halder, N., Sengupta, D., & Patra, S. K. (2015). Green synthesis of silver nanoparticles using fresh water green alga Pithophora oedogonia (Mont.) Wittrock and evaluation of their antibacterial activity. *Applied Nanoscience*, *5*(6), 703–709.

Sintubin, L., De Windt, W., Dick, J., Mast, J., Van Der Ha, D., Verstraete, W., & Boon, N. (2009). Lactic acid bacteria as reducing and capping agent for the fast and efficient production of silver nanoparticles. *Applied Microbiology and Biotechnology*, *84*(4), 741–749.

Solanki, M. K., Kashyap, P. L., Ansari, R. A., & Kumari, B. (2020). *Microbiomes and Plant Health: Panoply and Their Applications*, Academic Press, Elsevier, The Netherland. ISBN: 9780128197158. www.elsevier.com/books/microbiomes-and-plant-health/solanki/978-0-12-819715-8.

Solanki, M. K., Kashyap, P. L., Kumari, B., Ansari, R. A., Sumbul, A., Rizvi, R., & Mahmood, I. (2021). Mycorrhizal fungi and its importance in plant health amelioration. In: Solanki, M.K., Kashyap, P.L., Ansari, R.A. & KumariB. *Microbiomes and Plant Health* (pp. 205–223). Academic Press, Elsevier.

Sosan, A., Svistunenko, D., Straltsova, D., Tsiurkina, K., Smolich, I., Lawson, T., ... & Demidchik, V. (2016). Engineered silver nanoparticles are sensed at the plasma membrane and dramatically modify the physiology of Arabidopsis thaliana plants. *The Plant Journal*, *85*(2), 245–257.

Srinath, B. S., & Rai, V. R. (2015). Biosynthesis of highly monodispersed, spherical gold nanoparticles of size 4–10 nm from spent cultures of Klebsiella pneumoniae. *3 Biotech*, *5*(5), 671–676.

Srinath, B. S., Namratha, K., & Byrappa, K. (2017). Eco-friendly synthesis of gold nanoparticles by gold mine bacteria Brevibacillus formosus and their antibacterial and biocompatible studies. *IOSR Journal of Pharmacy*, *7*(8), 53–60.

Sudha, A., Jeyakanthan, J., & Srinivasan, P. (2017). Green synthesis of silver nanoparticles using Lippia nodiflora aerial extract and evaluation of their antioxidant, antibacterial and cytotoxic effects. *Resource-Efficient Technologies*, *3*(4), 506–515.

Suganya, K. U., Govindaraju, K., Kumar, V. G., Dhas, T. S., Karthick, V., Singaravelu, G., & Elanchezhiyan, M. (2015). Blue green alga mediated synthesis of gold nanoparticles and its antibacterial efficacy against Gram positive organisms. *Materials Science and Engineering: C*, *47*, 351–356.

Sumbul, A., Ansari, R. A., Rizvi, R., & Mahmood, I. (2020). Azotobacter: a potential bio-fertilizer for soil and plant health management. *Saudi Journal of Biological Sciences*, *27*(12), 3634–3640.

Sun, J. Z., Peter Kingori, G., Si, R. W., Zhai, D. D., Liao, Z. H., Sun, D. Z., ... & Yong, Y. C. (2015). Microbial fuel cell-based biosensors for environmental monitoring: a review. *Water Science and Technology*, *71*(6), 801–809.

Suresh, R., Vijayaraj, A., Giribabu, K., Manigandan, R., Prabu, R., Stephen, A., ... & Narayanan, V. (2013). Fabrication of iron oxide nanoparticles: magnetic and electrochemical sensing property. *Journal of Materials Science: Materials in Electronics*, *24*, 1256–1263.

Sutradhar, P., Saha, M., & Maiti, D. (2014). Microwave synthesis of copper oxide nanoparticles using tea leaf and coffee powder extracts and its antibacterial activity. *Journal of Nanostructure in Chemistry*, *4*(1), 86.

Tarafdar, J. C. (2015). Nanoparticle production, characterization and its application to horticultural crops. ICAR: Jodhpur, Rajasthan, India, pp. 222–229.

Tarafdar, J. C., Xiang, Y., Wang, W. N., Dong, Q., & Biswas, P. (2012). Nanoparticle synthesis characterization and application to solve some chronic agricultural problems. *Applied Biology Research*, *14*, 138–144.

Thakkar, K. N., Mhatre, S. S., & Parikh, R. Y. (2010). Biological synthesis of metallic nanoparticles. *Nanomedicine: Nanotechnology, Biology and Medicine*, *6*(2), 257–262.

Thema, F. T., Manikandan, E., Gurib-Fakim, A., & Maaza, M. (2016). Single phase Bunsenite NiO nanoparticles green synthesis by Agathosma betulina natural extract. *Journal of Alloys and Compounds*, *657*, 655–661.

Tomer, A. K., Rahi, T., Neelam, D. K., & Dadheech, P. K. (2019). Cyanobacterial extract-mediated synthesis of silver nanoparticles and their application in ammonia sensing. *International Microbiology*, *22*(1), 49–58.

Tong, Z. H., Bischoff, M., Nies, L. F., Carroll, N. J., Applegate, B., & Turco, R. F. (2016). Influence of fullerene (C_{60}) on soil bacterial communities: aqueous aggregate size and solvent co-introduction effects. *Scientific Reports*, *6*(1), 1–9.

Tripathy, S., Rodrigues, J., & Shimpi, N. G. (2023). Top-down and Bottom-up Approaches for Synthesis of Nanoparticles. *Nanobiomaterials: Perspectives for Medical Applications in the Diagnosis and Treatment of Diseases*, *145*, 92–130.

Tsekhmistrenko, S. I., Bityutskyy, V. S., Tsekhmistrenko, O. S., Horalskyi, L. P., Tymoshok, N. O., & Spivak, M. Y. (2020). Bacterial synthesis of nanoparticles: a green approach. *Biosystems Diversity*, *28*(1), 9–17.

Uddin, S., Safdar, L. B., Iqbal, J., Yaseen, T., Laila, S., Anwar, S., ... & Quraishi, U. M. (2021). Green synthesis of nickel oxide nanoparticles using leaf extract of Berberis balochistanica: Characterization, and diverse biological applications. *Microscopy Research and Technique*, *84*(9), 2004–2016.

Usman, M., Farooq, M., Wakeel, A., Nawaz, A., Cheema, S. A., ur Rehman, H., ... & Sanaullah, M. (2020). Nanotechnology in agriculture: current status, challenges and future opportunities. *Science of the Total Environment*, *721*, 137778.

Wei, H., & Wang, E. (2013). Nanomaterials with enzyme-like characteristics (nanozymes): next-generation artificial enzymes. *Chemical Society Reviews*, *42*(14), 6060–6093.

Wu, F., You, Y., Werner, D., Jiao, S., Hu, J., Zhang, X., ... & Wang, X. (2020). Carbon nanomaterials affect carbon cycle-related functions of the soil microbial community and the coupling of nutrient cycles. *Journal of Hazardous Materials*, *390*, 122144.

Yin, M., Wu, C. K., Lou, Y., Burda, C., Koberstein, J. T., Zhu, Y., & O'Brien, S. (2005). Copper oxide nanocrystals. *Journal of the American Chemical Society*, *127*(26), 9506–9511.

Ying, S., Guan, Z., Ofoegbu, P. C., Clubb, P., Rico, C., He, F., & Hong, J. (2022). Green synthesis of nanoparticles: current developments and limitations. *Environmental Technology & Innovation*, *26*, 102336.

Zhang, H., Zhou, H., Bai, J., Li, Y., Yang, J., Ma, Q., & Qu, Y. (2019). Biosynthesis of selenium nanoparticles mediated by fungus Mariannaea sp. HJ and their characterization. *Colloids and Surfaces A: Physicochemical and Engineering Aspects*, *571*, 9–16.

Zhang, W. (2018). Global pesticide use: profile, trend, cost/benefit and more. *Proceedings of the International Academy of Ecology and Environmental Sciences*, *8*(1), 1.

Zhao, S., Su, X., Wang, Y., Yang, X., Bi, M., He, Q., & Chen, Y. (2020). Copper oxide nanoparticles inhibited denitrifying enzymes and electron transport system activities to influence soil denitrification and N_2O emission. *Chemosphere*, *245*, 125394.

10 Nanoengineering for Sustainable Crop Health

Leveraging the Potential of Nanotechnology in Enhancing Disease Resistance and Yield Improvement

Manjunath Hubballi, Rizwan Ali Ansari, Rose Rizvi, Noor Fatima, Noah Totsline, Archana A. M., Charanpreet Kaur, and Archana T. S.

10.1 INTRODUCTION

Despite the fact that food production has risen multiple times in the recent past, it is surprising to know that still, half a billion people across the globe experience a shortage of nutritious food (Pinstrup-Andersen 1999). Malnutrition has been attributed to be the major cause of poor health in the population living in many parts of the world (Swinburn et al., 2019). Thus, the primary goal of the global food production system of the twenty-first century is to eradicate hunger and malnutrition. To meet the need for food and energy by 2050, agricultural production must increase by at least 60–100% from current levels in order to achieve the same (Sijpestijn et al., 2022). Global arable land is continuously shrinking with the least chance for horizontal expansion (Schneider et al., 2022). In addition, the crop production system is much more hindered by the increased scarcity of natural resources across the globe because they regenerate more quickly than they consume, especially when it comes to water (di Santo et al., 2022). Thus, achieving the goals of production is hindered by an arena of factors. Apart from the abiotic factors that come in the way of production, there are a huge number of microbes and insects that also contribute to hindering production (Ansari and Mahmood 2017a; Ansari and Mahmood 2019a; Ansari and Mahmood 2019b; Ansari et al., 2020; Ansari, Rizvi, et al., 2020; Ansari et al., 2019). It is estimated that nearly 70% of the total yield loss that occurs globally has been attributed to pest attacks, among which disease itself contributes 20–25% (Thakur et al., 2023, Tiwari et al., 2021). The losses due to diseases in different crops vary significantly with variety, location, climate and management practices being adopted (Mahmood et al., 2019; Ansari and Khan 2012a; Ansari and Khan 2012b; de Freitas Bueno et al., 2019). It is reported that the mean losses in food crops such as wheat, rice, maize, potato and soybean are: 21.50%, 30.30%, 22.60%, 17.20% and 21.40%, respectively. Apart from these, there is a considerable amount of loss which occurs at post-harvest stage (Tripathi et al., 2022). The exhaustive list of diseases and their reported yield losses have been listed (Table 10.1).

Nanoparticles can be engineered to encapsulate, adsorb, or chemically attach bioactive compounds like fungicides, antimicrobials, and other agrochemicals. Their small size allows them to penetrate

DOI: 10.1201/9781003256762-10

TABLE 10.1
Yield losses due to diseases in important crops

S. No	Crop	Disease	Causal organism	Percent loss	Reference
1.	Wheat	Black rust	*Puccinia graminis tritici*	70%	Rehman et al. (2022)
		Brown/leaf rust	*P. triticina*	>50%	Afzal et al. (2023)
		Yellow rust/Stripe Rust	*P. striiformis*	80–100%	Kumar et al. (2022)
2.	Rice	Blast of rice	*Magnaporthe oryzae*	30 to 61	Rambabu et al. (2022)
		Brown spot of rice	*Bipolaris oryzae*	6–90%	Priyadashani, et al. (2022)
		Bacterial blight	*Xanthomonas oryzae* pv. *oryzae*	80–90%	Kim et al. (2019)
3.	Maize	Southern corn blight of maize	*Cochliobolus heterostrophus*	>40%	Debnath and Biswas (2020)
4.	Potato	Late blight	*Phytophthora infestans*	>90%	Duarte et al. (2019)
		Early blight	*Alternaria solani*	80–100%	Singh et al. (2020)
5.	Tomato	Late blight	*Phytophthora infestans*	80%	Soresa (2022)
		Early Blight	*Alternaria solani*	70–90%	Prabhakar et al. (2020)
		Bacterial wilt	*Ralstonia solanacearum*	100%	Soresa (2022)
6.	Chilli	Anthracnose	*Colletotrichum capsici*	80–90%	Jatin et al. (2022)
7.	Onion	Purple blotch	*Alternaria porri.*	50–100%	Das et al. (2022)
8.	Sugarcane	Red rot	*Colletotrichum falcatum*	>70%	Hossain et al. (2020)
9.	Sunflower	Head rot	*Sclerotium rolfsii*	20–100%	Hahn (2002)
10.	Mustard	White rust	*Albugo candida*	23–54.5%	Bal and Kumar (2014)
11.	Soybean	Rust	*Phakopsora pachyrhizi,*	>90%	Godoy et al. (2016)
12.	Coffee	Rust	*Hemileia vastatrix*	>75%	Taju and Yesuf (2022)

plant tissues and cells (Panpatte et al., 2016). This targeted delivery helps concentrate the active ingredients where they are needed most, improving efficacy compared to conventional application methods (Yetisgin et al., 2020). Nanoparticle delivery systems can also prolong the release of active ingredients, providing longer-lasting protection. The precision delivery and slower release minimize the amount of chemicals that end up in the wider environment. This reduces impacts on non-target organisms and prevents groundwater contamination (Côa et al., 2020). The composition of various nanoparticles can be tailored for optimal delivery of each specific bioactive compound (Đorđević et al., 2015). Significant research is still needed to fully understand nanoparticle fate, transport, and ecotoxicity in agricultural systems. But overall, nanotechnology shows great potential for increasing the efficiency of agrochemical delivery while reducing environmental impacts. This chapter endeavours to elucidate the transformative role of nanoengineering in bolstering crop health sustainability, with a particular focus on augmenting disease resistance and optimizing yield enhancement. By delving into cutting-edge advancements in nanotechnology, it aims to unravel the intricate mechanisms through which nanoparticles exert their prowess in combating plant pathogens and fostering agricultural productivity. Additionally, it endeavours to dissect the inherent challenges and promising prospects inherent in the integration of nanotechnology within the agricultural domain, thereby charting a roadmap for innovative research endeavours and pragmatic applications in the field.

10.2 IMPORTANT PLANT DISEASES AND THEIR EFFECT ON FOOD AND NUTRITIONAL SECURITY

The current definition of one health includes plant health; however, the plant health component has not been properly discussed in one health approaches despite its vital importance in survival of life forms on the earth (Rizzo et al., 2021; Ibrahim et al., 2020; Devi et al., 2020; Ibrahim et al., 2022;

Keshari et al., 2024). The importance of plant health for public welfare prompted the United Nations to designate 2020 as the International Year of Plant Health (Fletcher et al., 2009). Since plants are the source of 80% of the food consumed maintaining the good health of the plant is important to achieve food security (FAO 2020). Food security is a fundamental aspect of the one health approach and serves as a cornerstone of the United Nations' Sustainable Development Goals (UNGA, 2019) In a broader sense, food security is the equilibrium between the amount of food produced by agriculture and the growing population's need for food (Smil, 2000). The imbalance in demand and output is not new, but of late it has increased. Food security is attained when all individuals within a population have equitable access to a safe, nutritious, and adequate supply of food to meet their dietary requirements for optimal health consistently (FAO 2019). Plant diseases significantly affect the four essential elements of food security: accessibility, availability, utilization, and stability. As a result, they pose a considerable threat to global food security (Serge et al., 2019). A few examples of the plant diseases affecting components of food security are given below with their possible impacts.

10.3 DISEASE MITIGATING STRATEGIES

The plant disease management aims in mitigating the yield losses caused by various plant diseases while maintaining and improving plant health for better productivity (Prasad et al., 2020; Solanki et al., 2020; Ansari et al., 2024; Sumbul et al., 2020). Disease management has traditionally relied heavily on the use of toxic chemicals. Undoubtedly, fungicides have contributed enormously to the effective management of many economically important diseases and even today there have been fungicides which are an integral part of production systems in agriculture (Chiu et al., 2022). However, the chemical fungicides resulted in documentation of their adverse impacts on environment in terms of toxicity in soil, residue in food crops and impact on growth and development of non-target organisms in the ecosystem (Rani et al., 2021; Ansari and Mahmood 2017b; Fabiyi et al., 2020; Patil et al., 2024; Solanki et al., 2021). Further, many strains of plant pathogens are reported to have resistance against a wide array of fungicides (El-Saadony et al., 2022). The increased consciousness on the ill effects of these fungicides has forced the development of an alternative arsenal for these. The other potential alternative methods of managing the diseases are the use of biological agents, either endophytes or ectophytes, or the soil borne beneficial microbes. These microbes exert a range of modes of action through which they can contribute to disease management effectively (Chaloner et al., 2021; Ansari et al., 2017; Ansari et al., 2023; Boyno et al., 2024; Fabiyi et al., 2024). Further, these agents are eco-friendly and nontoxic, having no residue in food and hence, are regarded as the better alternatives to fungicides (El-Saadony et al., 2022). The species of *Pseudomonas, Bacillus and Trichoderma* are the most widely used biological agents in disease management (Palmieri et al., 2022). Despite these inherent advantages with bioagents, there efficacy can't be equated with the chemical fungicides owing to their specificity, survivability under different niches, and the wider availability. Further, the stringent protocols associated with these agents in many countries have restricted their usage in agriculture.

Exploiting the inherited resistance in crops is one of the most important aspects of managing disease. It is one of the safest and cheapest means of managing diseases in many crop plants. Plants with pathogen resistance can reduce disease impact on crop development and yield, thus contributing to eliminate the world's hunger crisis. As long as there are no other diseases present in addition to the disease-resistant plants, additional efforts to reduce disease losses are not necessary. Furthermore, planting a disease-resistant variety has the additional advantage of reducing the probability of disease invasion (Vyska et al., 2016). The most traditional means of breeding for resistance involve introduction, selection, mutation, hybridization and somaclonal variation (Anil et al., 2018; Khan et al., 2020). However, recent development in sciences has led to discovery of many genes contributing to resistance against pathogens in wild as well as cultivated crops (Gururani et al., 2012). The developments have made it possible to transfer genes across the genus with quite satisfactory results.

10.4 NANOPARTICLES AND THEIR USES IN PLANT DISEASE MANAGEMENT

A nanoparticle as compared to bulk products, has more functionality, responsiveness, and efficiency (Sekhon, 2014). Currently, different types of nanoparticles are known to exist and they are classified based on several parameters, including size, shape, magnetic properties, crystalline nature, chemical nature, and physicochemical properties (Ahmad et al., 2022). Using only nanoparticles, plant seeds, leaves, or roots can be directly protected from insects, bacteria, fungi, viruses, and other pests and pathogens. Further, they can also be used as carrier material along with antifungal, antibacterial molecules to be used as pesticides in order to increase the uptake, shelf life and stability of pesticides. Thus, nanoparticles play a dual role in plant protection. Various researchers across the globe have reported the antibacterial, antifungal, and antiviral properties of metal nanoparticles, including titanium, zinc oxide, copper, and silver in the recent past (Ingle, 2021). Several morphological and physiological alterations in plants occur due to myriad NPs. The type, chemical composition, reactivity and most importantly, the effective nanoparticle dosage are vital in exploiting these particles for plant protection.

10.4.1 Synthesis

Various methodologies exist for the fabrication of nanoparticles, encompassing physical, chemical, and biological modalities. These methodologies can be broadly categorized into two main approaches: top-down and bottom-up methodologies. Diverse techniques such as electro-explosion, sputtering, etching, mechanical milling, and laser ablation are employed to effect size reduction at the atomic scale in a top-down fashion (Miu and Dinischiotu, 2022). The basic principle incorporates the use of biological organisms and plants for synthesizing the green nanoparticles.

Physical synthesis processes include ball milling, sputtering, laser ablation, evaporation-condensation, and plasma-based procedures. The majority of these methods have very slow rates of producing metal nanoparticles (Surender and Ajay, 2021). A ball milling technique, for example, results in a 50% or lower output of nanoparticles (Taghipour et al., 2022). Only 6–8% of the particles produced by sputtering are smaller than 100 nm, resulting in a wide particle size distribution (Seetharaman et al., 2018). Laser ablation offers a significant edge over alternative techniques for generating metal colloids due to its chemical reagent-free nature. This method facilitates the production of unadulterated and pristine metal colloids (Dolgaev et al., 2002). In employing this method, silver nanospheroids ranging in diameter from 20 to 50 nm were synthesized through laser ablation in water, utilizing femtosecond laser pulses with a wavelength of 800 nm (Hashim, 2012). However, substantial energy consumption is needed for plasma and laser ablation procedures. Physical processes are often deemed unsuitable for commercial applications due to their inherent drawbacks, including high costs, wide size distribution, slow production rates, generation of waste by-products, and excessive energy consumption (Rajput et al., 2023).

Numerous chemical methodologies have been developed for nanoparticle synthesis, and many of these techniques are commonly utilized in the production of nanostructured materials. Examples include pyrolysis, sol-gel processes, chemical reduction, polyol synthesis, chemical vapor deposition, microemulsion techniques, and hydrothermal synthesis (Hachem et al., 2022). In the chemical-mediated synthesis of nanoparticles, various reducing agents are frequently employed to reduce metal ions in both aqueous and non-aqueous solutions. These include the tollens reagent, poly (ethylene glycol)-block copolymers, elemental hydrogen, sodium citrate, ascorbate, sodium borohydride, the polyol method, and nitrogen. These reducing agents catalyse the breakdown of Ag^+ ions into metallic silver, facilitating the aggregation of these atoms into oligomeric clusters. Eventually, metallic silver particles in a colloidal form these clusters (Merga et al., 2007). Furthermore, using these dangerous reagents and chemicals throughout the synthesis process and the generation of byproducts pose significant risks to both the environment and human health. These specific NPs can therefore only be used in biological contexts. A developing field in green nanotechnology is innovative NP creation based on green chemistry. There is a novel approach to producing NPs with natural reducing and stabilizing agents through the use of biological synthesis techniques. Without the need for excessive energy or harmful chemicals, it offers a cost-effective and environmentally benign alternative to physiochemical synthesis (Miu and Dinischiotu, 2022). The biosynthesis of NPs involves biological units ranging from basic unicellular to intricate multicellular such as bacteria, actinomycetes, yeasts, algae, fungi and plants (Moosa et al., 2015). Microbial cultures' extracellular and intracellular filtrates serve as reducing agents in nanoparticle synthesis. Bacteria, fungi, yeast, and actinomycetes, typically tolerant to metals, thrive in harsh environments. Microbes utilize these inherent properties to accept, gather, and transform metals into metal ions. Common redox processes involving intracellular and extracellular routes convert metals into metal ions.

NADH and NADH-dependent nitrate reductase enzymes act as electron donors in the biosynthesis of silver nanoparticles by *Bacillus licheniformis* (Kalishwaralal et al., 2008). Silver nanoparticles were produced by *Fusarium oxysporum* by utilizing anthraquinones and nitrate reductase (Singh et al., 2022). NADPH nitrate reductase seems to be an important component in the production of metallic nanoparticles, based on the bacterial and fungal synthesis of metallic nanoparticles as described previously.

10.4.2 Characterization

In order to confirm the effectiveness of the nanoparticles after their synthesis, characterization is necessary before application. Characterization necessitates the use of a dependable and accurate set of protocols. A variety of techniques are utilized to characterize the size, shape, atomic composition, crystal structure, surface charge, and other physical, chemical, and biological properties of synthesized nanomaterials. Advanced microscopy methods, including transmission electron, scanning electron and atomic force microscopy, and scanning transmission X-ray microscopy, are employed to accurately determine the size and morphology of nanoparticles. Moreover, X-ray fluorescence microscopy is employed to analyse the atomic composition of the nanomaterials (Dutta et al., 2022). UV visible spectroscopy provides information on the optical characteristics of nanoparticles by detecting the nanoparticles' reflected light. Additionally, it assesses whether these nanoparticles are stable in the suspension of colloids or not.

Dynamic light scattering is a method employed to observe the Brownian motion of nanoparticles within a colloidal system. The surface charge of nanoparticles with bioactive substances determines their electrostatic interaction, and the surface charge can be measured using the zeta potential value. X-Ray Diffraction (XRD) is another vital method for characterizing nanoparticles. This technique unveils the crystalline structures of the nanoparticles with precision (Mourdikoudis et al., 2018).

10.5 VARIOUS TYPES OF NANOPARTICLES IN CROP PROTECTION

In today's world, the usage of nanoparticles propels traditional agricultural practices forward by reducing hazardous outputs of agriculture and elevating the management of diseases in crop plants. A myriad of nanoparticles is being utilized in the suppression of a diverse range of plant pathogens. This comprises carbon nanoparticles, non-metal nanoparticles, chitosan nanoparticles, quantum dots, liposomes and dendrimers in addition to metals and metal oxide nanoparticles (Elmer and White, 2018; Bayoumy et al., 2020). Among these nanoparticles, the majority do not yet have applications in plant pathology, although numerous studies have shown that using these nanoparticles can suppress the pathogens in the foliar, root, stem, and fruit.

10.5.1 Silver Nanoparticles and Their Effectiveness in Disease Suppression

Silver has the potential to disinfect about 650 microbes and is benign to humans, but it also modulates the metabolism function within the germs (Deshmukh et al., 2019). In comparison to synthetic fungicides, it could be utilized to manage various plant diseases in a very effective and safe manner (El-Baky et al., 2021). For controlling plant pathogens, Ag nanoparticles were the first to be incorporated into agriculture. Antimicrobial activity is well-studied, well-characterized, and well-known for this nanoparticle (Jogaiah et al., 2020). Silver nanoparticles have exhibited remarkable efficacy in suppressing pathogens such as *Phytophthora parasitica*, *Meloidogyne spp.*, and *Fusarium spp.* (Elmer and White, 2018; Banerjee et al., 2021). Additionally, silver nanoparticles have demonstrated potent antifungal properties against *Macrophomina phaseolina, Alternaria alternata, Sclerotinia sclerotiorum, Rhizoctonia solani, Curvularia lunata*, and *Botrytis cinerea* in well diffusion assays (Amna et al., 2021). Jagana et al. (2017) showed that green silver nanoparticles suppress the disease progression of the pathogen *Colletotrichum musae* by almost 70% in banana fruit. The defence strategies of plants such as increased lignin deposition in vascular bundles, are enhanced by the application of silver nanoparticles. It serves as a possible tool for managing agricultural diseases. It was reported that in *Arabidopsis thaliana,* 25% disease resistance was achieved by applying a mixture of AgNPs and silica on *Pseudomonas syringae* pv. *tomato* bacteria (Chu et al., 2012).

10.5.2 Copper Nanoparticles

A copper-based fungicide has the potential to induce substantial damage to various biomolecules, including proteins, lipids, and DNA. This damage occurs through the generation of highly reactive hydroxyl radicals, leading to widespread molecular and cellular damage.

A formulated fungicide from Cu nanoparticles can inhibit *Phytophthora infestans* progression in tomato plants (Giannousi et al., 2013). Polyphenol oxidase and other defence-related enzymes were enhanced in plants as a result of CuO nanoparticle application (Sarkar et al., 2020). Ponmurugan et al. (2016) revealed that biologically synthesized *Streptomyces griseus*-mediated CuNPs can reduce the root rot disease of tea. A Bordeaux mixture consisting of CuNPs was reported to prevent the nematode (*Xiphinema index*) in grapes (Elmer and White, 2018). Disease resistance in various crops is improved by using pesticides and fertilizers with nano Cu formulations.

10.5.3 Zinc-Nanoparticles

Zinc nanoparticles have demonstrated notable antimicrobial properties, making them valuable for various applications in plant disease management. The mechanism of action of nano-ZnO, derived from zinc nitrate, against significant pathogenic fungi is a topic of considerable interest and research (Rajeshkumar, 2019). Several studies on ZnNPs have shown that it has a high level of effectiveness against a wide range of plant pathogenic fungi, including *Alternaria alternata*, *Mucor plumbeus*, *Sclerotinia sclerotiorum*, *Rhizoctonia solani*, *Fusarium oxysporum* and *Rhizopus stolonifera* (Rehman et al., 2023). Derbalah et al. (2013) conducted a comprehensive assessment to evaluate the efficacy of foliar sprays containing 500 g/mL ZnO nanoparticles. Their study aimed to compare the performance of these nanoparticles against a range of treatments, including the conventional fungicide tetraconazole, nano silica, diatomaceous earth, and six bacterial biocontrol agents. Consistently two years later in the study, nano-Zn treatment was substantially more effective than tetraconazole in reducing disease severity and increasing dry weight, root yield and sugar content.

10.5.4 Carbon Nanoparticles

Carbon-based nanoparticles have many unique physical attributes and are used in different scientific fields. Several studies have demonstrated that carbon nanoparticles have antimicrobial as well as beneficial impacts on plant growth. The three primary types of carbon nanoparticles utilized in the treatment of disease are graphene oxides, carbon nanotubes, and fullerenes. With graphene oxides at 250 µg/mL, 95% of *Xanthomonas oryzae* pv. *oryzae* bacterial cells are killed as compared to 13.3% with bismerthiazol (Chen et al., 2013). Research findings illustrate that single-walled carbon nanotubes inhibit conidia production in both *Fusarium poae* and *Fusarium graminearum* (Wang et al., 2014).

10.5.5 Non-Metal, Metalloids, Metallic Oxides Nanoparticles

The most studied non-metal nanoparticles are nano Ti (titanium) and nano S (sulfur). Nano Ti is employed as a potential antifungal agent because of its autocatalytic properties (Boxi et al., 2016). Bacterial leaf spot on roses caused by *Xanthomonas* sp. is suppressed by a novel antibacterial light-activated nano-TiO_2/Zn nanoparticle combination. As the mixture also contains disease-suppressive nano-ZnO, the effects of TiO_2 and nano-ZnO cannot be distinguished (Paret et al., 2013). The sulfur nanoparticles are another category of nanoparticles that are effective in defending plants from the fungal diseases caused by *Fusarium solani*, *F. oxysporum*, *Aspergillus niger* and *Venturia inaequalis*. Metalloid Si nanoparticles have some effects on disease management. Powdery mildew was inhibited by using nano Si together with Ag (Park et al., 2006). When Si ions are continuously taken up by the plant, they have the greatest impact on the plant (Datnoff et al., 2007). Wilt disease in tomatoes, egg plant and watermelons could be treated using nano-MnO which is a metallic oxide nanoparticle. At 50 µg/mL, Nano-MgO inhibited the germination of conidia of *F. oxysporum*, *A. alternata*, *M. plumbeus* and *R. stolonifera* more than nano-ZnO (Wani and Shah, 2012).

10.5.6 Liposomes

A liposome is a spherical vesicle containing at least one phospholipid bilayer. Nanosized particles generally have an empty interior that may be coated with antimicrobial materials. Their sizes vary, but most fall within the nano range (Tziveleka et al., 2022). Liposomes have the ability to suppress gram-negative bacteria that were encapsulated with antimicrobial substances such as herbal extracts and chitosan which contain phenolics. Due to their high-water stability, liposomes are mostly employed for crop irrigation treatment to suppress diseases (Matouskova et al., 2016).

10.5.7 Dendrimers

These are tree-like nanoparticles that are mostly used as carriers for chemicals or DNA to induce disease resistance. Despite their use in human medicine, there are limited studies on their use in plant disease management (Xiao et al., 2022). Basically, the function of dendrimers is to enable the transportation of disease-inhibiting chemicals to those tissues which they are unable to reach (Elmer and White, 2018). The use of nanoparticles in plant disease management has lately been gaining momentum owing to its multifaceted uses. The list of most important fungal, bacterial and viral plant diseases controlled by using nanoparticles are given Table 10.2.

10.6 NANOPARTICLE INDUCED RESISTANCE IN PLANTS

Plants have developed a resistance mechanism which mimics the innate immune system of animals wherein it fights against the potential pathogens (El-Shetehy et al., 2021). The more important aspect of plant resistance is that it can be enhanced by use of microbial agents or a specific group of chemicals called resistance inducing chemicals (Ali et al., 2023). The special feature of induced resistance is that it is effective against a range plant pathogen. The induced resistance by beneficial rhizosphere microbial agents has been demonstrated in many agricultural crops against a range of pathogens. In the recent past there has been surge in interest in using nanoparticles for inducing disease resistance in plants against fungi, bacterial and viral disease (Dutta et al., 2022). The

TABLE 10.2
Application of different nanoparticles in the suppression of plant pathogens

S. No.	Target Pathogen	Host	Reference	
		Virus		
1.	AgNPs	*Tomato mosaic virus*	*Solanum lycopersicum*	El-Dougdoug et al. (2018)
2.	AgNPs	*Potato virus Y*	*Solanum lycopersicum*	El-Shazly et al. (2017)
3.	AgNPs	*Tomato Spotted Wilt Virus*	*Chenopodium amaranticolor*	Shafie et al. (2018)
4.	Au NPs	*Barley yellow dwarf virus*	*Hordeum vulgare*	Alkubaisi and Aref (2017)
5.	Carboxymethyl chitosan-Ti NPs	*Bean yellow mosaic virus*	*Vicia faba*	Sofy et al. (2020)
6.	CNTs	*Tomato mosaic virus*	*N. benthamiana*	Adeel et al. (2021)
7.	Curcumin-milk proteins NPs	*Potato virus Y*	*Solanum tubrosum*	Taha et al. (2019)
8.	Fe NPs	*Tobacco mosaic virus*	*N. benthamiana*	Cai et al. (2020)
9.	GO-AgNPs	*Tomato bushy stunt virus*	*Lactuca sativa*	Elazzazy et al. (2017)
10.	NiO NPs	*Cucumber mosaic virus*	*Cucumis sativus*	Derbalah and Elsharkawy, (2019)
11.	SiO_2 NPs	*Tomato yellow leaf curl virus*	*S. lycopersicum*	El-Sawy et al. (2017)
12.	TiO_2 NPs	*Broad bean stain virus*	*V. faba*	Lavanya and Arun (2021)
13.	ZnO NPs	*Cucumber mosaic virus*	*S. melongena*	El-Sawy et al. (2017)
14.	ZnO NPs	*Tobacco mosaic virus*	*N. benthamiana*	Cai et al. (2019)
15.	ZnO NPs	*Tobacco mosaic virus*	*S. lycopersicum*	Abdelkhalek and Al-Askar (2020)
16.		**Fungus**		
17.	AuNPs	*Candida albicans*	*N. benthamiana*	Aljabali et al. (2018)
18.	Calcium carbonate	*R. solani*	cucumber	Hatfaludi et al. (2004)
19.	Lecithin/Chitosan	*F. oxysporum*	Beans	Ilk et al. (2017)
20.	AgNPs	*Bipolaris sorokiniana* and *Magnaporthe Grisea*	*Lolium perenne* L.	Jo et al. (2009)
21.	CuO and Cu_2O NPs	*Phytophthora infestans*	*Solanum lycopersicum*,	Giannousi et al. (2013)
22.	Nano Si-Ag	*Pythium ultimum, Colletotrichum gloeosporioides, Magnaporthe grisea, Botrytis cineria* and *Rhizoctonia solani*	*Cucurbita pepo*	Park et al. (2006)
23.	Silica NPs	*Alternaria sp*	*Solanum lycopersicum*	Derbalah et al. (2018)
24.	TiO_2 NPs	*P. cubensis*	*Cucumis sativus*	Cui et al. (2009)
25.	TiO_2 NPs	*Cercospora beticola*	*Beta vulgaris* L.	Hamza et al. (2016)
26.	AgNPs	*Fusarium oxysporum* f. sp. *ciceri*	*Cicer arietinum*	Kaur et al. (2018)
27.		**Bacteria**		
28.	Cu composites	*Xanthomonas euvesicatoria*	*Solanum lycopersicum*	Fan et al. (2021)
29.	Nano Si-Ag	*Xanthomonas compestris* pv. *vesicatoria* and *Pseudomonas syringae*	*Cucurbita pepo*	Park et al. (2006)
30.	TiO_2 NPs	*P. syringae pv. lachrymans*	*Cucumis sativus*	Cui et al. (2009)

TABLE 10.2 (Continued)
Application of different nanoparticles in the suppression of plant pathogens

S. No.	Target Pathogen	Host	Reference	
31.	TiO_2 NPs	*Xanthomonas hortorum* pv. *pelargonii*, *X. axonopodis* pv. *Poinsettiicola*	bacterial blight of geranium (*Pelargonium hortorum*), Xanthomonas leaf spot of poinsettia (*Euphorbia pulcherrima*)	Norman and Chen (2011)
32.	ZnO NPs	*Pseudomonas aeruginosa*	*Aeromonas hydrophila*	Jayaseelan et al. (2012)
33.	ZnO NPs	*Ralstonia solanacearum*	*Solanum lycopersicum*	Khan et al. (2021)
34.	ZnO NPs	*Escherichia*	in vitro	Kairyte et al. (2013)
35.	ZnO/nanocopper composite	*Xanthomonas citri* subsp. *citri*	Citrus	Young et al. (2018)

nanoparticles induce resistance in plants through generating reactive oxygen species, production of pathogenesis related proteins and also by inducing some specific enzymes such as catalase, peroxidase, polyphenol oxidase, superoxide dismutase (Tan et al., 2018). The nano particles of ceric oxide, zinc oxide, titanium dioxide, cupric oxide, iron oxide, aluminium oxide, silver and nickel oxide are reported to induce bursts of oxidation in different crops (Georgakopoulos et al., 2018). Abdelrhim et al. (2021) reported the silicon dioxide nanoparticles could induce an innate immune response and thereby it could activate the antioxidant system in wheat plants against *Rhizoctonia solani*. The copper nanoparticles are reported to cause nanotoxicity in a few cases. However, their use at low dosages helps to upregulate the defence related genes in watermelons and tomatoes infected with the *Fusarium* (Borgatta et al., 2018; Ma et al., 2020). Du et al. (2022) reported the application of silicon dioxide nanoparticles by foliar spray in rice seedlings decreased the blast incidence by 70% by increasing the salicylic acid content. Further, application of copper nanoparticles (CuONPs (40 nm, 99.00% pure) in tobacco seedlings significantly enhanced the many defence related gene which helped the tobacco plants to reduce infection by soil-borne fungal pathogen *Phytophthora nicotianae* causing black shank disease. Buchman et al. (2019) reported that some Mesoporous Silica Nanoparticles (MSNs) and chitosan-coated mesoporous silica nanoparticles (CTS-MSNs) can suppress fusarium wilt (*Fusarium oxysporum* f. sp. *niveum*) in watermelons (*Citrullus lanatus*). The application of rice straw biochar and fly ash nanoparticles as soil amendments significantly improved defence against bacterial leaf spot on peppers caused by *Xanthomonas campestris* pv. *vesicatoria*. The X-ray diffractometry of both the nanoparticles showed unique composition of nanoforms which favoured plant defence against pathogen invasion (Aftab et al., 2022). The antibacterial activity of nanoparticles has been demonstrated well against many bacterial pathogens belonging to both gram positive and gram negative which infect plants (Tariq et al., 2022). Applying fly ash nanoparticles and rice straw biochar as soil amendments greatly increased the peppers' resistance to bacterial leaf spot brought on by *Xanthomonas campestris* pv. *vesicatoria*. According to Aftab et al. (2022), the X-ray diffractometry analysis of the two nanoparticles revealed a distinct nanoform composition that supported plant defence against pathogen invasion. Reactive oxygen species production in plants has been linked to the bactericidal action of silver nitrate nano formulations.

According to Derbalah and Elsharkawy (2019), after four days of cucumber mosaic virus inoculation, cucumber plants treated with nickel oxide nanoparticles boosted the expression of the peroxidase gene in cucumber, which in turn decreased the incidence of CMV. Likewise, iron-induced production of reactive oxygen species in tobacco inhibits TMV population.

10.7 NANOPARTICLES MEDIATED GENE TRANSFER AND CROP PROTECTION: A MOLECULAR APPROACH OR DEVELOPMENT OF RESISTANT VARIETIES/HYBRIDS THROUGH NANOPARTICLES

Disease-resistant crop varieties/hybrids provide an effective and safe way to protect the global food supply from plant pathogens and reduce dependency on environmentally harmful chemical treatments (Dong & Ronald, 2019). Pathogens quickly adapt to overcome plant resistance and pathogens are unintentionally introduced to new geographic ranges, infecting populations of susceptible plant hosts. Speed and efficiency in the production of disease resistant plants is necessary to keep pace with the spread and evolution of plant pathogens (Dong and Ronald, 2019; van Esse et al., 2020). Plant resistance is genetically complex and the development of stable and broad-spectrum resistance in plants at the genetic level is a challenging and time-consuming task using traditional breeding methods (Nelson et al., 2018; Dong and Ronald, 2019). Genetic editing techniques like CRISPR-Cas9, zinc-finger nucleases, and Transcription Activator-Like Effectors (TALENs) are powerful tools for the insertion of resistance transgenes, preforming resistance-associated gene knockouts, or producing resistance gene overexpression (Cunningham et al., 2018; Zhu et al., 2020). Unlike animals, plants possess a rigid and multi-layered cell wall limiting delivery of exogenous genetic material. Traditional methods of gene delivery must bypass the cell wall barrier using external force and each have unique limitations. Use of ballistic bombardment by gene guns risks damage to tissue and to the genome sequence while have poor efficiency and being highly imprecise (Cunningham et al., 2018). *Agrobacterium*-mediated delivery of genetic material always introduces sequences of foreign DNA and has a limited host range of infectible plants, with several important leguminous and monocot crops being resistant to transformation (Hwang et al., 2017). Only the plant nuclear genome can be modified using *Agrobacterium* and specific organelles cannot be targeted. PolyEthyleneGlycol (PEG)-mediated gene transfer is limited to a few plant cell-types and protoplasts. Viral-delivery vectors have limited space for genetic cargo within the viral capsid and can elicit immunogenic host responses (Cunningham et al., 2018). These techniques all involve lengthy and laborious regeneration of plants from tissue cultures using complex mixtures of plant hormones specific to the host plant species (Cunningham et al., 2018). Some plant species such as cotton are resistant to such tissue culture regeneration techniques (Zhao et al., 2017). As such, novel technologies are needed to expedite resistant crop production and improve their efficiency to keep pace with ever-adapting plant pathogens which threaten food security by reducing yield. Engineered NanoParticles (NPs) represent a promising new platform for delivery of genetic material to plants. NPs are particles of organic or inorganic matter, generally having a diameter of 1 to 100 nm in at least a single dimension (Khan et al., 2019). NPs are imperceptible using light-based microscopes, and electron or atomic-force microscopy is required for visualization (Grobelny et al., 2011). NPs vary greatly in material composition and morphology. They are generally classified according to their composition and unique physiochemical properties. Common categories include biologicals (liposomes, chitosan, peptides), carbon-based, silicas, metalloids, polymerics, and ceramics (Khan et al., 2019). NPs occur widely in nature and can also be synthesized via techniques including self-assembly, lithography, or surface deposition (Hasan, 2015; Khan et al., 2019). They have demonstrated diverse usage in industries including cosmetics, manufacturing, and medical sciences (Hasan, 2015). NP-delivery of genetic material can be mediated by a variety of physical means including formation of membrane pores via electrical fields, light waves, or sound waves (Cunningham et al., 2018). Magneto infection under the force of a magnetic field can also be used to infiltrate cells (Zhao et al., 2017). Passive bioabsorbable mechanisms of delivery include cationic carriers, envelope penetration, or direct uptake by the cell through bio-permeability or natural openings such at stomata (Ansari and Rehman, 2021). NPs have delivered functionally and structurally diverse biomolecular cargo (DNA, mRNA, siRNA, endonucleases, recombinases, peptides) to plants and animal (Cunningham et al., 2018; Mujtaba et al., 2021). The use of NPs in the production of disease-resistant transgenic plants has only recently been explored. Hassan and colleagues employed silicon carbide whisker nanoparticles for the incorporation of a rice chitinase gene into peanut plants (*Arachis hypogaea*) to impart increased resistance against peanut leaf spot, induced by a fungal pathogen. RT-PCR confirmed higher levels of chitinase mRNA transcripts in transgenic peanuts compared to control plants, with transgenic plants displaying greater resistance to the pathogen (Hassan et al., 2016). Potato plants were given CRISPR-Cas9 endonuclease complexes via chitosan NPs, which allowed them to alter the coilin gene and overproduce Cajal protein bodies. According to Makhotenko et al. (2019), transgenic plants that produced more Cajal bodies showed increased resistance to the potato virus Y (Potyviride). Chitosan nanoparticles were utilized to introduce *Arabidopsis thaliana* thionin genes into potato plants (*Solanum tuberosum*), resulting in heightened resistance against fungal pathogens such as *Rhizoctonia solani* and *Alternaria alternata* (Abdel-Razik et al., 2017). Tawfik et al. (2022) employed a chitosan NP system akin to this one to introduce *A. thaliana* thionin genes into *Allium cepa*, which resulted in increased resistance to the fungal disease *Aspergillus niger*. These studies together suggest the potential field application of transgenic disease resistant crops produced by NP-mediated gene delivery as an effective means of reducing disease-associated yield loss.

10.8 BENEFITS OF USING NANOPARTICLES FOR PRODUCTION OF DISEASE RESISTANT PLANTS

NP-mediated gene delivery for the production of disease resistant crop varieties offers multiple advantages over traditional techniques. Plant-cell penetration is enhanced while damage to plant tissue and the genome by mechanical force or cell-wall degrading enzymes is avoided. Chang et al. demonstrated mesoporous silica nanoparticles which bypassed the *Arabidopsis thaliana*

TABLE 10.3
Impact of different nanoparticles on plant resistance against a range of plant pathogens

S. No.	Nanoparticles	Host Plant	Pathogen/Disease	Resistance Gene Delivered/Edited	Reference
1.	Silicon carbide whiskers	Peanut (*Arachis hypogaea*)	Peanut Leaf Spot by *Cercosporadium personatum*	Rice chitinase controlled by CaMV 35S promoter delivered as transgene	Hassan et al. (2016)
2.	Chitosan and gold nanoparticles bound to Cas9 endonuclease and a short guide RNA	Potato (*Solanum tuberosum*)	Potato Virus Y (Potyviride)	Potato coilin gene edited for enhanced production of Cajal bodies	Makhotenko et al. (2019)
3.	Chitosan nanoparticles	Potato (*Solanum tuberosum*)	*Alternaria alternata* and *Rhizoctonia solani*	*Arabidopsis thaliana* thionin genes *AT1G12660* and *AT1G12663* delivered as transgenes	Abdel-Razik et al. (2017)
4.	Chitosan nanoparticles	Onion (*Allium cepa*)	*Aspergillus niger*	*Arabidopsis thaliana* thionin gene (t*hio-60*) delivered as transgene	Tawfik et al. (2022)

root epidermal layer and reached the deeper cortex and endoderm layers using a simple co-culture technique, yielding transient transgene expression (Chang et al., 2013). NPs were also passively delivered to watermelons through stomatal openings and systemically distributed via phloem vasculature (Raliya et al., 2016). NPs such as carbon nanotubes also provide protection for the genetic cargo against degradation by plant endogenous nucleases (Demirer et al., 2018). It is possible to precisely tailor the physicochemical characteristics of NPs to target subcellular targets such as organelles, in addition to certain plant tissue types such as leaves, shoots, and roots (Serag et al., 2011; Cunningham et al., 2018). The cationic charge of gold NPs could be adjusted to determine NP accumulation fate *in planta*, with cations traveling in vasculature to roots and anions accumulating in shoots and leaves (Zhu et al., 2012). Depending on their unique weight, multiwalled carbon nanotubes have also been shown to concentrate in some organelles, such as the nucleus and plastids, eschewing endosome destruction (Serag et al., 2011). Additionally, a range of distinct dicot species' chloroplasts might be precisely targeted by single-walled carbon nanotubes covered in a lipid sheath (Kwak et al., 2019). Focusing on particular organelles by NPs as carriers of gene-editing technologies like CRISPR-Cas9 and TALENs provide for transient expression of heritable edits without the need to integrate transgenes into the nuclear genome features, appealing from a regulatory standpoint regarding Genetically Modified Organisms (GMOs) (Demirer et al., 2018). Many NPs like chitosan are non-toxic and biodegradable with a safe environmental fate (Cunningham et al., 2018).

10.9 DRAWBACKS OF USING NANOPARTICLES FOR RESISTANT PLANTS

While NP-mediated gene delivery has successfully generated disease-resistant plants in several studies, the technology has limitations and aspects which require further study to ensure safety. The ecological impact of NPs requires further study, especially given the great diversity of NP shape, size, and chemical properties (Kumar et al., 2022). Clarifying the safety of NPs for human health is also necessary (Demirer et al., 2021). It can be difficult to achieve stable genetic transformation in plants using NPs since many methods that have been documented only produce temporary changes that are not heritable across generations (Cunningham et al., 2018). However, a study that used magnet infection of pollen with iron-based nanoparticles to confer resistance to insect pests resulted in persistent transgenic cotton plants (Zhao et al., 2017). Seeds produced from transformed pollen produced adult plants with stable expression of the transgenes. Insect resistance was significant in the T3 generation and PCR confirmed the heritability of transgene material. The authors also successfully transformed other dicots using the same methodology (pumpkin, pepper, zucchini) and had some lesser success with a monocot (*Lilium brownii*) (Zhao et al., 2017). A follow-up investigation revealed a restricted host range by failing to change multiple significant monocots (maize, sorghum, and lily) utilizing the same previously published methods of NP-magnet pollen infection (Vejlupkova et al., 2020). Plants transformed via NP-mediated means represent a novel technological frontier in plant research with a set of advantages and disadvantages specific to the target host plant and to the technology in general.

10.10 CONCLUSIONS AND FUTURE PROSPECTS

The chapter elucidates cutting-edge advancements in nanotechnology and its pivotal role in mitigating food scarcity and epidemics. Nanoparticles serve as a potent armamentarium against diverse pathogens, bolstering crop resilience and mitigating yield losses, thus ameliorating food shortages. While multifarious methodologies exist to combat these challenges, nanoscience emerges as a quintessential approach by engineering precision-engineered nanoparticles. Each nanoparticle variant embodies distinct functionalities, augmenting plant fortitude and fostering the genesis of nutrient-rich cultivars. Furthermore, nanotechnology transcends disease mitigation, offering avenues for enhancing food nutritive profiles, prolonging product shelf lives, and fostering the cultivation of agriculturally advantageous commodities. Moreover, the purview of nanotechnology transcends mere disease mitigation, heralding a new era of agricultural innovation wherein food nutritive profiles are enhanced, product shelf lives prolonged, and agriculturally advantageous commodities fostered. By harnessing the transformative potential of nanoscience, we stand poised to revolutionize agricultural paradigms and fortify global food security

REFERENCES

Abdelkhalek, A., & Al-Askar, A. A. (2020). Green synthesized ZnO nanoparticles mediated by Mentha spicata extract induce plant systemic resistance against Tobacco mosaic virus. *Applied Science*, 10(15), 5054.

Abdel-Razik, A. B., Hammad, I. A., & Tawfik, E. (2017). Transformation of thionin genes using chitosan nanoparticle into potato plant to be resistant to fungal infection. *Journal of Biotechnology and Biochemistry*, 03(3), 01–13.

Abdelrhim, A. S., Mazrou, Y. S. A., Nehela, Y., Atallah, O. O., El-Ashmony, R. M., & Dawood, M. F. A. (2021). Silicon dioxide nanoparticles induce innate immune responses and activate antioxidant machinery in wheat against *Rhizoctonia solani*. *Plants*, *10*(12), Art. 12.

Adeel, M., Farooq, T., White, J. C., Hao, Y., He, Z., & Rui, Y. (2021). Carbon-based nanomaterials suppress *tobacco mosaic virus* (TMV) infection and induce resistance in *Nicotiana benthamiana*. *Journal of Hazardous Materials*, *404*, 124167.

Aftab, S., Iqbal, M. Z., & Iqbal, M. W. (2022). Programmable photo-induced doping in 2D materials. *Advanced Materials Interfaces*, *9*(32), 2201219.

Afzal, S., Tariq, M. A., Khanum, A., Haroon, A., Arshad Hussain, M., Chaudary, A.-U.-R., Batool, M., Bashir, M. A., Alajmi, R., Essa, M., Ejaz, S., & Samreen, S. (2023). Grasp of wheat leaf rust through plant leaves extract and bioagent as an eco-friendly measure. *Journal of King Saud University – Science*, *35*(2), 102452.

Ahmad, E. A., Farrag, A. A., Kheder, A. A., & Shaaban, A. (2022). Effect of phytoplasma associated with sesame phyllody on ultrastructural modification, physio-biochemical traits, productivity and oil quality. *Plants*, *11*(4), Art. 4.

Ali, S., Tyagi, A., & Bae, H. (2023). Plant microbiome: An ocean of possibilities for improving disease resistance in plants. *Microorganisms*, *11*(2), Art. 2.

Aljabali, A. A. A., Akkam, Y., Al Zoubi, M. S., Al-Batayneh, K. M., Al-Trad, B., Abo Alrob, O., Alkilany, A. M., Benamara, M., & Evans, D. J. (2018). Synthesis of gold nanoparticles using leaf extract of *Ziziphus zizyphus* and their antimicrobial activity. *Nanomaterials*, *8*(3), Art. 3.

Alkubaisi, N. A., & Aref, N. M. A. (2017). Dispersed gold nanoparticles potentially ruin gold barley yellow dwarf virus and eliminate virus infectivity hazards. *Applied Nanoscience*, *7*(1), 31–40.

Amna, M., T., Khan, U. N., Amin, B., Javed, M. T., Mehmood, S., Farooq, M. A., Sultan, T., Munis, M. F. H., & Chaudhary, H. J. (2021). Characterization of bio-fabricated silver nanoparticles for distinct anti-fungal activity against sugarcane phytopathogens. *Microscopy Research and Technique*, *84*(7), 1522–1530.

Anil, V. S., Lobo, S., & Bennur, S. (2018). Somaclonal variations for crop improvement: Selection for disease resistant variants in vitro. *Plant Science Today*, *5*(2), Art. 2.

Ansari, M. A., & Rehman, S. (2021). *Microbial Nanotechnology: Green Synthesis and Applications*. Springer.

Ansari, R. A., & Khan, T. A. (2012a). Parasitic association of root-knot nematode, *Meloidogyne incognita* on guava. *e-Journal of Science Technology*, *5*(12), 65–67.

Ansari, R. A., & Khan, T. A. (2012b) Diversity and community structure of phytonematodes associated with guava in and around Aligarh, Uttar Pradesh, India. *Trends in Biosciences*, *5*(3), 202–204.

Ansari, R. A., & Mahmood, I. (2017a). Determination of disease incidence caused by *Meloidogyne* spp. and or *Fusarium udum* on pigeonpea in Aligarh district: A survey. *Trends in Biosciences*, *10*(24), 5239–5243.

Ansari, R. A., & Mahmood, I. (2017b). Optimization of organic and bio-organic fertilizers on soil properties and growth of pigeon pea. *Scientia Horticulturae*, *226*, 1–9.

Ansari, R. A., & Mahmood, I. (2019a). *Plant Health Under Biotic Stress – Volume II, Microbial Interactions*, Springer Nature, Singapore. ISBN: 978-981-13-6040-4

Ansari, R. A., & Mahmood, I. (2019b). *Plant Health Under Biotic Stress – Volume I, Organic Strategies*, Springer Nature, Singapore. ISBN: 978-981-13-6043-5

Ansari, R. A., Mahmood, I., Rizvi, R., Sumbul, A., & Safiuddin. (2017). Siderophores: Augmentation of soil health and crop productivity. *Probiotics in Agroecosystem*, 291–312.

Ansari, R. A., Rizvi, R., & Mahmood, I. (2023). *Mycorrhizal Symbiosis and Agroecosystem Restoration*, Springer.

Ansari, R. A., Rizvi, R., & Mahmood, I. (2020). *Management of Phytonematodes: Management of Phytonematodes: Recent Advances and Future Challenges*, Springer Nature, Singapore. ISBN: 978-981-15-4087-5

Ansari, R. A., Rizvi, R., Sumbul, A., & Mahmood, I. (2020). Plant-growth-promoting rhizobacteria (PGPR)-based sustainable management of phytoparasitic nematodes: Current understandings and future challenges. *Management of Phytonematodes: Recent Advances and Future Challenges*, 51–71.

Ansari, R. A., Sumbul, A., Rizvi, R., & Mahmood, I. (2019). Organic soil amendments: Potential tool for soil and plant health management. In Ansari RA, Mahmood I (eds), *Plant Health Under Biotic Stress: Volume 1: Organic Strategies* (pp. 1–35).

Ansari, R. A., Sumbul, A., Rizvi, R., & Mahmood, I. (2024). Impact of environmental gases on mycorrhizal symbiosis and its influence on ecosystem functioning under the current climate change scenario. In *Mycorrhizal Symbiosis and Agroecosystem Restoration* (pp. 51–76). Singapore: Springer Nature Singapore.

Bal, R. S., & Kumar, A. (2014). Studies on the epidemiology of white rust and Alternaria leaf blight and their effect on the yield of Indian mustard. *African Journal of Agricultural Research*, 9(2), 302–306.

Banerjee, A., Sarkar, A., Acharya, K., & Chakraborty, N. (2021). Nanotechnology: An emerging hope in crop improvement. *Letters in Applied Nano BioScience*, *10*, 2784–2803.

Bayoumy, A. M., Elhaes, H., Osman, O., Kholmurodov, K. T., Hussein, T., & Ibrahim, M. A. (2020). Effect of nano metal oxides on heme molecule: Molecular and biomolecular approaches. *Biointerface Research in Applied Chemistry*, 10, 4837–4845.

Borgatta, J., Ma, C., Hudson-Smith, N., Elmer, W., Pérez, C., Torre-Roche, R., Zuverza-Mena, N., Haynes, C., White, J., & Hamers, R. (2018). Copper based nanomaterials suppress root fungal disease in watermelon (*Citrullus lanatus*): Role of particle morphology, composition and dissolution behavior. *ACS Sustainable Chemistry & Engineering*, *6*(11), 14847–14856.

Boxi, S. S., Mukherjee, K., & Paria, S. (2016). Ag doped hollow TiO_2 nanoparticles as an effective green fungicide against *Fusarium solani* and *Venturia inaequalis* phytopathogens. *Nanotechnology*, *27*(8), 085103.

Boyno, G., Ansari, R. A., Durak, E. D., Güneş, H., Çevik, R., & Demir, S. (2024). Arbuscular mycorrhizal technology in sustainable agriculture: Current knowledge and challenges in agroforestry. In: Ansari, R.A., Rizvi, R., Mahmood, I. (eds), *Mycorrhizal Symbiosis and Agroecosystem Restoration*, 173–195.

Buchman, J., Elmer, W., Ma, C., Landy, K., White, J., & Haynes, C. (2019). Chitosan-Coated *Mesoporous Silica* nanoparticle treatment of *Citrullus lanatus* (Watermelon): Enhanced fungal disease suppression and modulated expression of stress-related genes. *ACS Sustainable Chemistry & Engineering, XXXX*.

Cai, L., Cai, L., Jia, H., Liu, C., Wang, D., & Sun, X. (2020). Foliar exposure of Fe3O4 nanoparticles on *Nicotiana benthamiana*: Evidence for nanoparticles uptake, plant growth promoter and defence response elicitor against plant virus. *Journal of Hazardous Materials.*, *393*, 122415.

Cai, L., Liu, C., Fan, G., Liu, C., & Sun, X. (2019). Preventing viral disease by ZnONPs through directly deactivating TMV and activating plant immunity in *Nicotiana benthamiana*. *Environmental Science: Nano*, *6*(12), 3653–3669.

Chaloner, T. M., Gurr, S. J., & Bebber, D. P. (2021). Plant pathogen infection risk tracks global crop yields under climate change. *Nature Climate Change*, *11*(8), 710–715.

Chang, F.-P., Kuang, L.-Y., Huang, C.-A., Jane, W.-N., Hung, Y., Hsing, Y. C., & Mou, C.-Y. (2013). A simple plant gene delivery system using mesoporous silica nanoparticles as carriers. *Journal of Materials Chemistry B*, *1*(39), 5279–5287.

Chen, J., Wang, X., & Han, H. (2013). A new function of graphene oxide emerges: Inactivating phytopathogenic bacterium *Xanthomonas oryzae* pv. *oryzae*. *Journal of Nanoparticle Research*, *15*(5), 1658.

Chiu, T., Poucet, T., & Li, Y. (2022). The potential of plant proteins as antifungal agents for agricultural applications. *Synthetic and Systems Biotechnology*, *7*(4), 1075–1083.

Chu, H., Kim, H.-J., Su Kim, J., Kim, M.-S., Yoon, B.-D., Park, H.-J., & Kim, C. Y. (2012). A nanosized Ag–silica hybrid complex prepared by γ-irradiation activates the defence response in Arabidopsis. *Radiation Physics and Chemistry*, *81*(2), 180–184.

Coa, F., Bortolozzo, L., Petry, R., Da Silva, G., Martins, C., Zigiotto, A., Sabino, C., Costa, R., Khan, L., Delite, F., & Martinez, D. (2020). *Environmental Toxicity of Nanopesticides Against Non-Target Organisms: The State of the Art* In: Fraceto, L.F., S.S. de Castro, V.L., Grillo, R., Ávila, D., Caixeta Oliveira, H., Lima, R. (eds), *Nanopesticides* (pp. 227–279). Springer, Cham.

Cui, H., Zhang, P., Gu, W., & Jiang, J. (2009). Application of anatase TiO2 sol derived from peroxotitannic acid in crop diseases control and growth regulation. *Technical Proceedings of the 2009 NSTI Nanotechnology Conference and Expo, NSTI-Nanotech 2009, 2*, 286–289.

Cunningham, F. J., Goh, N. S., Demirer, G. S., Matos, J. L., & Landry, M. P. (2018). Nanoparticle-mediated delivery towards advancing plant genetic engineering. *Trends in Biotechnology, 36*(9), 882–897.

Das, R., Deb, P., Pandey, H., Shyam, P., & Singh, D. (2022). Botanical synthesis of silver nanoparticles (AgNPs) and its antifungal effect against *Alternaria porri* causing purple blotch of onion: An *in vitro* and natural epiphytic study. *Journal of Agriculture and Food Research, 10*, 100390.

Datnoff, L. E., Rodrigues, F. A., & Seebold, K. W. (2007). Silicon and plant disease. *Mineral Nutrition and Plant Disease, 27*, 233–246.

de Freitas Bueno, R. C. O., Ansari, R. A., Lima, G. P. P., & Sakate, R. K. (2019). Phytosanitation: A novel approach toward disease management. In: Ansari, R., Mahmood, I. (eds), *Plant Health Under Biotic Stress: Volume 1: Organic Strategies*, 73–90.

Debnath, S., & Biswas, S. (2020). Evaluation of some inbred lines against southern corn leaf blight of maize under natural condition during kharif season of West Bengal. *Journal of Pharmacognosy and Phytochemistry, 9*(5), 2526–2528.

Demirer, G. S., Silva, T. N., Jackson, C. T., Thomas, J. B., W. Ehrhardt, D., Rhee, S. Y., Mortimer, J. C., & Landry, M. P. (2021). Nanotechnology to advance CRISPR–Cas genetic engineering of plants. *Nature Nanotechnology, 16*(3), 243–250.

Demirer, G. S., Zhang, H., Matos, J. L., Goh, N., Cunningham, F., Sung, Y., Chang, R., Aditham, A. J., Chio, L., Cho, M.-J., Staskawicz, B., & Landry, M. P. (2018). High aspect ratio nanomaterials enable delivery of functional genetic material without dna integration in mature plants (p. 179549). Biorx IV.

Derbalah, A., Shenashen, M., Hamza, A., Mohamed, A., & El Safty, S. (2018). Antifungal activity of fabricated mesoporous silica nanoparticles against early blight of tomato. *Egyptian Journal of Basic and Applied Sciences, 5*(2), 145–150.

Derbalah, A. S., El-Moghazy, S. M., & Godah, M. I. (2013). Alternative control methods of sugar-beet leaf spot disease caused by the fungus *Cercospora beticola* (Sacc). *Egyptian Journal of Biological Pest control, 23*(2), 247–254.

Derbalah, A. S. H., & M. M. Elsharkawy, A new strategy to control *Cucumber mosaic virus* using fabricated NiO-nanostructures, *Journal of Biotechnology, 306* (2019), 134–141.

Deshmukh, S. P., Patil, S. M., Mullani, S. B., & Delekar, S. D. (2019). Silver nanoparticles as an effective disinfectant: A review. *Materials Science and Engineering: C, 97*, 954–965.

Devi, T. S., Das, D., Ansari, R. A., Rizvi, R., Sumbul, A., & Mahmood, I. (2020). Role of organic additives in the sustainable management of phytoparasitic nematodes. *Management of Phytonematodes: Recent Advances and Future Challenges*, 279–295.

di Santo, N., Russo, I., & Sisto, R. (2022). Climate change and natural resource scarcity: A literature review on dry farming. *Land, 11*(12), Art. 12.

Dolgaev, S. I., Simakin, A. V., Voronov, V. V., Shafeev, G. A., & Bozon-Verduraz, F. (2002). Nanoparticles produced by laser ablation of solids in liquid environment. *Applied Surface Science, 186*(1), 546–551.

Dong, O. X., & Ronald, P. C. (2019). Genetic engineering for disease resistance in plants: Recent progress and future perspectives. *Plant Physiology, 180*(1), 26–38.

Đorđević, V., Balanč, B., Belščak-Cvitanović, A., Lević, S., Trifković, K., Kalušević, A., & Nedović, V. (2015). Trends in encapsulation technologies for delivery of food bioactive compounds. *Food Engineering Reviews, 7*, 452–490.

Du, J., Liu, B., Zhao, T., Xu, X., Lin, H., Ji, Y., Li, Y., Li, Z., Lu, C., Li, P., Zhao, H., Li, Y., Yin, Z., & Ding, X. (2022). Silica nanoparticles protect rice against biotic and abiotic stresses. *Journal of Nanobiotechnology, 20*(1), 197.

Duarte, H., Zambolim, L., Machado, F., Porto, H., & Rodrigues, F. (2019). Comparative epidemiology of late blight and early blight of potato under different environmental conditions and fungicide application programs. *Semina: Ciências Agrárias, 40*, 1805.

Durán, N., Marcato, P. D., Alves, O. L., De Souza, G. I., & Esposito, E. (2005). Mechanistic aspects of biosynthesis of silver nanoparticles by several *Fusarium oxysporum* strains. *Journal of Nanobiotechnology*, *3*(1), 8.

Dutta, P., Kumari, A., Mahanta, M., Biswas, K. K., Dudkiewicz, A., Thakuria, D., Abdelrhim, A. S., Mandal, M. K., & Mazumdar, N. (2022). Advances in nanotechnology as a potential alternative for plant viral disease management. *Frontiers in Microbiology*, *13*, 1–12.

Elazzazy, A. M., Elbeshehy, E. K., & Betiha, M. A. (2017). *In vitro* assessment of activity of graphene silver composite sheets against multidrug-resistant bacteria and *Tomato Bushy Stunt Virus*, Trop. *Journal of Pharma Research*, 16(11), 2705–2711.

El-Baky, N. A., & Amara, A. A. A. F. (2021). Recent approaches towards control of fungal diseases in plants: An updated review. *Journal of Fungi*, *7*(11), 900.

El-Dougdoug, N. K., Bondok, A. M., & El-Dougdoug, K. A. (2018). Evaluation of silver nanoparticles as antiviral agent against TOMV and PVY in tomato plants. *Middle East Journal of Applied Sciences*, *8*(1), 100–111.

Elmer, W., & White, J. C. (2018). The future of nanotechnology in plant pathology. *Annual Review of Phytopathology*, *56*, 111–133.

El-Saadony, M. T., Saad, A. M., Soliman, S. M., Ahmed, A. I., El-Tahan, A. M., Abd El-Mageed, T. A., Elrys, A. S., El-Tarabily, K. A., & AbuQamar, S. F. (2022). Plant growth-promoting microorganisms as biocontrol agents of plant diseases: Mechanisms, challenges, and future perspectives. *Frontiers in Plant Science*, *13*, 1–19.

El-Sawy, M., Elsharkawy, M., Abass, J., & Hagag, E. (2017). Inhibition of *Tomato Yellow Leaf Curl Virus* by *Zingiber officinale* and *Mentha longifolia* extracts and silica nanoparticles and its reflection on tomato growth and yield. *International Journal of Antivirals & Antiretrovirology*, *1*, 1–6.

El-shazly, M., Attia, Y., Kabil, F., Anis, E., & Hazman, M. (2017). Inhibitory effects of salicylic acid and silver nanoparticles on potato virus Y-infected potato plants in Egypt. *Middle East Journal of Agriculture Research*, *6*, 835–848.

El-Shetehy, M., Moradi, A., Maceroni, M., Reinhardt, D., Petri-Fink, A., Rothen-Rutishauser, B., Mauch, F., & Schwab, F. (2021). Silica nanoparticles enhance disease resistance in *Arabidopsis* plants. *Nature Nanotechnology*, *16*(3), 344–353.

Fabiyi, O. A., Alabi, R. O., & Ansari, R. A. (2020). Nanoparticles' synthesis and their application in the management of phytonematodes: An overview. In: Ansari, R., Rizvi, R., Mahmood, I. (eds), *Management of Phytonematodes: Recent Advances and Future Challenges* (pp. 125–140). Springer, Singapore.

Fabiyi, O. A., Ansari, R. A., Rizvi, R., Sumbul, A., & Mahmood, I. (2024). Influence of arbuscular mycorrhizal fungi on soil health amelioration and plant fitness under hostile environment. In Ansari, R. A., Rizvi, R., Mahmood, I. (eds), *Mycorrhizal Symbiosis and Agroecosystem Restoration* (pp. 227–248). Springer Nature Singapore .

Fan, Q., Liao, Y.-Y., Kunwar, S., Da Silva, S., Young, M., Santra, S., Minsavage, G. V., Freeman, J. H., Jones, J. B., & Paret, M. L. (2021). Antibacterial effect of copper composites against *Xanthomonas euvesicatoria*. *Crop Protection*, *139*, 105366.

Fletcher, J., Franz, D., & Leclerc, J. E. (2009). Healthy plants: Necessary for a balanced "One health" concept. *Veterinaria Italiana*, *45*(1), 79–95.

Food and Agriculture Organization of the United Nations (FAO). (2020). International year of plant health – protecting plants, protecting life.

Food and Agriculture Organization of the United Nations. (2019). *The State of Food and Agriculture 2019: Moving Forward on Food Loss and Waste Reduction*. Food and Agricultural Organization, Rome, Italy.

Georgakopoulos-Soares, I., Morganella, S., Jain, N., Hemberg, M., & Nik-Zainal, S. (2018). Noncanonical secondary structures arising from non-B DNA motifs are determinants of mutagenesis. *Genome Research*, *28*(9), 1264–1271.

Giannousi, K., Avramidis, I., & Dendrinou-Samara, C. (2013). Synthesis, characterization, and evaluation of copper-based nanoparticles as agrochemicals against *Phytophthora infestans*. *RSC Advances*, *3*, 21743.

Godoy, C., Seixas, C., Seixas, S., Soares, R., Marcelino-Guimaraes, F., Meyer, M., Costamilan, L., Soja, E., Carlos, R., Strass, J., & Amaral, A. (2016). Asian soybean rust in Brazil: Past, present, and future. *Pesquisa Agropecuária Brasileira*, *51*, 407–421.

Grobelny, J., DelRio, F. W., Pradeep, N., Kim, D.-I., Hackley, V. A., & Cook, R. F. (2011). Size measurement of nanoparticles using atomic force microscopy. *Methods in Molecular Biology*, *697*, 71–82.

Gururani, M. A., Venkatesh, J., Upadhyaya, C. P., Nookaraju, A., Pandey, S. K., & Park, S. W. (2012). Plant disease resistance genes: Current status and future directions. *Physiological and Molecular Plant Pathology*, *78*, 51–65.

Hachem, K., Ansari, M. J., Saleh, R. O., Kzar, H. H., Al-Gazally, M. E., Altimari, U. S., Hussein, S. A., Mohammed, H. T., Hammid, A. T., & Kianfar, E. (2022). Methods of chemical synthesis in the synthesis of nanomaterial and nanoparticles by the chemical deposition method: A review. *BioNanoScience*, *12*(3), 1032–1057.

Hahn, V. (2002). Genetic variation for resistance to *Sclerotinia* head rot in sunflower inbred lines. *Field Crops Research*, *77*(2), 153–159.

Hamza, A., El-Mogazy, S., & Derbalah, A. (2016). Fenton reagent and titanium dioxide nanoparticles as antifungal agents to control leaf spot of sugar beet under field conditions. *Journal of Plant Protection Research*, *56*.

Hasan, S. (2015). A review on nanoparticles: Their synthesis and types. *Research Journal of Recent Sciences*, *4*(ISC-2014), 1–3.

Hashim, A. A. (2012). *The delivery of nanoparticles*. InTech. ISBN 978-953-51-0615-9. www.issp.ac.ru/ebooks/books/open/The_Delivery_of_Nanoparticles.pdf

Hassan, M. ul, Akram, Z., Ali, S., Ali, G. M., Zafar, Y., Shah, Z. H., & Alghabari, F. (2016). Whisker-mediated transformation of peanut with chitinase gene enhances resistance to leaf spot disease. *Crop Breeding and Applied Biotechnology*, *16*, 108–114.

Hatfaludi, T., Liska, M., Zellinger, D., Ousman, J. P., Szostak, M., Ambrus, A., Jalava, K., & Lubitz, W. (2004). Bacterial ghost technology for pesticide delivery. *Journal of Agricultural and Food Chemistry*, *52*(18), 5627–5634.

Hossain, M. I., Ahmad, K., Siddiqui, Y., Saad, N., Rahman, Z., Haruna, A. O., & Bejo, S. K. (2020). Current and prospective strategies on detecting and managing *Colletotrichum falcatum* causing red rot of sugarcane. *Agronomy*, *10*(9), Art. 9.

Hwang, H.-H., Yu, M., & Lai, E.-M. (2017). Agrobacterium-mediated plant transformation: Biology and applications. *The Arabidopsis Book*, *15*, e0186.

Ibrahim, D. S. S., Elderiny, M. M., Ansari, R. A., Rizvi, R., Sumbul, A., & Mahmood, I. (2020). Role of *Trichoderma* spp. in the management of plant-parasitic nematodes infesting important crops. In Ansari, R., Rizvi, R., Mahmood, I. (eds), *Management of Phytonematodes: Recent Advances and Future Challenges* (pp. 259–278). Springer, Singapore.

Ibrahim, D. S. S., Mostafa, M., Abdellatif, A. A., & Ansari, R. A. (2022). Sustainable management of soil-borne plant pathogens. In *Earth Systems Protection and Sustainability: Volume 1* (pp. 35–56). Cham: Springer International Publishing.

Ingle, A. P. (2021). *Nanotechnology in Plant Growth Promotion and Protection*. John Wiley & Sons, Incorporated.

Ilk, S., Saglam, N., & Özgen M. (2017). Kaempferol loaded lecithin/chitosan nanoparticles: preparation, characterization, and their potential applications as a sustainable antifungal agent. *Artificial Cells, Nanomedicine, and Biotechnology*, Aug; *45*(5), 907–916.

Jagana, D., Hegde, Y., & Lella, R. (2017). Green nanoparticles—A novel approach for the management of banana anthracnose caused by *Colletotrichum musae*. *International Journal of Current Microbiology and Applied Sciences*, *6*, 1749–1756.

Jatin, C., Sachitanand, D., & Dipika, M. (2022). Anthracnose of chilli: A review. *The Pharma Innovation Journal*, *SP-11*(6), 1811–1815.

Jayaseelan, C., Rahuman, A. A., Kirthi, A. V., Marimuthu, S., Santhoshkumar, T., Bagavan, A., Gaurav, K., Karthik, L., & Rao, K. V. B. (2012). Novel microbial route to synthesize ZnO nanoparticles using *Aeromonas hydrophila* and their activity against pathogenic bacteria and fungi. *Spectrochimica Acta. Part A, Molecular and Biomolecular Spectroscopy*, *90*, 78–84.

Jo, Y.-K., Kim, B. H., & Jung, G. (2009). Antifungal activity of silver ions and nanoparticles on phytopathogenic fungi. *Plant Disease*, *93*(10), 1037–1043.

Jogaiah, S., Singh, H. B., Fraceto, L. F., & De Lima, R. (2020). *Advances in Nano-Fertilizers and Nano-Pesticides in Agriculture: A Smart Delivery System for Crop Improvement*. Woodhead Publishing.

Kah, M., & Hofmann, T. (2014). Nanopesticide research: Current trends and future priorities. *Environment International*, *63*, 224–235.

Kairyte, K., Kadys, A., & Luksiene, Z. (2013). Antibacterial and antifungal activity of photoactivated ZnO nanoparticles in suspension. *Journal of Photochemistry and Photobiology. B, Biology*, *128*, 78–84.

Kalishwaralal, K., Deepak, V., Ramkumarpandian, S., Nellaiah, H., & Sangiliyandi, G. (2008). Extracellular biosynthesis of silver nanoparticles by the culture supernatant of *Bacillus licheniformis*. *Materials Letters*, *62*(29), 4411–4413.

Kaur, P., Thakur, R., Duhan, J., & Chaudhury, A. (2018). Management of wilt disease of chickpea in vivo by silver nanoparticles; biosynthesized by rhizospheric microflora of chickpea (*Cicer arietinum* L.). *Journal of Chemical Technology & Biotechnology*, *93*, 3233–3243.

Keshari, N., Kranti, K. V. V. S. K., Gunda, N. K., & Ansari, R. A. (2024). Arbuscular mycorrhizal fungi: A potential agent for phytonematodes management in diverse agro-climatic zones. In *Mycorrhizal Symbiosis and Agroecosystem Restoration* (pp. 147–169). Singapore: Springer Nature Singapore.

Keshvadi, M., Karimi, F., Valizadeh, S., & Valizadeh, A. (2019). Comparative study of antibacterial inhibitory effect of silver nanoparticles and garlic oil nanoemulsion with their combination. *Biointerface Research in Applied Chemistry*, *9*, 4560–4566.

Khan H, A., Hassan, M., & Khan, M. (2020). *Conventional Plant Breeding Program for Disease Resistance* (pp. 27–51).

Khan, I., Saeed, K., & Khan, I. (2019). Nanoparticles: Properties, applications, and toxicities. *Arabian Journal of Chemistry*, *12*(7), 908–931.

Khan, R. A. A., Tang, Y., Naz, I., Alam, S. S., Wang, W., Ahmad, M., Najeeb, S., Rao, C., Li, Y., Xie, B., & Li, Y. (2021). Management of *Ralstonia solanacearum* in tomato using ZnO nanoparticles synthesized through *Matricaria chamomilla*. *Plant Disease*, *105*(10), 3224–3230.

Kim, D.-Y., Kadam, A., Shinde, S., Saratale, R. G., Patra, J., & Ghodake, G. (2018). Recent developments in nanotechnology transforming the agricultural sector: A transition replete with opportunities. *Journal of the Science of Food and Agriculture*, *98*(3), 849–864.

Kim, S.-M., & Reinke, R. F. (2019). A novel resistance gene for bacterial blight in rice, Xa43(t) identified by GWAS, confirmed by QTL mapping using a bi-parental population. *PLOS ONE*, *14*(2), e0211775.

Kumar, A., Choudhary, A., Kaur, H., Guha, S., Mehta, S., & Husen, A. (2022). Potential applications of engineered nanoparticles in plant disease management: A Critical update. *Chemosphere*, *295*, 133798.

Kumar, S., Singh, R., Khilari, K., Singh, D. V., & Mishra, P. (2022). To study the effect of different environment factor on disease development and progression of wheat stripe rust under field condition. *Pharma Innovation Journal*, *SP-11*(7), 436–442.

Kwak, S.-Y., Lew, T. T. S., Sweeney, C. J., Koman, V. B., Wong, M. H., Bohmert-Tatarev, K., Snell, K. D., Seo, J. S., Chua, N.-H., & Strano, M. S. (2019). Chloroplast-selective gene delivery and expression in planta using chitosan-complexed single-walled carbon nanotube carriers. *Nature Nanotechnology*, *14*(5), 447–455.

Lavanya, R., & Arun, V. (2021). Detection of Begomovirus in chilli and tomato plants using functionalized gold nanoparticles. *Scientific Reports*, *11*(1), 14203.

Ma, C., Borgatta, J., Hudson, B. G., Tamijani, A. A., De La Torre-Roche, R., Zuverza-Mena, N., Shen, Y., Elmer, W., Xing, B., Mason, S. E., Hamers, R. J., & White, J. C. (2020). Advanced material modulation of nutritional and phytohormone status alleviates damage from soybean sudden death syndrome. *Nature Nanotechnology*, *15*(12), 1033–1042.

Mahmood, I., Rizvi, R., Sumbul, A., & Ansari, R. A. (2019). Potential role of plant growth promoting rhizobacteria in alleviation of biotic stress. In: Ansari, R. A., Mahmood, I. (eds), *Plant Health Under Biotic Stress: Volume 2: Microbial Interactions* (pp. 177–188).

Makhotenko, A. V., Khromov, A. V., Snigir, E. A., Makarova, S. S., Makarov, V. V., Suprunova, T. P., Kalinina, N. O., & Taliansky, M. E. (2019). Functional analysis of coilin in virus resistance and stress tolerance of potato *Solanum tuberosum* using CRISPR-Cas9 Editing. *Doklady. Biochemistry and Biophysics*, *484*(1), 88–91.

Matouskova, P., Marova, I., Bokrova, J., & Benesova, P. (2016). Effect of encapsulation on antimicrobial activity of herbal extracts with lysozyme. *Food Technology and Biotechnology*, *54*(3), 304–316.

Merga, G., Wilson, R., Lynn, G., Milosavljevic, B., & Meisel, D. (2007). Redox catalysis on "Naked" silver nanoparticles. *Journal of Physical Chemistry C, 111*(3), 12220–12226.

Miu, B. A., & Dinischiotu, A. (2022). New green approaches in nanoparticles synthesis: An overview. *Molecules (Basel, Switzerland)*, *27*(19), 6472.

Mohamed, T., El-Motlak, M. H., Mamdouh, S., Ashour, M., Ahmed, H., Qayyum, H., & Mahmoud, A. (2022). Excitation wavelength and colloids concentration-dependent nonlinear optical properties of silver nanoparticles synthesized by laser ablation. *Materials*, *15*(20), Art. 20.

Moosa, A. A., Ridha, A. M., & Al-kaser, M. (2015). Process parameters for green synthesis of silver nanoparticles using leaves extract of aloe vera plant process parameters for green synthesis of silver nanoparticles using leaves extract of aloe vera plant. *International Journal of Multidisciplinary and Current Research*, *3*, 966–975.

Mourdikoudis, S., Pallares, R. M., & Thanh, N. T. K. (2018). Characterization techniques for nanoparticles: Comparison and complementarity upon studying nanoparticle properties. *Nanoscale*, *10*(27), 12871–12934.

Mujtaba, M., Wang, D., Carvalho, L. B., Oliveira, J. L., Espirito Santo Pereira, A. do, Sharif, R., Jogaiah, S., Paidi, M. K., Wang, L., Ali, Q., & Fraceto, L. F. (2021). Nanocarrier-mediated delivery of miRNA, RNAi, and CRISPR-Cas for plant protection: Current trends and future directions. *ACS Agricultural Science & Technology*, *1*(5), 417–435.

Nelson, R., Wiesner-Hanks, T., Wisser, R., & Balint-Kurti, P. (2018). Navigating complexity to breed disease-resistant crops. *Nature Reviews Genetics*, *19*(1), 21–33.

Norman, D. (2011). Effect of foliar application of titanium dioxide on bacterial blight of geranium and *Xanthomonas* leaf spot of poinsettia. *Horticulture Science*, *44*, 426–428.

Palmieri, D., Ianiri, G., Del Grosso, C., Barone, G., De Curtis, F., Castoria, R., & Lima, G. (2022). Advances and perspectives in the use of biocontrol agents against fungal plant diseases. *Horticulturae*, *8*(7), Art. 7.

Panpatte, D. G., & Jhala, Y. K. (2019). *Nanotechnology for Agriculture: Crop Production & Protection*. Springer Nature.

Panpatte, D., Jhala, Y. G., Shelat, H., & Vyas, R. V. (2016). *Nanoparticles: The Next Generation Technology for Sustainable Agriculture* (pp. 289–300).

Paret, M. L., Vallad, G. E., Averett, D. R., Jones, J. B., & Olson, S. M. (2013). Photocatalysis: Effect of light-activated nanoscale formulations of TiO_2 on *Xanthomonas perforans* and control of bacterial spot of tomato. *Phytopathology*, *103*(3), 228–236.

Park, H. J., Kim, S. H., Kim, H. J., & Choi, S. H. (2006). A new composition of nanosized silica-silver for control of various plant diseases. *Plant Pathology Journal*, *22*, 295–302.

Patil, S., Ansari, R. A., Gurikar, C., Bhavan, S., & Hariprasad, K. (2024). Propagation of mycorrhizal fungi and their strength in plant health ameliorations. In *Mycorrhizal Symbiosis and Agroecosystem Restoration* (pp. 77–98). Singapore: Springer Nature Singapore.

Pinstrup-Andersen, P., Pandya-Lorch, R., & Rosegrant, M. W. (1999). World food prospects: Critical issues for the early twenty-first century. Food Policy report International Food Policy Research Institute Washington, D.C. October 1999.

Ponmurugan, P., Manjukarunambika, K., Elango, V., & Gnanamangai, B. M. (2016). Antifungal activity of biosynthesized copper nanoparticles evaluated against red root-rot disease in tea plants. *Journal of Experimental Nanoscience*, *11*, 1019–1031.

Prabhakar, M., Raja, P., & Awasthi, D. (2020). Deep learning-based assessment of disease severity for early blight in tomato crop. *Multimedia Tools and Applications*, *79*, 39–40.

Prasad, B. N., Subramanyam, B., Lakshmipathi, R. N., Ansari, R. A., Rizvi, R., Sumbul, A., ... & Rachmi, C. M. (2020). Utilization of beneficial microorganisms in sustainable control of phytonematodes. In: Ansari, R., Rizvi, R., Mahmood, I. (eds), *Management of phytonematodes: Recent Advances and Future Challenges*, 317–337.

Priyadarshani, C., Wickramasinghe, D., Egodawatta, C., Benaragama, D., Aruni, P., & Devasinghe, U. (2022). Effect of rates and sources of N fertilizer application on dynamics of rice brown leaf spot disease (*Bipolaris oryzae*) incidences in the dry zone of SriLanka. *Journal of Tropical Crop Science*, *9* (3), 165–173.

Rajeshkumar, S. (2019). Antifungal impact of nanoparticles against different plant pathogenic fungi. In D. K. Tripathi, P. Ahmad, S. Sharma, D. K. Chauhan, & N. K. Dubey (eds), *Nanomaterials in Plants, Algae and Microorganisms* (pp. 197–217). Academic Press.

Rajput, V. D., El-Ramady, H., Upadhyay, S. K., Minkina, T., Ahmed, B., & Mandzhieva, S. (2023). *Nano-Biofortification for Human and Environmental Health*. Springer Nature.

Raliya, R., Franke, C., Chavalmane, S., Nair, R., Reed, N., & Biswas, P. (2016). Quantitative understanding of nanoparticle uptake in watermelon plants. *Frontiers in Plant Science*, 7, 1–10.

Rambabu, B., Lingaiah, N., Chandra, B., Ravikanti, S., & Yadla, H. (2022). Genetic variability studies in advanced backcross derived lines for brown plant hopper and blast resistance in rice (*oryza sativa* l.). *Biological Forum – An International Journal, 14*(2a), 386–388.

Rani, L., Thapa, K., Kanojia, N., Sharma, N., Singh, S., Grewal, A. S., Srivastav, A. L., & Kaushal, J. (2021). An extensive review on the consequences of chemical pesticides on human health and environment. *Journal of Cleaner Production, 283*, 124657.

Rehman, A., Naeem, M. S., Naseer, M. F., & Khan, M. A. (2022). Pursuit of the stripe rust (*Puccinia striiformis* f. sp. *tritici* (pst) disease resistance and its relation with morphological attributes in wheat (*Triticum aestivum* l.). *Pakistan Journal of Phytopathology, 34*(1), Art. 1.

Rehman, F., Pakar, N., Ur Rehman, S., Javed, M. T., Munis, M., & Chaudhary, H. (2023). Zinc oxide nanoparticles: Biogenesis and applications against phytopathogens. *Journal of Plant Pathology, 106*, 1–45.

Rizzo, D. M., Lichtveld, M., Mazet, J. A. K., Togami, E., & Miller, S. A. (2021). Plant health and its effects on food safety and security in a One Health framework: Four case studies. *One Health Outlook, 3*(1), 6.

Sarkar, J., Chakraborty, N., Chatterjee, A., Bhattacharjee, A., Dasgupta, D., & Acharya, K. (2020). Green synthesized copper oxide nanoparticles ameliorate defence and antioxidant enzymes in *Lens culinaris. Nanomaterials, 10*, 312–339.

Schneider, J. M., Zabel, F., Schünemann, F., Delzeit R, & Mauser, W. (2022). Global cropland could be almost halved: Assessment of land saving potentials under different strategies and implications for agricultural markets. *PLOS ONE,* 17(2): e0263063.

Seetharaman, P. K., Chandrasekaran, R., Gnanasekar, S., Chandrakasan, G., Gupta, M., Manikandan, D. B., & Sivaperumal, S. (2018). Antimicrobial and larvicidal activity of eco-friendly silver nanoparticles synthesized from endophytic fungi *Phomopsis liquidambaris. Biocatalysis and Agricultural Biotechnology, 16*, 22–30.

Sekhon, B. S. (2014). Nanotechnology in agri-food production: An overview. *Nanotechnology, Science and Applications*, I, 31–53.

Serag, M. F., Kaji, N., Gaillard, C., Okamoto, Y., Terasaka, K., Jabasini, M., Tokeshi, M., Mizukami, H., Bianco, A., & Baba, Y. (2011). Trafficking and subcellular localization of multiwalled carbon nanotubes in plant cells. *ACS Nano*, 5(1), 493–499.

Serge Savary, Laetitia W., Sarah J. P., Paul E., Neil McRoberts & Andy Nelson. (2019). The global burden of pathogens and pests on major food crops. *Nature Ecology & Evolution, 3*, 430–439.

Shafie, R. M., Salama, A. M., & Farroh, K (2018). In silver nanoparticles activity against *Tomato Spotted Wilt Virus. Middle East Journal of Applied Sciences, 7*, 1251–1267.

Sijpestijn, G. F., Wezel, A., & Chriki, S. (2022). Can agroecology help in meeting our 2050 protein requirements? *Livestock Science, 256*, 104822.

Singh, A., Gaurav, S., Shukla, G., Rani, P., Kumar, B., & Kumar, A. (2020). Evaluation of mycogenic silver and zinc oxide nanoparticles as potential control agent against early blight (*Alternaria solani*) of potato (*Solanum tuberosum* L.). *Journal of Advanced Scientific Research, 11*(2), 164–172.

Singh, A., Singh, N. B., Hussain, I., Singh, H., & Singh, S. (2015). Plant-nanoparticle interaction: An approach to improve agricultural practices and plant productivity. *International Journal of Pharmaceutical Science Invention (IJPSI), 4*, 2319–6718.

Singh, P., Kumar, V., Bakshi, M., Sillanpää, M., & Hussain, C. M. (2022). Environmental applications of microbial nanotechnology: Emerging trends in environmental remediation (Edition 1). Elsevier Science, 424p.

Smil, V. (2000). *Feeding the World: A Challenge for the Twenty-First Century*. Cambridge: The Massachusetts Institute of Technology Press.

Sofy, A. R., Hmed, A. A., Alnaggar, A. E.-A. M., Dawoud, R. A., Elshaarawy, R. F. M., & Sofy, M. R. (2020). Mitigating effects of *Bean yellow mosaic virus* infection in faba bean using new carboxymethyl chitosan-titania nanobiocomposites. *International Journal of Biological Macromolecules, 163*, 1261–1275.

Solanki, M. K., Kashyap, P. L., Kumari, B., Ansari, R. A., Sumbul, A., Rizvi, R., & Mahmood, I. (2021). Mycorrhizal fungi and its importance in plant health amelioration. In *Microbiomes and Plant Health* (pp. 205–223). Academic Press.

Solanki, M. K., Kashyap, P. L., Ansari, R. A., & Kumari, B. (2020). *Microbiomes and Plant Health: Panoply and Their Applications*. The Netherland: Academic Press, Elsevier. ISBN: 9780128197158

Soresa, D., (2022). Estimating combining ability and response of tomato (*Solanum Lycopersicum* l.) genotypes to late blight (*Phytophthora Infestans*) disease. *Journal of Genetic Engineering and Biotechnology Research*, *4*(3), 328–339.

Sumbul, A., Ansari, R. A., Rizvi, R., & Mahmood, I. (2020). Azotobacter: A potential bio-fertilizer for soil and plant health management. *Saudi Journal of Biological Sciences*, *27*(12), 3634–3640.

Surender and Ajay Gahlot. (2021). Physical methods of nanoparticles preparation – An Overview. *International Journal of Advances in Engineering and Management* (*IJAEM*), *3*(12), 812–816

Swinburn, B. A., Kraak, V. I., Allender, S., Atkins, V. J., Baker, P. I., Bogard, J. R., Brinsden, H., Calvillo, A., De Schutter, O., Devarajan, R., Ezzati, M., Friel, S., Goenka, S., Hammond, R. A., Hastings, G., Hawkes, C., Herrero, M., Hovmand, P. S., Howden, M., & Dietz, W. H. (2019). The global syndemic of obesity, undernutrition, and climate change: The Lancet Commission report. *The Lancet*, *393*(10173), 791–846.

Taghipour, S., Yeung, K.-L., & Ataie-Ashtiani, B. (2022). Efficiency of mechanochemical ball milling technique in the preparation of Fe/TiO2 Photocatalysts. *Chemical Engineering, 6*, 77.

Taha, S. H., El-Sherbiny, I. M., Salem, A. S., Abdel-Hamid, M., Hamed, A. H., & Ahmed, G. A. (2019). Antiviral activity of curcumin loaded milk proteins nanoparticles on *Potato Virus Y*. *Pakistan Journal of Biological Sciences: PJBS*, *22*(12), 614–622.

Taju, M., & Yesuf N. (2022). Review on coffee leaf rust (*Hemileia vastatrix*) and its management in Ethiopia. *International Journal of Research Studies in Science, Engineering and Technology*, *8*(4), 5–13.

Tan, B. L., Norhaizan, M. E., Liew, W.-P.-P., & Sulaiman Rahman, H. (2018). Antioxidant and oxidative stress: A mutual interplay in age-related diseases. *Frontiers in Pharmacology*, *9*, 1162.

Tariq, M., Mohammad, K. N., Ahmed, B., Siddiqui, M. A., & Lee, J. (2022). Biological synthesis of silver nanoparticles and prospects in plant disease management. *Molecules (Basel, Switzerland)*, *27*(15), 4754.

Tawfik, E., Hammad, I., & Bakry, A. (2022). Production of transgenic *Allium cepa* by nanoparticles to resist *Aspergillus niger* infection. *Molecular Biology Reports*, *49*(3), 1783–1790.

Thakur, P. S., Sheorey, T., & Ojha, A. (2023). VGG-ICNN: A lightweight CNN model for crop disease identification. *Multimedia Tools and Applications*, *82*(1), 497–520.

Tiwari, V., Joshi, R. C., & Dutta, M. K. (2021). Dense convolutional neural networks based multiclass plant disease detection and classification using leaf images. *Ecological Informatics*, *63*, 101289.

Tripathi, A. N., Tiwari, S. K., Behera, T. K., Tripathi, A. N., Tiwari, S. K., & Behera, T. K. (2022). Postharvest diseases of vegetable crops and their management. In *Postharvest Technology—Recent Advances, New Perspectives and Applications*. IntechOpen.

Tziveleka, L.-A.; Pippa, N.; Ioannou, E.; Demetzos, C.; Roussis, V. (2022). Development of Ulvan-containing liposomes as antibacterial drug delivery platforms. *Journal of Functional Biomaterials, 13*, 186.

United Nations General Assembly (UNGA). (2019). Resolution adopted by the general assembly on December 20th, 2019.

Van Esse, H. P., Reuber, T. L., & van der Does, D. (2020). Genetic modification to improve disease resistance in crops. *The New Phytologist*, 225(1), 70–86.

Vejlupkova, Z., Warman, C., Sharma, R., Scheller, H. V., Mortimer, J. C., & Fowler, J. E. (2020). No evidence for transient transformation via pollen magnetofection in several monocot species. *Nature Plants*, *6*(11), Art. 11.

Vyska, M., Cunniffe, N., Gilligan, C. (2016). Trade-off between disease resistance and crop yield: A landscape-scale mathematical modelling perspective. *Journal of Royal Society of Interface,* 13, 0451.

Wang, X., Liu, X., Chen, J., Han, H., & Yuan, Z. (2014). Evaluation and mechanism of antifungal effects of carbon nanomaterials in controlling plant fungal pathogen. *Carbon*, *68*, 798–806.

Wani, A. H., & Shah, M. A. (2012). A unique and profound effect of MgO and ZnO nanoparticles on some plant pathogenic fungi. *Journal of Applied Pharmaceutical Science*, *2*(3), 40–44.

Xiao, X., Teng, F., Shi, C., Chen, J., Wu, S., Wang, B., Meng, X., Essiet Imeh, A., & Li, W. (2022). Polymeric nanoparticles—Promising carriers for cancer therapy. *Frontiers in Bioengineering and Biotechnology Frontiers in Bioengineering and Biotechnology*, *10*, 1024143.

Yetisgin, A. A., Cetinel, S., Zuvin, M., Kosar, A., & Kutlu, O. (2020). Therapeutic nanoparticles and their targeted delivery applications. *Molecules*, *25*(9), Art. 9.

Young, M., Ozcan, A., Myers, M. E., Johnson, E. G., Graham, J. H., & Santra, S. (2018). Multimodal Generally Recognized as Safe ZnO/Nanocopper Composite: A novel antimicrobial material for the management of citrus phytopathogens. *Journal of Agricultural and Food Chemistry*, *66*(26), 6604–6608.

Zhao, X., Meng, Z., Wang, Y., Chen, W., Sun, C., Cui, B., Cui, J., Yu, M., Zeng, Z., Guo, S., Luo, D., Cheng, J. Q., Zhang, R., & Cui, H. (2017). Pollen magnetofection for genetic modification with magnetic nanoparticles as gene carriers. *Nature Plants*, 3(12), Art. 12.

Zhu, H., Li, C., & Gao, C. (2020). Applications of CRISPR–Cas in agriculture and plant biotechnology. *Nature Reviews Molecular Cell Biology*, *21*(11), Art. 11.

Zhu, Z.-J., Wang, H., Yan, B., Zheng, H., Jiang, Y., Miranda, O. R., Rotello, V. M., Xing, B., & Vachet, R. W. (2012). Effect of surface charge on the uptake and distribution of gold nanoparticles in four plant species. *Environmental Science & Technology*, 46(22), 12391–12398.

11 Management of Plant Disease Using Nanotechnology

A Paradigm Shift in Crop Protection

Tesleem Taye Bello, Abdulmujib Yusuf, Taofeek T. Adegboyega, and Oluwatoyin Adenike Fabiyi

11.1 INTRODUCTION

Agricultural pathogens constitute a major avenue for yield loss in crop production worldwide. Crop diseases caused by agricultural pathogens constitute a major avenue for substantial economic losses to farmers, agricultural industries, and national economies through reduced yields, increased production costs, and decreased marketability of affected crops (Ansari and Mahmood 2017a; Ansari and Mahmood 2019a; Ansari and Mahmood 2019b; Ansari et al., 2020; Ansari et al., 2020). Aside from this, many crop diseases can also affect wild plant species, leading to ecosystem disruptions and loss of biodiversity (de Freitas Bueno et al., 2019; Ibrahim et al., 2020; Devi et al., 2020; Ibrahim et al., 2022; Keshari et al., 2024; Prasad et al., 2020; Solanki et al., 2020). International trade in agricultural commodities is subject to strict phytosanitary regulations aimed at preventing the spread of plant pests and diseases. Failure to manage crop diseases can result in trade restrictions, market access barriers, and economic losses for exporting countries. Conventional management methods of these economically important pathogens over time rely essentially on chemical pesticides. Continuous use of agricultural chemicals for managing agricultural pathogens has been exploited to prevent farmer's crops from attack by a wide variety of pathogenic organisms causing infections and reducing yield. However, these chemicals are characterized by persistent active ingredients that have been found to have devastating effects on soil and water composition, non-target organisms, fruits, vegetables as well as altering the soil microbial balance (Fabiyi and Olatunji, 2021; Fabiyi et al., 2023a; Ansari et al., 2019; Mahmood et al., 2019; Ansari and Khan 2012). Therefore, in view of the high costs and detrimental action of chemicals there is a general call for reduction in chemical pesticides application in agricultural pests and diseases management. The advent of nanotechnology has however, opened new doors to new concepts and has also advanced with significant impact in different fields such as medicine, engineering, pharmacology and the likes. NanoParticles (NPs) are particles that are extremely small with size typically ranging between 1–100 nm (Holister et al., 2003). NPs can be produced from different sources such as metals, ceramics, clay or polymers. As a result of their minute sizes, nanoparticles are known to often exhibit distinct features and behaviours which confers on them special abilities when compared with bulk particles (Auffan et al., 2009; Fabiyi, 2023a;b; Anees et al., 2024). Nanotechnology has been successfully applied in disease diagnosis, imaging and in targeted drug delivery. In electronics, nanoparticles are used to enhance the performance of electronic devices and improve energy efficiency. Nanoparticles offer a more ecologically sound option in agricultural pest management due to their capacity to release their active ingredients slowly, thus leading to improved efficiency at an optimum application rate, thereby

DOI: 10.1201/9781003256762-11

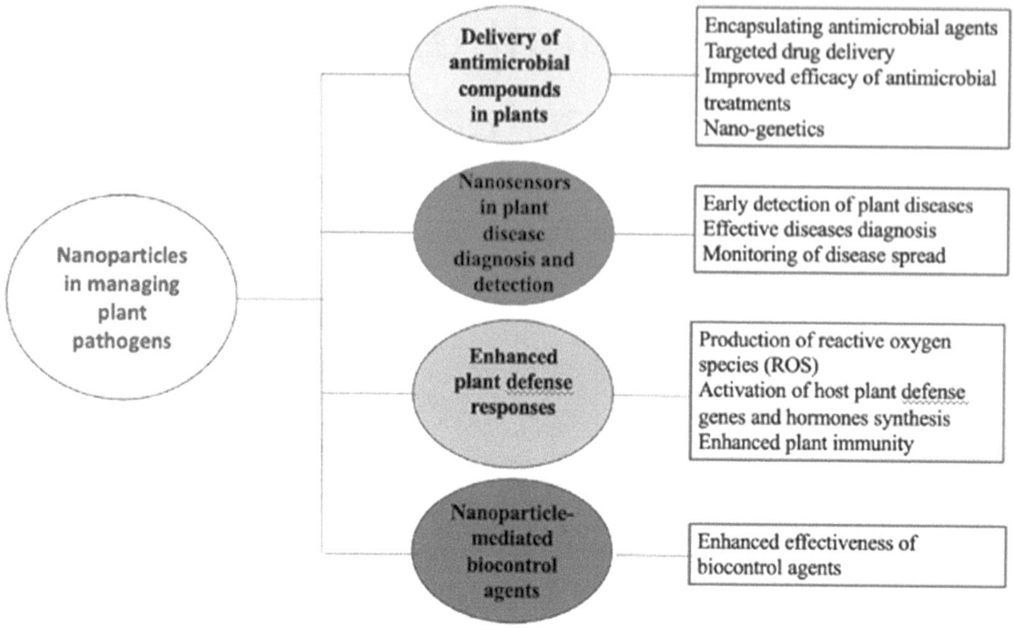

FIGURE 11.1 Illustration depicting the diverse applications of nanotechnology in combating agricultural plant diseases. The diagram highlights various sectors where nanotechnology plays a crucial role, including disease detection, targeted delivery of antimicrobial agents, enhancing plant defence system, and increasing biocontrol agent potential in knocking down the pathogens population through nanomaterial-based strategies.

reducing costs while ensuring effective delivery (Fabiyi et al., 2020; Kumar et al., 2021; Fabiyi et al., 2021; Fabiyi et al., 2023b; Bello and Fabiyi 2024). The inherent potential of nanotechnology has been successfully applied in different aspects of agriculture and food production worldwide but its potential in managing agricultural pests and pathogens is yet to be fully harnessed. Figure 11.1 highlights some applications of nanotechnology in crop protection which can be viewed in terms of the different mechanisms of action:

1. Nanoparticles in direct crop protection utilised as nano fertilizers, pesticide formulations, and so forth
2. Nanosensors for detecting plant disease
3. Nanoparticles to enhance plant defence systems
4. Nanoparticles as carriers in pesticide formulations or genetic materials such as the double-stranded RNA (dsRNA) (Worrall et al., 2018). As carriers, NPs offer a number of advantages, including:
   (i) Increased targeted uptake by the intended pest
   (ii) Enhanced solubility for pesticides that are poorly soluble in water
   (iii) Decreased toxicity
   (iv) Longer shelf life (Hyles et al., 2017). Furthermore, when exposed to environmental elements like UV rays and rain, nanocarriers may help boost the stability and efficacy of nanopesticides, which could result in a significant decrease in the frequency of applications leading to an overall reduction in toxicity. This chapter highlight the various mechanisms through which nanomaterials are employed in managing agriculturally important plant diseases.

11.2 DELIVERY OF ANTIMICROBIAL COMPOUNDS IN PLANTS

Nanoparticles have been effectively utilized to serve as carriers for delivering antimicrobial compounds to plants (Ivanova et al., 2018). This is usually achieved by encapsulating antimicrobial agents within nanoparticles, so that their stability and controlled release can be enhanced. This targeted delivery system has helped to improve the effectiveness of antimicrobial treatments, reducing the amount of chemicals needed and minimizing potential negative impacts on the environment (Hu and Xianyu, 2021).

11.3 APPLICATION OF NANOTECHNOLOGY (NANOSENSORS) IN PLANT DISEASE DIAGNOSIS AND DETECTION

Another notable area of application of nanotechnology in managing bacterial diseases of plants is in the use of nanosensors for the early detection and monitoring of bacterial diseases in plants (Kashyap et al., 2019; John et al., 2022). Nanosensors are tiny devices that can detect and measure changes in properties at the nanoscale, whether they are chemical, physical, or biological. These sensors, which consist of nanomaterials such as nanowires or nanoparticles, are designed to detect specific molecules or chemicals with high specificity and sensitivity (Javaid et al., 2021). Applications for nanosensors are numerous and span a range of industries, including food safety, healthcare, and environmental monitoring, and security (Deshpande et al., 2018; Adam and Gopinath, 2022). For example, they have been used to detect pollutants in water or air, monitor glucose levels in diabetic patients, detect pathogens in food, or identify explosives in security screening (Deshpande et al., 2018). Nanosensors work by utilizing various sensing mechanisms, such as changes in electrical conductivity, optical properties, or mass (Adam and Gopinath, 2022). Due to the great potential of nanosensor, they have been used to effectively detect specific biomarkers or signals associated with some bacterial infections, providing rapid and accurate detection even before visible symptoms appear (Yusof and Isha, 2020). Early detection of several bacterial pathogens in plants has enabled timely intervention and management strategies, reducing the spread and severity of bacterial diseases in plants (Kaushal and Wani, 2017; John et al., 2022).

11.4 ENHANCED PLANT DEFENCE RESPONSES

To strengthen plant defence mechanisms against various bacterial diseases, a variety of nanoparticles have been employed (Ahmed et al., 2022). Different NPs have been linked to the activation of defence responses and the expression of genes relevant to plant defence, which increases plant resistance to bacterial infections (Siddiqui et al., 2019). This strategy can strengthen plant defences against bacterial infections and lessens that susceptibility. Plants have been shown to respond defensively to nanoparticles in a number of ways (Elmer et al., 2018; El-Shetehy et al., 2021). Literature reports have demonstrated that plants can engage defence systems, in response to nanoparticle exposure, that are akin to those triggered in reaction to infections or environmental challenges (Cai et al., 2020; Iqbal et al., 2020). One common defence-related reaction in plants is the production of Reactive Oxygen Species (ROS) (Abdal Dayem et al., 2017).

Although ROS are known to be highly reactive chemicals that can damage cells, they are also necessary for plants to signal defence responses (O'Brien et al., 2012; Nanda et al. 2010). Research has demonstrated that exposure to specific nanoparticles can cause plants to produce Reactive Oxygen Species (ROS), which can activate defence genes and cause the manufacture of chemicals relevant to defence (Segal and Wilson, 2018). Moreover, interactions between NPs and plant cell walls have been reported to activate defence responses through a series of signalling cascades (O'Brien et al., 2012; Molnár et al., 2020). Defence hormones like ethylene, Salicylic Acid (SA), and Jasmonic Acid (JA), which are involved in the regulation of several defence pathways in plants, are expected to be synthesised to complete this process (Kasote et al., 2019; Kurczyńska et al.,

2021; Sonkar et al., 2021). Furthermore, nanoparticles have the ability to disrupt gene expression during defensive reactions. It has been shown through various studies that exposure to nanoparticles can upregulate Pathogenesis-Related (PR) genes that have roles in plant defence against pathogens (García-Sánchez et al., 2015). Nanoparticles have been utilized to transmit genetic material like antimicrobial genes or RNA interference (RNAi) molecules, into plants (Meena et al., 2012; Lv et al., 2020). This approach, known as nanogenetics, allows for targeted and efficient delivery of genetic material to plant cells, enabling the plants to produce antimicrobial compounds or silence specific genes involved in bacterial infection (Worrall et al., 2018). Nanoparticle-based delivery systems have shown great potential for delivering genetic material into cells (Li et al., 2006). These systems secure and deliver genetic material, such as DNA or RNA, to the target cells by using nanoparticles as carriers (Cai et al., 2019; 2020). Genetic materials have been transferred via a variety of nanoparticle types, including liposomes, polymeric nanoparticles, and inorganic nanoparticles (Kashyap et al., 2015; Demirer et al., 2019). To improve these nanoparticles' stability, cellular absorption, and release of the genetic material, their size, surface charge, and surface changes were altered (Demirer and Landry, 2017). One of the main advantages of the delivery systems utilizing nanomaterial is their ability to protect the genetic material from degradation and enzymatic degradation in the plant system. The nanoparticles can encapsulate the genetic material within their core or attach it to their surface, providing a protective barrier and increasing its stability during transport (Worrall et al., 2018).

11.5 NANOPARTICLE-MEDIATED BIOCONTROL AGENTS

Biocontrol products are crucial in the treatment of plant diseases because they use living organisms to lower plant pathogen populations (Ansari et al., 2017; Ansari and Mahmood 2017b; Raymaekers et al., 2020; Sumbul et al., 2020; Solanki et al., 2021;Ansari et al., 2023; 2024; Patil et al., 2024). By directly opposing plant pathogens through competition for nutrients and space, the synthesis of antimicrobial chemicals, or the stimulation of plant defence systems, certain bacteria, fungus, and other microbes can function as biocontrol agents. Examples include *Bacillus subtilis, Trichoderma* spp., and *Pseudomonas fluorescens* (Poveda and Eugui, 2022). Some fungi parasitize plant pathogens, either by infecting and killing them directly or by competing for resources. Examples include species of the genus *Trichoderma*, which parasitizes other fungi and suppress their growth, thereby reducing disease incidence (Poveda and Eugui, 2022; Boyno et al., 2024). Endophytic microbes live within plant tissues and do not cause illness. They can benefit their host in a number of ways, including increased pathogen resistance (Latz et al., 2018; De Silva et al., 2019). Secondary metabolites produced by endophytic bacteria and fungi can either promote plant defences or impede the establishment of plant diseases (De Silva et al., 2019). Plant development and health are enhanced by certain rhizosphere bacteria that produce phytohormones, improve nutrient uptake, and build systemic resistance to diseases (Wang et al., 2021). Examples include strains of *Pseudomonas* and *Bacillus* that colonize plant roots and provide protection against soil-borne pathogens (Wang et al., 2021). Nanoparticles have been utilized to enhance the efficacy of biocontrol. Many of these biocontrol agents, such as beneficial bacteria or bacteriophages, are used for managing bacterial diseases (Ogunyemi et al., 2023). By attaching or encapsulating these biocontrol agents with nanoparticles, their stability, survival, and targeted delivery to the plants was improved, increasing their effectiveness in suppressing bacterial pathogens (Wagemans et al., 2022; Ogunyemi et al., 2023). Several nanoparticles can be engineered to possess antimicrobial properties, which can help control bacterial pathogens. As an example, it has been found that silver nanoparticles exhibit broad-spectrum antibacterial activity against bacteria and other plant diseases. By incorporating these nanoparticles into sprays or coatings, it is possible to specifically target and limit the growth of bacterial infections on plant surfaces. As a result, numerous plant diseases have demonstrated the immense potential that nanotechnology holds. The antiviral (Singh et al., 2022) and antibacterial (Kim et al., 2012) properties of metal nanoparticles have been the subject of extensive research.

11.6 NANOPARTICLES FOR MANAGING BACTERIAL DISEASES OF PLANTS

In the management of plant disease using NPs, silver NPs are the most widely studied for their antimicrobial properties against plant pathogens (Kim et al., 2012; Parveen and Siddiqui, 2021). They can inhibit the growth and activity of bacterial cells by disrupting their cell membranes and interfering with cellular processes. Many bacterial pathogens of plants such as *Erwinia amylovora* causing fire blight, *Pseodomonas syringea* causing canker, and *Xanthomonas campestris* causing bacterial leaf spot have been effectively controlled using differently synthesized AgNPs worldwide (Parveen and Siddiqui, 2021). For example, a bacterial disease, known as fire blight, which is caused by the bacterium *Erwinia amylovora* is responsible for substantial yield losses in plants especially within the Rosaceae family, such as apple and pear trees. Green synthesized AgNPs show inhibitory effects on bacterial growth when applied as sprays or coatings to control the disease (Tarighi and Nejad, 2023). A number of plant bacterial strains including *Xanthomonas citri* subsp. *citri*, which causes citrus canker, bacterial leaf spot, soft rot, and bacterial wilt, have been the subject of research on the antibacterial capabilities of copper nanoparticles (CuNPs) (Gkanatsiou et al., 2019; Varympopi et al., 2020; Ma et al., 2022). CuNPs can emit copper ions, which can be poisonous to bacteria and cause cell death. CuNPs have also shown the ability to reduce bacterial growth and suppress the disease (Atiq et al., 2022). Zinc oxide NPs (ZnONPs) have attracted wide interest for their antimicrobial activity against bacterial pathogens causing bacterial blight, bacterial spot, crown gall and leaf spot diseases (Parveen and Siddiqui, 2021; Shantharaj et al., 2023). ZnONPs have shown effectiveness in managing these pathogens by inducing oxidative stress and disrupting bacterial cell membranes thereby inhibiting the bacterial growth with the resultant effect of reducing the disease symptoms (Fan et al., 2023). Chitosan nanoparticles derived from chitin, a natural polymer found in crustacean shells, have antimicrobial properties. They can disrupt bacterial cell membranes and inhibit bacterial growth. Chitosan nanoparticles have been investigated and found effective for controlling several bacterial diseases of agricultural importance such as the citrus canker, bacterial soft rot and leaf blight (Kashyap et al., 2015; Ahmed et al., 2022). Carbon-based NPs, such as carbon nanotubes and graphene oxide, have shown antimicrobial activity against plant bacterial pathogens. They can trigger cell death through their interactions with the bacterial cells and the damage they cause to the membrane (Ahmed et al., 2022). Carbon-based nanoparticles have been studied and found effective for controlling bacterial diseases like bacterial leaf spot and bacterial wilt. A list of common fungal pathogens of plants that have been controlled with different kinds of nanoparticles are given in Table 11.1.

11.7 NANOPARTICLES FOR MANAGING FUNGAL DISEASES OF PLANTS

Several metal nanoparticles have been explored and proved effective in managing different fungal pathogens of different crops worldwide (Kim et al., 2012). Below is a brief summary of some plant fungal pathogens that have been effectively controlled using different kinds of metal nanoparticles. AgNPs have been extensively studied with respect to their applications in managing different kinds of pathogenic agents. AgNPs are known to produce antifungal effects on many common and economically important plant fungal pathogens such as *Rhizoctonia solani* causing root rot in many ornamental crops, *Alternaria alternata* causing brown rot in many fruits and vegetables, *Botrytis cinerea,* causing soft rot in many horticultural crops, Rice blast disease caused by *Magnaporthe oryzae* and *Curvularia lunata* causing leaf spot of maize and other cereals and so forth (Kim et al., 2012). Copper nanoparticles (CuNPs) have been found to have antifungal properties against late blight caused by the oomycete pathogen *Phytophthora infestans*, by inhibiting the growth of

TABLE 11.1
Nanoparticles in managing bacterial, fungal and viral pathogens of infecting wide range of crops

S. No.	NPs	Pathogens	Diseases/agents	References
		Bacterial plant pathogens		
1.	AgNPs	*Xanthomonas oryzae pv. oryzae*	Bacterial blight of rice	Ibrahim et al. (2019) Ahmed et al. (2020)
2.		*Erwinia cacticida*	Soft rot diseases of cactus	Paulkumar et al. (2014)
3.		*Dickeya dadantii*	Soft rot diseases of crops	Hossain et al. (2019)
4.		*Ralstonia solanacearum*	Wilt diseases of crops	Wang et al. (2022)
5.		*Agrobacterium tumefaciens*	Crown gall diseases	Mohammad and Abd El-Rahman (2015)
6.		*Pseudomonas syringae*	Bacterial dieback diseases	Jiang et al. (2022)
7.		*Erwinia carotovora, Ralstonia solanacearum*	Soft rot diseases	Bibi et al. (2023)
8.	ZnONPs	*Xanthomonas oryzae pv. oryzae*	Bacterial blight of rice	Ogunyemi et al. (2023)
9.		*Ralstonia solanacearum*	Wilt diseases of crops	Khan et al. (2021)
10.		*Pseudomonas syringae, Xanthomonas campestris pv. vesicatoria, Pectobacterium carotovorum* sub sp. *carotovorum, Ralstonia solanacearum*	Bacterial dieback diseases, Bacterial blight diseases, wilt diseases	Parveen and Siddiqui (2021)
11.		*Pseudomonas syringae*	Bacterial dieback diseases	Fan et al. (2023)
12.		*Xylella fastidiosa*	Pierce's diseases of grapevine	Shantharaj et al. (2023)
13.		*Xanthomonas axonopodis* pv. *phaseoli*	Bacterial blight of beans	Siddiqui et al. (2018)
14.	CuONPs	*Pseudomonas syringae* pv tomato	Bacterial leaf spot	Li et al. (2017)
15.		*Ralstonia solanacearum*	Wilt diseases of crops	Chen et al. (2019)
16.	CuNPs	*Pseudomonas syringae pv tabaci* *Xanthomonas citri*	Brown spot diseases of tobacco; Bacterial blight of citrus	Ma et al. (2022)
17.		*Agrobacterium tumefaciens*	Crown gall diseases	Varympopi et al. (2020)
18.		*Erwinia amylovora, Xanthomonas campestris, Pseudomonas syringae*	Soft rot; Bacterial blight; Leaf spot diseases	Gkanatsiou et al. (2019)
19.	MgONPs	*Ralstonia solanacearum*	Wilt diseases of crops	Cia et al. (2018)
20.	FeONPs	*Ralstonia solanacearum*		Alam et al. (2019)
		Fungal plant pathogens		
21.	AgNPs	*Aspegillus flavus, Aspergillus niger,, Fusarium oxysporum, Curvularia lunata, Rhizopus arrizus*	Ear rot disease; Post-harvest decay of fruits; Leaf spot; fusarium wilt.	Savithramma et al. (2011)
22.		*Alternaria alternata, Botrytis cinerea, Sclerotinia aclerotium, Rhizoctonia solani*	Fungal spots rots and blights; White mold disease; Damping off, root and stem rot diseases.	Krishnaraj et al. (2012)

TABLE 11.1 (Continued)
Nanoparticles in managing bacterial, fungal and viral pathogens of infecting wide range of crops

S. No.	NPs	Pathogens	Diseases/agents	References
23.	CuNPs	*Aspergillus niger, Aspegillus flavus.*	Ear rot disease; Post-harvest decay of fruits	Rajesh et al. (2018)
24.		*Altrnaria alternata, Curvularia lunat,, Fusarium oxysporum, Phoma destructive*	Fungal spots rots and blights diseases; Fusarium wilt	Kanhed et al (2014)
25.		*Fusarium oxysporum,* Fusarium *culmorum* Fusarium *germaniearum.*	Fusarium wilt diseases	Shende et al. (2018)
26.	CuONPs	*Alternaria alternata Fusarium oxysporum, Phytium ultinum,*	Fungal spots rots and blights diseases; Fusarium wilt	Hassan et al. (2018)
27.	ZnONPs	*Aspergillus niger, Aspegillus flavus, Fusarium germaniearum. Aspergillus niger*	Ear rot disease; Post-harvest decay of fruits	Rajiv et al. (2013)
28.		*Fusarium germaniearum Botryosphaeria dothidea Diploda seriata, Alternaria mali*	Fusarium wilt Rot diseases complex	Ahmed et al. (2020) Ramachandrappa et al. (2019)
29.		*Fusarium graminearum*	Fusarium wilt diseases	Ibrahim et al. (2020)
30.		*Fusarium oxysporum* fsp lycopersici, *Alternaria solani*		Parveen and Siddiqui, (2021)
31.		*Botryosphaeria dothidea*		
32.	AuNPs	*Aspergillus niger, Alternaria solani, Puccina graminis,*	Ear rot disease, rust	Jayaseelan et al. (2013); Islam et al., (2019)
33.		*Sclerotium rolfsii*	Rot diseases of vegetables	Thanighaiarassu et al. (2014)
34.	MgONPs	*Fusarium oxysporum*	Fusarium wilt diseases	Abdel-Aziz et al. (2020)
35.		*Fusarium solani, Fusarium oxysporum*		Abdallah et al. (2022)
		Viral plant pathogens		
36.	AgNPs	The sunhemp rossette virus		Jain and Kothari (2014)
37.		The new banana bunchy-top virus		Mahfouze et al. (2020)
38.		Potato virus, Tomato mosaic virus		Noha et al. (2018)
39.		Tobacco mosaic virus		Wang et al. (2016)
40.		Tomato spotted wilt virus		Shafie et al. (2018)
41.	ZnONPs	Cucumber mosaic virus		Hao et al. (2017)
42.		Tobacco mosaic virus		Cai et al. (2019)
43.	AuNPs	Barley yellow mosaic virus, Cotton leaf curl virus		Amudha et al. (2011)
44.		Barley yellow dwarf virus		Briddon and Markham (2000)
45.	SiO_2NPs	Tomato yellow leaf curl virus		Mofiat (1999)
46.		Papaya ring spot virus		Tan et al. (2018)
47.		Tobacco mosaic virus		Cai et al. (2019)

the pathogen thereby reducing disease severity. Green synthesized AgNPs were used to manage common economically important fungal pests of crops such as *Fusarium oxysporum*, *Rhizopus arrizus*, *Curvularia lunata*, *Aspergillus flavus*, and *Aspergillus niger* (Savithramma et al., 2012). Many bio-synthesized gold nanoparticles (AuNPs) have shown efficacy in controlling several economically important agricultural plant pathogens. For example, fungi pathogens such as *Sclerotium rolfsii* causing the southern blight diseases in vegetables and melons, *Alternaria solani* causing early blight diseases, *Puccina graminis* which causes stem rot disease in cereals and other crops (Krishnaraj et al., 2012). Additionally, zinc nanoparticles have shown promise in combating a number of fungal plant pathogens, including *Diploda seriata*, which causes grapevine canker disease, *Fusarium graminearum*, which causes the head blight disease of wheat and barley, and *Botryosphaeria dothidea*, which causes cankers on numerous trees and shrubs (Ramachandrappa et al., 2019; Ahmed et al., 2020).

11.8 NANOPARTICLES FOR MANAGING VIRAL DISEASES OF PLANTS

Plant viruses account for a substantial amount of yield losses in agricultural production worldwide. This is due to their genome diversity, rapid mutation and evolution. Several nanoparticles have shown great promise in the management of plant viruses (Vargas-Hernandez et al., 2020; Farooq et al., 2021; El Gamal et al., 2022). As against the large volumes of available literature regarding other plant pathogens such as bacteria and fungi, few reports have addressed the use of nanotechnology in managing viral diseases of agricultural crops worldwide. Some nanoparticles which have proved effective in managing several viral pathogens of plants are highlighted in Table 11.1.

11.9 CONCLUSIONS AND FUTURE INSIGHTS

Crop diseases can significantly reduce yields and quality, leading to food shortages and insecurity, especially in regions heavily dependent on agriculture for food supply. Effective disease management helps safeguard food production and availability. The application of nanoparticles has proved to be one environmentally sound alternative to this problem. Nanoparticles may influence the efficacy of conventional pesticides by enhancing their solubility, stability, and targeted delivery to specific plant organs or pathogens, reducing the amount of pesticide needed and minimizing environmental impact. Certain nanoparticles can trigger plant defence mechanisms, enhancing the plants' innate abilities to resist pathogen attacks. Nanotechnology no doubt has a great potential for addressing disease problems affecting agricultural production. It offers an unprecedented range of applications with positive attributes such as having enhanced absorption rate and improved bioavailability, toxic selectivity and opportunities for the early detection of pathogens. However, it is crucial to know that the success of nanoparticles in controlling crop diseases can vary depending on factors including the concentration, size of the particles and formulation. Further research is needed to optimize their use and assess their residual impacts on plants and the environment as well.

REFERENCES

Abdal Dayem, A., Hossain, M. K., Lee, S. B., Kim, K., Saha, S. K., Yang, G. M., ... & Cho, S. G. (2017). The role of reactive oxygen species (ROS) in the biological activities of metallic nanoparticles. *International Journal of Molecular Sciences*, *18*(1), 120.

Abdel-Aziz, M. M., Emam, T. M., & Elsherbiny, E. A. (2020). Bioactivity of magnesium oxide nanoparticles synthesized from cell filtrate of endobacterium *Burkholderia rinojensis* against Fusarium oxysporum. *Materials Science and Engineering: C*, *109*, 110617.

Adam, T., & Gopinath, S. C. (2022). Nanosensors: Recent perspectives on attainments and future promise of downstream applications. *Process Biochemistry*, *117*, 153–173.

Ahmed, T., Noman, M., Jiang, H., Shahid, M., Ma, C., Wu, Z., ... & Li, B. (2022). Bioengineered chitosan-iron nanocomposite controls bacterial leaf blight disease by modulating plant defense response and nutritional status of rice (*Oryza sativa* L.). *Nano Today*, *45*, 101547.

Ahmed, T., Shahid, M., Noman, M., Niazi, M. B. K., Mahmood, F., Manzoor, I., ... & Chen, J. (2020). Silver nanoparticles synthesized by using *Bacillus cereus* SZT1 ameliorated the damage of bacterial leaf blight pathogen in rice. *Pathogens*, *9*(3), 160.

Alam, T., Khan, R. A. A., Ali, A., Sher, H., Ullah, Z., & Ali, M. (2019). Biogenic synthesis of iron oxide nanoparticles via *Skimmia laureola* and their antibacterial efficacy against bacterial wilt pathogen *Ralstonia solanacearum*. *Materials Science and Engineering: C*, *98*, 101–108.

Amudha, J., Balasubramani, G., Malathi, V. G., Monga, D., & Kranthi, K. R. (2011). Cotton leaf curl virus resistance transgenics with antisense coat protein gene (AV1). *Current Science*, *101*(3), 300–307.

Anees, A., Gupta, R., Phanindra, P. V., Fabiyi, O. A., Thera, U. K., Bello, T. T., & Ahmad, F. (2024). Green Synthesis of Nanoparticles and Applications. In *Advanced Nanotechnology in Plants* (pp. 89 –122) CRC Press.

Ansari, R. A., & Khan, T. A. (2012). Parasitic association of root-knot nematode, *Meloidogyne incognita* on guava. *e-Journal of Science Technology*, *5*(12), 65–67.

Ansari, R. A., & Mahmood, I. (2017a). Determination of disease incidence caused by *Meloidogyne* spp. and or *Fusarium udum* on pigeonpea in Aligarh district: A survey. *Trends in Biosciences, 10*(24), 5239–5243.

Ansari, R. A., & Mahmood, I. (2017b). Optimization of organic and bio-organic fertilizers on soil properties and growth of pigeon pea. *Scientia Horticulturae*, *226*, 1–9.

Ansari, R. A., Mahmood, I., Rizvi, R., Sumbul, A., & Safiuddin. (2017). Siderophores: Augmentation of soil health and crop productivity. In Kumar, V., Kumar, M., Sharma, S., Prasad, R. (eds.), *Probiotics in Agroecosystem*. Springer, Singapore.

Ansari, R. A., Rizvi, R., & Mahmood, I. (2023). *Mycorrhizal Symbiosis and Agroecosystem Restoration*, Springer.

Ansari, R. A., Rizvi, R., Sumbul, A., & Mahmood, I. (2020). Plant-growth-promoting rhizobacteria (PGPR)-based sustainable management of phytoparasitic nematodes: Current understandings and future challenges. In Ansari, R., Rizvi, R., Mahmood, I. (eds), *Management of Phytonematodes: Recent Advances and Future Challenges* (pp. 51–71). Singapore: Springer.

Ansari, R. A., Sumbul, A., Rizvi, R., & Mahmood, I. (2019). Organic soil amendments: potential tool for soil and plant health management. In Ansari, R., Mahmood, I. (eds.), *Plant Health Under Biotic Stress*. Singapore: Springer.

Ansari, R. A., Sumbul, A., Rizvi, R., & Mahmood, I. (2024). Impact of environmental gases on mycorrhizal symbiosis and its influence on ecosystem functioning under the current climate change scenario. In *Mycorrhizal Symbiosis and Agroecosystem Restoration* (pp. 51–76). Singapore: Springer Nature Singapore.

Ansari, R. A., & Mahmood, I. (2019a). *Plant Health Under Biotic Stress – Volume II, Microbial Interactions*. Singapore: Springer Nature. ISBN: 978-981-13-6040-4.

Ansari, R. A., & Mahmood, I. (2019b). *Plant Health Under Biotic Stress – Volume I, Organic Strategies*. Singapore: Springer Nature. ISBN: 978-981-13-6043-5.

Ansari, R. A., Rizvi, R., & Mahmood, I. (2020). *Management of Phytonematodes: Management of Phytonematodes: Recent Advances and Future Challenges*. Singapore: Springer Nature. ISBN: 978-981-15-4087-5.

Atiq, M., Mazhar, H. M. R., Rajput, N. A., Ahmad, U., Hameed, A., & Lodhi, A. (2022). Green synthesis of silver and copper nanoparticles from leaves of *Eucalyptus globulus* and assessment of its antibacterial potential towards *Xanthomonas citri* pv. *citri* causing citrus canker. *Applied Ecology and Environmental Research*, *20*(3), 2205–2213.

Auffan, M., Rose, J., Bottero, J. Y., Lowry, G. V., Jolivet, J. P., & Wiesner, M. R. (2009). Towards a definition of inorganic nanoparticles from an environmental, health and safety perspective. *Nature Nanotechnology*, *4*(10), 634–641.

Bello, T. T., & Fabiyi, O. A. (2024). Facile Biosynthesis of Gold Nanoparticles Using Microbes and Their Applications in the Management of Economically Important Plant Pathogens. In *Nanotechnology in Plant Health* (pp. 164–179). CRC Press.

Bibi, Z., Ali, M., Abohashrh, M., Ahmad, I., Khan, H., Ali, M., ... & Muhmood, T. (2023). Biologically synthesized silver nanoparticles efficiently control plant pathogenic Bacteria-*Erwinia carotovora* and *Ralstonia solanacearum. Inorganics*, *11*(7), 309.

Boyno, G., Ansari, R. A., Durak, E. D., Güneş, H., Çevik, R., & Demir, S. (2024). Arbuscular mycorrhizal technology in sustainable agriculture: Current knowledge and challenges in agroforestry. *Mycorrhizal Symbiosis and Agroecosystem Restoration*, *9*, 173–195.

Briddon, R. W., & Markham, P. G. (2000). Cotton leaf curl virus disease. *Virus Research*, *71*(1–2), 151–159.

Cai, L., Cai, L., Jia, H., Liu, C., Wang, D., & Sun, X. (2020). Foliar exposure of Fe_3O_4 nanoparticles on *Nicotiana benthamiana*: Evidence for nanoparticles uptake, plant growth promoter and defense response elicitor against plant virus. *Journal of Hazardous Materials*, *393*, 122415.

Cai, L., Chen, J., Liu, Z., Wang, H., Yang, H., & Ding, W. (2018). Magnesium oxide nanoparticles: Effective agricultural antibacterial agent against *Ralstonia solanacearum. Frontiers in Microbiology*, *9*, 790.

Cai, L., Liu, C., Fan, G., Liu, C., & Sun, X. (2019). Preventing viral disease by ZnONPs through directly deactivating TMV and activating plant immunity in *Nicotiana benthamiana. Environmental Science: Nano*, *6*(12), 3653–3669.

Cai, Q., He, B., Weiberg, A., Buck, A. H., & Jin, H. (2019). Small RNAs and extracellular vesicles: New mechanisms of cross-species communication and innovative tools for disease control. *PLoS Pathogens*, *15*(12), e1008090.

Chen, J., Mao, S., Xu, Z., & Ding, W. (2019). Various antibacterial mechanisms of biosynthesized copper oxide nanoparticles against soilborne *Ralstonia solanacearum. RSC Advances*, *9*(7), 3788–3799.

de Freitas Bueno, R. C. O., Ansari, R. A., Lima, G. P. P., & Sakate, R. K. (2019). Phytosanitation: A novel approach toward disease management. In Ansari, R., Mahmood, I. (eds.), *Plant Health Under Biotic Stress: Volume 1: Organic Strategies* (pp. 73–90). Springer, Singapore.

De Silva, N. I., Brooks, S., Lumyong, S., & Hyde, K. D. (2019). Use of endophytes as biocontrol agents. *Fungal Biology Reviews*, *33*(2), 133–148.

Demirer, G. S., & Landry, M. P. (2017). Delivering genes to plants. *Chemical Engineering Progress*, *113*(4), 40–45.

Demirer, G. S., Zhang, H., Matos, J. L., Goh, N. S., Cunningham, F. J., Sung, Y., ... & Landry, M. P. (2019). High aspect ratio nanomaterials enable delivery of functional genetic material without DNA integration in mature plants. *Nature Nanotechnology*, *14*(5), 456–464.

Deshpande, A. H., Weldode, J. M., & Pise, J. S. (2018). Applications of nano-sensors in various fields: A review. *International Journal of Management, Technology and Engineering* (8), 1403–1408.

Devi, T. S., Das, D., Ansari, R. A., Rizvi, R., Sumbul, A., & Mahmood, I. (2020). Role of organic additives in the sustainable management of phytoparasitic nematodes. *Management of Phytonematodes: Recent Advances and Future Challenges*, 279–295.

El Gamal, A. Y., Atia, M. M., Sayed, T. E., Abou-Zaid, M. I., & Tohamy, M. R. (2022). Antiviral activity of chitosan nanoparticles for controlling plant-infecting viruses. *South African Journal of Science*, *118*(1–2), 1–9.

Elmer, W., Ma, C., & White, J. (2018). Nanoparticles for plant disease management. *Current Opinion in Environmental Science & Health*, *6*, 66–70.

El-Shetehy, M., Moradi, A., Maceroni, M., Reinhardt, D., Petri-Fink, A., Rothen-Rutishauser, B., ... & Schwab, F. (2021). Silica nanoparticles enhance disease resistance in *Arabidopsis* plants. *Nature Nanotechnology*, *16*(3), 344–353.

Fabiyi O. A., Olatunji, G. A., Atolani, O., & Olawuyi, R. O. (2020). Preparation of bio-nematicidal nanoparticles of *Eucalyptus officinalis* for the control of cyst nematode (*Heterodera sacchari*). *The Journal of Animal and Plant Science*, *30*(5), 1172–1177.

Fabiyi, O. A., Lateef, A., Gueguim-Kana, E. B., Beukes, L. S., Matyumza, N., Bello, T. T., & Olatunji, G. A. (2023b). Correction: Characterization and nematicidal potential of copper, iron and zinc nanoparticles synthesized from *Tridax procumbens* L. extract on *Meloidogyne incognita* infected cabbage plants. *European Journal of Plant Pathology*, *168*(2), 419.

Fabiyi, O. A., & Olatunji, G. A. (2021). Environmental sustainability: Bioactivity of *Leucaena leucocephala* leaves and pesticide residue analysis in tomato fruits. *Acta Universitatis Agriculturae et Silviculturae Mendelianae Brunensis*, *69*(4), 473–480.

Fabiyi, O. A., Adebisi, O. O., Falore, S. O., & Claudius-Cole, A. O. (2023a). *Correction: Potential Inhibition of Entomopathogenic Nematodes and Plant Growth-Promotingbacteria with Exposure to Selected Herbicides and Insecticides*. Vegetos.

Fabiyi, O. A. (2023a). The Perspective of Nanotechnology in Plant Protection. In Avila-Quezada, G. D., & Rai, M. (eds.), *Nanophytopathology* (pp. 23–38).

Fabiyi, O. A. (2023b). Nano clay materials and their role in plant disease management. In Avila-Quezada, G. D., & Rai, M. (eds.), *Nanophytopathology* (pp. 145–156).

Fabiyi, O. A., Claudius-Cole, A. O., Olatunji, G. A., Abubakar, D. O., & Adejumo, O. A. (2021). Evaluation of the *in vitro* Response of *Meloidogyne incognita* to silver nano particle liquid from Agricultural wastes. *Agrivita Journal of Agricultural Sciences, 43*(3), 524–534.

Fan, G., Xiao, Q., Li, Q., Xia, Y., Feng, H., Ma, X., ... & Sun, X. (2023). Antimicrobial mechanisms of ZnO nanoparticles to phytopathogen *Pseudomonas syringae*: Damage of cell envelope, suppression of metabolism, biofilm and motility, and stimulation of stomatal immunity on host plant. *Pesticide Biochemistry and Physiology, 194*, 105455.

Farooq, T., Adeel, M., He, Z., Umar, M., Shakoor, N., da Silva, W., ... & Rui, Y. (2021). Nanotechnology and plant viruses: An emerging disease management approach for resistant pathogens. *ACS Nano, 15*(4), 6030–6037.

García-Sánchez, S., Bernales, I., & Cristobal, S. (2015). Early response to nanoparticles in the Arabidopsis transcriptome compromises plant defence and root-hair development through salicylic acid signalling. *BMC Genomics, 16*(1), 1–17.

Gkanatsiou, C., Karamanoli, K., Menkissoglu-Spiroudi, U., & Dendrinou-Samara, C. (2019). Composition effect of Cu-based nanoparticles on phytopathogenic bacteria. Antibacterial studies and phytotoxicity evaluation. *Polyhedron, 170*, 395–403.

Hao, Y., Cao, X., Ma, C., Zhang, Z., Zhao, N., Ali, A., ... & Rui, Y. (2017). Potential applications and antifungal activities of engineered nanomaterials against grey mold disease agent *Botrytis cinerea* on rose petals. *Frontiers in Plant Science, 8*, 1332.

Hayles, J., Johnson, L., Worthley, C., & Losic, D. (2017). Nanopesticides: A review of current research and perspectives. *New Pesticides and Soil Sensors*, 193–225.

Holister, P., Weener, J. W., Roman, C., & Harper, T. (2003). Nanoparticles. *Technology White Papers, 3*, 1–11.

Hossain, A., Hong, X., Ibrahim, E., Li, B., Sun, G., Meng, Y., ... & An, Q. (2019). Green synthesis of silver nanoparticles with culture supernatant of a bacterium *Pseudomonas rhodesiae* and their antibacterial activity against soft rot pathogen *Dickeya dadantii. Molecules, 24*(12), 2303.

Hu, J., & Xianyu, Y. (2021). When nano meets plants: A review on the interplay between nanoparticles and plants. *Nano Today, 38*, 101143.

Ibrahim, D. S. S., Elderiny, M. M., Ansari, R. A., Rizvi, R., Sumbul, A., & Mahmood, I. (2020). Role of *Trichoderma* spp. in the management of plant-parasitic nematodes infesting important crops. In Ansari, R., Rizvi, R., Mahmood, I. (eds.), *Management of Phytonematodes: Recent Advances and Future Challenges* (pp. 259–278). Singapore: Springer. https://doi.org/10.1007/978-981-15-4087-5_11

Ibrahim, D. S. S., Mostafa, M., Abdellatif, A. A., & Ansari, R. A. (2022). Sustainable Management of Soil-Borne Plant Pathogens. In Furze, J. N., Eslamian, S., Raafat, S. M., Swing, K. (eds), *Earth Systems Protection and Sustainability: Volume 1* (pp. 35–56). Cham: Springer International Publishing.

Ibrahim, E., Zhang, M., Zhang, Y., Hossain, A., Qiu, W., Chen, Y., ... & Li, B. (2020). Green-synthesization of silver nanoparticles using endophytic bacteria isolated from garlic and its antifungal activity against wheat Fusarium head blight pathogen *Fusarium graminearum. Nanomaterials, 10*(2), 219.

Ibrahim, E., Fouad, H., Zhang, M., Zhang, Y., Qiu, W., Yan, C., ... & Chen, J. (2019). Biosynthesis of silver nanoparticles using endophytic bacteria and their role in inhibition of rice pathogenic bacteria and plant growth promotion. *RSC Advances, 9*(50), 29293–29299.

Iqbal, M. S., Singh, A. K., Singh, S. P., & Ansari, M. I. (2020). Nanoparticles and plant interaction with respect to stress response. In Bhushan, I., Singh, V., Tripathi, D. (eds.), *Nanomaterials and Environmental Biotechnology. Nanotechnology in the Life Sciences* (pp. 1–15.). Springer, Cham. https://doi.org/10.1007/978-3-030-34544-0_1

Ivanova, N., Gugleva, V., Dobreva, M., Pehlivanov, I., Stefanov, S., & Andonova, V. (2018). *Silver Nanoparticles as Multi-Functional Drug Delivery Systems* (pp. 71–91). IntechOpen, London, UK. Nanoparticles for managing bacterial diseases of plants.

Jain, D., & Kothari, S. L. (2014). Green synthesis of silver nanoparticles and their application in plant virus inhibition. *Journal of Mycology and Plant Pathology, 44*(1), 21–24.

Javaid, M., Haleem, A., Singh, R. P., Rab, S., & Suman, R. (2021). Exploring the potential of nanosensors: A brief overview. *Sensors International, 2*, 100130.

Jiang, L., Xiang, S., Lv, X., Wang, X., Li, F., Liu, W., ... & Sun, X. (2022). Biosynthesized silver nanoparticles inhibit *Pseudomonas syringae* pv. *tabaci* by directly destroying bacteria and inducing plant resistance in *Nicotiana benthamiana*. *Phytopathology Research*, *4*(1), 43.

John, S. A., Chattree, A., Ramteke, P. W., Shanthy, P., Nguyen, T. A., & Rajendran, S. (2022). Nanosensors for Plant Health Monitoring. In Denizli, A., Anh Nguyen, T., Rajendran, S., Yasin, G., Kumar Nadda, A. (eds), *Nanosensors for Smart Agriculture* (pp. 449–461). Elsevier.

Kashyap, P. L., Kumar, S., Jasrotia, P., Singh, D. P., & Singh, G. P. (2019). Nanosensors for plant disease diagnosis: Current understanding and future perspectives. In Pudake, R., Chauhan, N., Kole, C. (eds.), *Nanoscience for Sustainable Agriculture* (pp. 189–205). Cham: Springer. https://doi.org/10.1007/978-3-319-97852-9_9

Kashyap, P. L., Xiang, X., & Heiden, P. (2015). Chitosan nanoparticle based delivery systems for sustainable agriculture. *International Journal of Biological Macromolecules*, *77*, 36–51.

Kasote, D. M., Lee, J. H., Jayaprakasha, G. K., & Patil, B. S. (2019). Seed priming with iron oxide nanoparticles modulate antioxidant potential and defense-linked hormones in watermelon seedlings. *ACS Sustainable Chemistry & Engineering*, *7*(5), 5142–5151.

Kaushal, M., & Wani, S. P. (2017). Nanosensors: Frontiers in precision agriculture. In Prasad, R., Kumar, M., Kumar, V. (eds), *Nanotechnology: An Agricultural Paradigm* (pp. 279–291). Singapore: Springer.

Keshari, N., Kranti, K. V. V. S. K., Gunda, N. K., & Ansari, R. A. (2024). Arbuscular mycorrhizal fungi: A potential agent for phytonematodes management in diverse agro-climatic zones. In *Mycorrhizal Symbiosis and Agroecosystem Restoration* (pp. 147–169). Singapore: Springer Nature Singapore.

Khan, R. A. A., Tang, Y., Naz, I., Alam, S. S., Wang, W., Ahmad, M., ... & Li, Y. (2021). Management of *Ralstonia solanacearum* in tomato using ZnO nanoparticles synthesized through Matricaria chamomilla. *Plant Disease*, *105*(10), 3224–3230.

Khan, Y., Sadia, H., Ali Shah, S. Z., Khan, M. N., Shah, A. A., Ullah, N., ... & Khan, M. I. (2022). Classification, synthetic, and characterization approaches to nanoparticles, and their applications in various fields of nanotechnology: A review. *Catalysts*, *12*(11), 1386.

Kim, S. W., Jung, J. H., Lamsal, K., Kim, Y. S., Min, J. S., & Lee, Y. S. (2012). Antifungal effects of silver nanoparticles (AgNPs) against various plant pathogenic fungi. *Mycobiology*, *40*(1), 53–58.

Krishnaraj, C., Ramachandran, R., Mohan, K., & Kalaichelvan, P. T. (2012). Optimization for rapid synthesis of silver nanoparticles and its effect on phytopathogenic fungi. *Spectrochimica Acta Part A: Molecular and Biomolecular Spectroscopy*, *93*, 95–99.

Kumar, S., Basumatary, I. B., Sudhani, H. P., Bajpai, V. K., Chen, L., Shukla, S., & Mukherjee, A. (2021). Plant extract mediated silver nanoparticles and their applications as antimicrobials and in sustainable food packaging: A state-of-the-art review. *Trends in Food Science & Technology*, *112*, 651–666.

Kurczyńska, E., Godel-Jędrychowska, K., Sala, K., & Milewska-Hendel, A. (2021). Nanoparticles—plant interaction: What we know, where we are? *Applied Sciences*, *11*(12), 5473.

Latz, M. A., Jensen, B., Collinge, D. B., & Jørgensen, H. J. (2018). Endophytic fungi as biocontrol agents: Elucidating mechanisms in disease suppression. *Plant Ecology & Diversity*, *11*(5–6), 555–567.

Li, C. X., Parker, A., Menocal, E., Xiang, S., Borodyansky, L., & Fruehauf, J. H. (2006). Delivery of RNA interference. *Cell Cycle*, *5*(18), 2103–2109.

Li, Y., Yang, D., & Cui, J. (2017). Graphene oxide loaded with copper oxide nanoparticles as an antibacterial agent against Pseudomonas syringae pv. tomato. *RSC Advances*, *7*(62), 38853–38860.

Lv, Z., Jiang, R., Chen, J., & Chen, W. (2020). Nanoparticle-mediated gene transformation strategies for plant genetic engineering. *The Plant Journal*, *104*(4), 880–891.

Ma, X., Zhu, X., Qu, S., Cai, L., Ma, G., Fan, G., & Sun, X. (2022). Fabrication of copper nanoparticle composite nanogel for high-efficiency management of *Pseudomonas syringae* pv. tabaci on tobacco. *Pest Management Science*, *78*(5), 2074–2085.

Mahfouze, H. A., El-Dougdoug, N. K., & Mahfouze, S. A. (2020). Virucidal activity of silver nanoparticles against Banana bunchy top virus (BBTV) in banana plants. *Bulletin of the National Research Centre*, *44*(1), 1–11.

Mahmood, I., Rizvi, R., Sumbul, A., & Ansari, R. A. (2019). Potential role of plant growth promoting rhizobacteria in alleviation of biotic stress. In Ansari, R., Mahmood, I. (eds.), *Plant Health Under Biotic Stress* (pp. 177–188). Springer, Singapore. https://doi.org/10.1007/978-981-13-6040-4_9

Meena, M., Zehra, A., Swapnil, P., Harish, M. A., Yadav, G., & Sonigra, P. (2021). Endophytic nanotechnology: An approach to study scope and potential applications. *Frontiers in Chemistry*, *9*, 613343.

Moffat, A. S. (1999). Geminiviruses emerge as serious crop threat. *Science, 286*(5446), 1835–1835.

Mohammad, T. G., & Abd El-Rahman, A. F. (2015). Environmentally friendly synthesis of silver nanoparticles using *Moringa oleifera* (Lam) leaf extract and their antibacterial activity against some important pathogenic bacteria. *Mycopath, 13*(1), 1–6.

Molnár, Á., Rónavári, A., Bélteky, P., Szőllősi, R., Valyon, E., Oláh, D., ... & Kolbert, Z. (2020). ZnO nanoparticles induce cell wall remodeling and modify ROS/RNS signalling in roots of Brassica seedlings. *Ecotoxicology and Environmental Safety, 206*, 111158.

Nanda, A. K., Andrio, E., Marino, D., Pauly, N., & Dunand, C. (2010). Reactive oxygen species during plant-microorganism early interactions. *Journal of Integrative Plant Biology, 52*(2), 195–204.

Noha, K., Bondok, A. M., & El-Dougdoug, K. A. (2018). Evaluation of silver nanoparticles as antiviral agent against ToMV and PVY in tomato plants. *Sciences, 8*(01), 100–111.

O'Brien, J. A., Daudi, A., Butt, V. S., & Paul Bolwell, G. (2012). Reactive oxygen species and their role in plant defence and cell wall metabolism. *Planta, 236*, 765–779.

Ogunyemi, S. O., Abdallah, Y., Ibrahim, E., Zhang, Y., Bi, J. A., Wang, F., ... & Xu, L. (2023). Bacteriophage-mediated biosynthesis of MnO2NPs and MgONPs and their role in the protection of plants from bacterial pathogens. *Frontiers in Microbiology, 14*, 1193206.

Parveen, A., & Siddiqui, Z. A. (2021). Zinc oxide nanoparticles affect growth, photosynthetic pigments, proline content and bacterial and fungal diseases of tomato. *Archives of Phytopathology and Plant Protection, 54*(17–18), 1519–1538.

Patil, S., Ansari, R. A., Gurikar, C., Bhavan, S., & Hariprasad, K. (2024). Propagation of mycorrhizal fungi and their strength in plant health ameliorations. In *Mycorrhizal Symbiosis and Agroecosystem Restoration* (pp. 77–98). Singapore: Springer Nature Singapore.

Poveda, J., & Eugui, D. (2022). Combined use of Trichoderma and beneficial bacteria (mainly Bacillus and Pseudomonas): Development of microbial synergistic bio-inoculants in sustainable agriculture. *Biological Control*, 105100.

Prasad, B. N., Subramanyam, B., Lakshmipathi, R. N., Ansari, R. A., Rizvi, R., Sumbul, A., ... & Rachmi, C. M. (2020). Utilization of beneficial microorganisms in sustainable control of phytonematodes. *Management of Phytonematodes: Recent Advances and Future challenges*, 317–337.

Ramachandrappa, L. T., Kalagatur, N. K., Mudili, V., Mohan, C. D., Rangappa, S., Prasad, B. D., Ashwini, B. S., Hashem, A., Alqarawi, A. A., Malik, J. A., et al. (2019). Biofabrication of zinc oxide nanoparticles with Syzygium aromaticum flower buds extract and finding its novel application in controlling the growth and mycotoxins of Fusarium graminearum. *Frontiers in Microbiology, 10*, 1244–1213.

Raymaekers, K., Ponet, L., Holtappels, D., Berckmans, B., & Cammue, B. P. (2020). Screening for novel biocontrol agents applicable in plant disease management–a review. *Biological Control, 144*, 104240.

Savithramma, N., Linga Rao, M., Ankanna, S., & Venkateswarlu, P. (2012). Screening of medicinal plants for effective biogenesis of silver nanoparticles and efficient antimicrobial activity. *International Journal of Pharmaceutical Sciences and Research, 3*(4), 1141–1148.

Segal, L. M., & Wilson, R. A. (2018). Reactive oxygen species metabolism and plant-fungal interactions. *Fungal Genetics and Biology, 110*, 1–9.

Shafie, R. M., Salama, A. M., & Farroh, K. Y. (2018). Silver nanoparticles activity against Tomato spotted wilt virus. *Middle East J. Agric. Res, 7*, 1251–1267.

Shafiq, T., Yasmin, H., Shah, Z. A., Nosheen, A., Ahmad, P., Kaushik, P., & Ahmad, A. (2022). Titanium oxide and zinc oxide nanoparticles in combination with cadmium tolerant *Bacillus pumilus* ameliorates the cadmium toxicity in maize. *Antioxidants, 11*(11), 2156.

Shantharaj, D., Naranjo, E., Merfa, M. V., Cobine, P. A., Santra, S., & De La Fuente, L. (2023). Zinc Oxide-based nanoformulation zinkicide mitigates the xylem-limited pathogen *Xylella fastidiosa* in tobacco and southern highbush blueberry. *Plant Disease, 107*(4), 1096–1106.

Siddiqui, Z. A., Khan, A., Khan, M. R., & Abd-Allah, E. F. (2018). Effects of zinc oxide nanoparticles (ZnO NPs) and some plant pathogens on the growth and nodulation of lentil (Lens culinaris Medik.). *Acta Phytopathologica et Entomologica Hungarica, 53*(2), 195–211.

Siddiqui, Z. A., Khan, M. R., Abd-Allah, E. F., & Parveen, A. (2019). Titanium dioxide and zinc oxide nanoparticles affect some bacterial diseases, and growth and physiological changes of beetroot. *International Journal of Vegetable Science, 25*(5), 409–430.

Singh, R., Kuddus, M., Singh, P. K., & Choden, D. (2022). Nanotechnology for nanophytopathogens: From detection to the management of plant viruses. *BioMed Research International, 2022*.

Solanki, M. K., Kashyap, P. L., Kumari, B., Ansari, R. A., Sumbul, A., Rizvi, R., & Mahmood, I. (2021). Mycorrhizal Fungi and Its Importance in Plant Health Amelioration. In *Microbiomes and Plant Health* (pp. 205–223). Academic Press.

Solanki, M. K., Kashyap, P. L., Ansari, R. A., & Kumari, B. (2020). *Microbiomes and Plant Health: Panoply and Their Applications*, Academic Press, Elsevier, The Netherland. ISBN: 9780128197158. www.elsevier.com/books/microbiomes-and-plant-health/solanki/978-0-12-819715-8.

Sonkar, S., Sharma, L., Singh, R. K., Pandey, B., Rathore, S. S., Singh, A. K., ... & Singh, S. P. (2021). Plant stress hormones nanobiotechnology. In *Nanobiotechnology: Mitigation of Abiotic Stress in Plants*, 349–373.

Sumbul, A., Ansari, R. A., Rizvi, R., & Mahmood, I. (2020). Azotobacter: A potential bio-fertilizer for soil and plant health management. *Saudi Journal of Biological Sciences*, *27*(12), 3634–3640.

Tan, B. L., Norhaizan, M. E., Liew, W. P. P., & Sulaiman Rahman, H. (2018). Antioxidant and oxidative stress: A mutual interplay in age-related diseases. *Frontiers in Pharmacology*, *9*, 1162.

Tarighi, S., & Nejad, M. S. (2023). Ecofriendly fabrication of silver nanoparticles using quince petal extract and its antibacterial properties against fire blight disease. *Journal of Natural Pesticide Research*, *4*, 100026.

Vargas-Hernandez, M., Macias-Bobadilla, I., Guevara-Gonzalez, R. G., Rico-Garcia, E., Ocampo-Velazquez, R. V., Avila-Juarez, L., & Torres-Pacheco, I. (2020). Nanoparticles as potential antivirals in agriculture. *Agriculture*, *10*(10), 444.

Varympopi, A., Dimopoulou, A., Theologidis, I., Karamanidou, T., Kaldeli Kerou, A., Vlachou, A., ... & Skandalis, N. (2020). Bactericides based on copper nanoparticles restrain growth of important plant pathogens. *Pathogens*, *9*(12), 1024.

Wagemans, J., Holtappels, D., Vainio, E., Rabiey, M., Marzachì, C., Herrero, S., ... & Turina, M. (2022). Going viral: Virus-based biological control agents for plant protection. *Annual Review of Phytopathology*, *60*, 21–42.

Wang, H., Liu, R., You, M. P., Barbetti, M. J., & Chen, Y. (2021). Pathogen biocontrol using plant growth-promoting bacteria (PGPR): Role of bacterial diversity. *Microorganisms*, *9*(9), 1988.

Wang, L., Pan, T., Gao, X., An, J., Ning, C., Li, S., & Cai, K. (2022). Silica nanoparticles activate defense responses by reducing reactive oxygen species under *Ralstonia solanacearum* infection in tomato plants. *NanoImpact*, *28*, 100418.

Wang, Y., Sun, C., Xu, C., Wang, Z., Zhao, M., Wang, C., ... & Chen, F. (2016). Preliminary experiments on nano-silver against tobacco mosaic virus and its mechanism. *Yancao Keji*, *49*, 22–30.

Worrall, E. A., Hamid, A., Mody, K. T., Mitter, N., & Pappu, H. R. (2018). Nanotechnology for plant disease management. *Agronomy*, *8*(12), 285.

Yusof, N. A., & Isha, A. (2020). Nanosensors for early detection of plant diseases. In *Nanomaterials for Agriculture and Forestry Applications* (pp. 407–419). Elsevier.

FSC
www.fsc.org
MIX
Papier | Fördert gute Waldnutzung
FSC® C083411

Kolibri 360 GmbH
Ferdinand-Jühlke-Straße 7
99095 Erfurt, Deutschland
produktsicherheit@kolibri360.de